Quantum Chemistry of Polymers –
Solid State Aspects

NATO ASI Series

Advanced Science Institutes Series

A series presenting the results of activities sponsored by the NATO Science Committee, which aims at the dissemination of advanced scientific and technological knowledge, with a view to strengthening links between scientific communities.

The series is published by an international board of publishers in conjunction with the NATO Scientific Affairs Division

A	Life Sciences	Plenum Publishing Corporation
B	Physics	London and New York
C	Mathematical and Physical Sciences	D. Reidel Publishing Company Dordrecht, Boston and Lancaster
D	Behavioural and Social Sciences	Martinus Nijhoff Publishers
E	Engineering and Materials Sciences	The Hague, Boston and Lancaster
F	Computer and Systems Sciences	Springer-Verlag
G	Ecological Sciences	Berlin, Heidelberg, New York and Tokyo

Series C: Mathematical and Physical Sciences Vol. 123

Quantum Chemistry of Polymers – Solid State Aspects

edited by

Janos Ladik

Lehrstuhl für theoretische Chemie,
Friedrich-Alexander-Universität Erlangen-Nürnberg, Erlangen, F.R.G.

and

Jean-Marie André

Laboratoire de Chimie Théorique Appliquée,
Facultés Universitaires de Namur, Namur, Belgium

with the cooperation of

Max Seel

Lehrstuhl für theoretische Chemie,
Friedrich-Alexander-Universität Erlangen-Nürnberg, Erlangen, F.R.G.

D. Reidel Publishing Company

Dordrecht / Boston / Lancaster

Published in cooperation with NATO Scientific Affairs Division

Proceedings of the NATO Advanced Study Institute on
Quantum Chemistry of Polymers – Solid State Aspects
Braunlage/Harz, F.R.G.
July 25-August 5, 1983

Library of Congress Cataloging in Publication Data

NATO Advanced Study Institute on Quantum Chemistry of Polymers—Solid State Aspects (1983: Braunlage, Germany)
Quantum chemistry of polymers—solid state aspects.

(NATO ASI series. Series C, Mathematical and physical sciences ; v. 123)
Proceedings of the NATO Advanced Study Institute on Quantum Chemistry of Polymers—Solid State Aspects, Braunlage/Harz, July 25-August 5, 1983.
Sponsored by the NATO Scientific Affairs Division and the National Foundation for Cancer Research.
Published in cooperation with NATO Scientific Affairs Division.
Includes indexes.
1. Polymers and polymerization—Congresses. 2. Quantum chemistry—Congresses. 3. Solid state chemistry—Congresses. I. Ladik, János. II. André, Jean-Marie, 1944- . III. Seel, Max. IV. North Atlantic Treaty Organization. Scientific Affairs Division. V. Title. VI. Series.
QD380.N38 1983 547.7′0448 84-3367
ISBN 90-277-1741-9

Published by D. Reidel Publishing Company
P.O. Box 17, 3300 AA Dordrecht, Holland

Sold and distributed in the U.S.A. and Canada
by Kluwer Academic Publishers,
190 Old Derby Street, Hingham, MA 02043, U.S.A.

In all other countries, sold and distributed
by Kluwer Academic Publishers Group,
P.O. Box 322, 3300 AH Dordrecht, Holland

D. Reidel Publishing Company is a member of the Kluwer Academic Publishers Group

Printed in The Netherlands.

TABLE OF CONTENTS

PREFACE

The NATO Advanced Study Institute on "Quantum Chemistry of Polymers; Solid State Aspects" was held at the MARITIM Congress Hotel Braunlage/Harz in the Federal Republic of Germany from July 25 - August 5, 1983.

We wish to express our deep gratitude to the NATO Scientific Affairs Division, the main sponsor of the Institute, and to the National Foundation for Cancer Research, Bethesda, Maryland for their substantial support. We sincerely thank Dr. Craig Sinclair, Director of the NATO Advanced Study Institutes program as well as the whole Advanced Study Institute/Advanced Research Workshop Advisory Board of the NATO Scientific Affairs Division, who have honored us by holding their external annual meeting during this School in Braunlage. We are very much indebted also to Dr. Mario Di Lullo, Director of the Advanced Research Workshop program of the NATO Scientific Affairs Division who together with Dr. Sinclair has given a very informative lecture about the NATO ASI/ARW programs. Special thanks are due to Mr. Franklin Salisbury, Executive Director of the National Foundation for Cancer Research, to Mrs. Tamara Salisbury, Deputy Director of the National Foundation for Cancer Research and to Dr. Mary Hennen Aldridge, President of the National Foundation for Cancer Research, who also honored the School with their presence. Last but not least we should like to express our gratitude to Professor Nikolaus Fiebiger, President of the Friedrich-Alexander-University Erlangen-Nürnberg, who also visited the School and has given a most thoughtprovoking lecture about the situation of research in the Federal Republic of Germany.

This volume contains the main lectures of the Institute. It is our great pleasure to thank all the lecturers for the most excellent and interesting lectures and for their well prepared manuscripts.

The School started with a review by André and Delhalle of the SCF LCAO crystal orbital theory in its <u>ab initio</u> form with special emphasis on fast computational methods for polymers with

J. Ladik et al. (eds.), Quantum Chemistry of Polymers - Solid State Aspects, vii–ix.

larger unit cells and on the problems of lattice summations in infinite chains. Delhalle has included in his lectures also the theoretical interpretation of ESCA spectra based on the energy band structure of polymers. Karpfen has reviewed the theoretical determination of phonon spectra of polymers with smaller unit cells based again on their band structure.

The next step was the discussion of electronic correlation both in larger molecules and in polymers. In his lectures Collins developed a general theory using the Green's function formalism for the excitons in polymers and solids as well as for the possibility of excitonic superconductivity in CuCl and CdS. The lectures of Kunz have covered both the problems of ground state correlation and the calculation of excitons including correlation. In the one hour lecture of Suhai the question of the excitonic spectra with correlation was also discussed and he has presented results for polyethylene.

Besides the theoretical lectures there were several series of lectures given by experimentalists. In the lecture course of Hanack synthesis methods and characterization of conducting polymers of macrocyclic metal complexes were presented. Pope discussed in his lecture course electronic charge and energy transport in organic solids assuming different mechanisms. The same topic was treated in the series of Roth, first of all in the case of polyacetylene, which was supplemented by the presentation of measurements for magnetic properties of these polymers. In the series of Bloor the chemical and physical properties of single crystal polydiacetylenes were discussed in detail. The lecture series of Chance contained different experiments on doped conjugated polymers (like polyacetylene, polypyrrole and polyparaphenylene) which were supplemented also by theoretical considerations. Finally, Heidberg has given a lecture about vibrational pre-evaporation and -dissorption on the surface of solids and Swenberg discussed time-dependence of fluorescence of polymers.

Returning to the theory a further topic of the School was the theory of disordered systems and of local perturbations. Martino in his series treated the effect of disorder on electronic transport in organic polymers using for the hopping of electrons in these systems a random walk theory. Del Re treated with the help of a Green's matrix formalism the effect of local perturbations on the electronic structure of polymers. Duke discussed in detail electron localization and delocalization in different organic metals and organic semiconductors. The problem of disorder and correlation in biopolymers and highly conducting polymers was treated in the series of Ladik who discussed also different possible physical mechanisms for the activation of oncogenes. In the series of Otto large scale *ab initio* computations on the energy band structures of polynucleotides, polypeptides and on their interactions were presented.

The lectures of the Institute were kept on a rather high level and they actually covered nearly all important aspects of the quantum theory of extended systems and their most representative applications to polymers.

The secretarial burden fell on Ms. S. Patzak who worked very efficiently and smilingly throughout the Institute and during the final presentation of these proceedings. We wish to thank her most heartedly. Dr. B. Gazdy, Dr. C. Liegener, Mr. W. Förner and Mr. M. Gies did a wonderful job in arranging many practical details. We are very much indebted also to them.

Last but not least we gratefully acknowledge the help and effective cooperation of Mr. E. Döring, Director of the MARITIM Congress Hotel, and Mr. G. Möller who together with their staff contributed very much to the success of the School and who made our stay in Braunlage so pleasant.

Erlangen, Namur, September 1983

Janos Ladik
Jean-Marie André
Max Seel

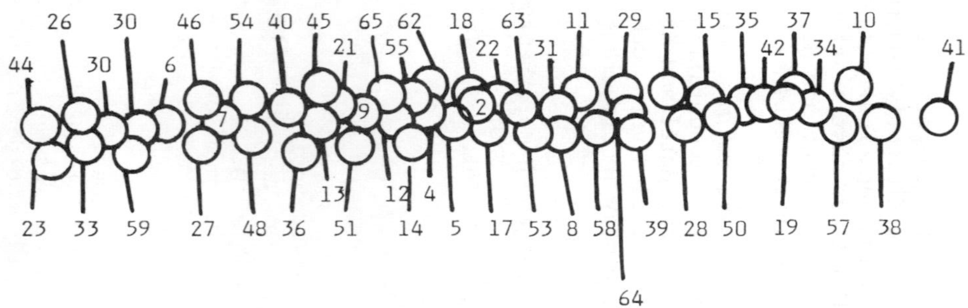
26 30 46 54 40 45 65 62 18 63 11 29 1 15 35 37 10
44 30 6 21 55 22 31 42 34 41
7 9 2
13 12 4
23 33 59 27 48 36 51 14 5 17 53 8 58 39 28 50 19 57 38
64

LIST OF PARTICIPANTS

1. Prof. J.-M. André
   Laboratoire de Chimie Théorique Appliquée
   Facultés Universitaires de Namur
   61, Rue de Bruxelles, B-5000 Namur, Belgium

2. Dr. M. Baldo
   Istituto di Fisica dell' Universita
   Corso Italia 57, I-95129 Catania, Italy

3. Dr. R. Bannehr
   ELTECH
   3, Route de Troinex, CH-1227 Carouge/Geneve, Switzerland

4. Dr. H. Baur
   ZKM 6201, BASF AG
   D-6700 Ludwigshafen, FRG

5. Prof. W.A. Bingel
   Lehrstuhl für Theoretische Chemie
   Universität Göttingen
   Tammannstr. 6, D-3400 Göttingen, FRG

6. Prof. D. Bloor
   Department of Physics
   Queen Mary College
   Mile End Road, London E1 4SN, England

7. V. Bodart
   Laboratoire de Chimie Théorique Appliquée
   Facultés Universitaires de Namur
   61, Rue de Bruxelles, B-5000 Namur, Belgium

8. Prof. V. Bonacic-Koutecky
   Institut für Physikalische Chemie und Quantenchemie
   Freie Universität Berlin
   Takustraße 3, D-1000 Berlin 33, FRG

9. Dr. J.-L. Brédas
Laboratoire de Chimie Théorique Appliquée
Facultés Universitaires de Namur
61, Rue de Bruxelles, B-5000 Namur, Belgium

10. Dr. R.R. Chance
Corporate Research Center
Allied Corporation
Morristown, New Jersey 07960, USA

11. Prof. J. Cizek
Department of Applied Mathematics
University of Waterloo
Waterloo, Ontario N2L 3G1, Canada

12. J.A. Cogordan
Quantum Chemistry Group
Uppsala University
Box 518, S-751 20 Uppsala, Sweden

13. Prof. T.C. Collins
Associate Vice President for Academic Affairs
University of Missouri
309 University Hall, Columbia, Missouri 65211, USA

14. L. Cruzeiro
Lab. Biomecanica
Instituto Gulbenkian de Ciencia
Rua da Quinta Grande, Ap. 14, P-2781 Oeiras, Portugal

15. Prof. J. Delhalle
Laboratoire de Chimie Théorique Appliquée
Facultés Universitaires de Namur
61, Rue de Bruxelles, B-5000 Namur, Belgium

16. Prof. G. Del Re
Cattedra di Chimica Teorica
Università di Napoli
Via Mezzocannone 4, I-80134 Napoli, Italy

17. Dr. K.J. Donovan
Department of Physics
Queen Mary College
Mile End Road, London E1 4SN, England

18. Prof. K. Dransfeld
Fakultät für Physik
Universität Konstanz
Postfach 5560, D-7750 Konstanz, FRG

19. Dr. C.B. Duke
Xerox Corporation
800 Phillips Road, Bldg. 114
Webster, New York 14580, USA

20. Dr. N. Ferrer-Anglada
Facultat de Fisica
Diagonal 645, Barcelona - 28, Spain

21. W. Förner
Lehrstuhl für Theoretische Chemie
Universität Erlangen-Nürnberg
Egerlandstraße 3, D-8520 Erlangen, FRG

22. Dr. B. Gazdy
Lehrstuhl für Theoretische Chemie
Universität Erlangen-Nürnberg
Egerlandstraße 3, D-8520 Erlangen, FRG

23. M. Gies
Lehrstuhl für Theoretische Chemie
Universität Erlangen-Nürnberg
Egerlandstraße 3, D-8520 Erlangen, FRG

24. Prof. M. Hanack
Institut für Organische Chemie
Universität Tübingen
Auf der Morgenstelle 18, D-7400 Tübingen, FRG

25. Prof. J. Heidberg
Institut für Physikalische Chemie und Elektrochemie
Universtät Hannover
Callinstrße 3-3a, D-3000 Hannover, FRG

26. A. Jahn
Institut für Organische und Markomolekulare Chemie
Universität Bremen
Postfach 330440, D-2800 Bremen, FRG

27. Dr. A. Karpfen
Institut für Theoretische Chemie und Strahelenchemie
Universität Wien
Währinger Straße 17, A-1090 Wien, Austria

28. Prof. J. Kaufman
Department of Chemistry
Johns Hopkins University
Baltimore, Maryland 21218, USA

29. Prof. J. Koutecky
Institut für Physikalische Chemie und Quantenchemie
Freie Universität Berlin
Takustraße 3, D-1000 Berlin, FRG

30. Prof. A.B. Kunz
Department of Physics and Materials Research Laboratory
University of Illinois at Urbana-Champaign
Urbana, Illinois 61801, USA

31. Prof. J. Ladik
Lehrstuhl für Theoretische Chemie
Universität Erlangen-Nürnberg
Egerlandstraße 3, D-8520 Erlangen, FRG

32. Prof. C. Levin
Department of Chemistry
Smith College
Northampton, MA 01063, USA

33. Dr. C. Liegener
Lehrstuhl für Theoretische Chemie
Universität Erlangen-Nürnberg
Egerlandstraße 3, D-8520 Erlangen, FRG

34. Prof. S.Z. Liu
Lehrstuhl für Theoretische Chemie
Universität Erlangen-Nürnberg
Egerlandstraße 3, D-8520 Erlangen, FRG

35. D. MacDonaill
Chemistry Department
Trinity College
Dublin 2, Ireland

36. Prof. G.L. Malli
Department of Chemistry
Simon Fraser University
Burnaby, B.C. Canada V5A 1S6

37. Prof. F. Martino
Physics Department
City College at the City University of New York
Convent Avenue at 138th Street, New York, N.Y. 10031, USA

38. Dr. D. Morton-Blake
Chemistry Department
Trinity College
Dublin 2, Ireland

39. Dr. H.B. Özgün
Balgat Baglarüstu 5 Cad.
6 Sok. N. 8
Ankara, Turkey

40. Prof. P. Otto
Lehrstuhl für Theoretische Chemie
Universität Erlangen-Nürnberg
Egerlandstraße 3, D-8520 Erlangen, FRG

41. Dr. S. Paksoy
Yildiz Universite
Kimya Bölümü, Yildiz-Istanbul, Turkey

42. J.C. Peng
Department of Physics
Queen Mary College
Mile End Road, London E1 4NS, England

43. Prof. M. Pope
Chemistry Department
New York University
4 Washington Place, Room 812
New York, New York 10003, USA

44. Dr. S. Roth
Max-Planck-Institut für Festkörperforschung
Heisenbergstraße 1, D-7000 Stuttgart 80, FRG

45. Dr. M. Seel
Lehrstuhl für Theoretische Chemie
Universität Erlangen-Nürnberg
Egerlandstraße 3, D-8520 Erlangen, FRG

46. Dr. M. Springborg
Max-Planck-Institut für Festkörperforschung
Heisenbergstraße 1, D-7000 Stuttgart 80, FRG

48. Dr. C. Swenberg
Armed Forces Radiobiology Research Institute
Bethesda, Maryland 20814, USA

49. Dr. S. Suhai
Deutsches Krebsforschungszentrum
Im Neuenheimer Feld 280, D-6900 Heidelberg, FRG

50. Dr. M. Talu
Abidin Daver Sok. Özgür Apt. No. 24/12
Cankaya, Ankara, Turkey

51. B. Thémans
Laboratoire de Chimie Théorique Appliquée
Facultés Universitaires de Manur
61, Rue de Bruxelles, B-5000 Namur, Belgium

52. I. Ülkem
Chemistry Department
Middle East Technical University
Ankara, Turkey

53. C. Ünaleroglu
Organic Chemistry Section
Department of Chemsitry
Hacettepe University
Ankara, Turkey

54. M. Wirth
Institute of Physical Chemistry
Chalmers University of Technology
S-412 96 Göteborg, Sweden

55. Dr. G. Zannoni
Dip. di Chimica Industriale e Ing. Chmica
Politecnico di Milano
Piazza Leonardo da Vinci, I-20133 Milano, Italy

56. Prof. G. Zerbi
Dip. di Chimica Industriale e Ing. Chimica
Politecnico di Milano
Piazza Leonardo da Vinci, I-20133 Milano, Italy

57. J.L. Zhu
Department of Physics and Measurement Technology
Linköping University
S-581 83 Linköping, Sweden

58. B. Zümreoglu
Department of Chemistry
Inorganic Chemistry Section
Hacettepe University
Ankara, Turkey

Special Guests

59. Dr. F. Salisbury
Executive Director
National Foundation for Cancer Research
7315 Wisconsin Avenue, Suite 332W
Bethèsda, Maryland 20814, USA

60. T. Salisbury
Deputy Director
National Foundation for Cancer Research
7315 Wisconsin Avenue, Suite 332W
Bethesda, Maryland 20814, USA

61. Dr. M.H. Aldridge
President
National Foundation for Cancer Research
7315 Wisconsin Avenue, Suite 332W
Bethesda, Maryland 20814, USA

62. Dr. C. Sinclair
Scientific Affairs Division
NATO
B-1110 Bruxelles, Belgium

63. Dr. M. Di Lullo
Scientific Affairs Division
NATO
B-1110 Bruxelles, Belgium

64. Prof. N. Fiebiger
President
Friedrich-Alexander-Universität Erlangen-Nürnberg
Schloßplatz 4, D-8520 Erlangen, FRG

65. Mrs. I. Fiebiger
Friedrich-Alexander-Universität Erlangen-Nürnberg
Schloßplatz 4, D-8520 Erlangen, FRG

TOWARDS SPECIFIC AB INITIO PROGRAMS FOR POLYMER CALCULATIONS

J.M. André[a*], V.P. Bodart[+], J.L. Brédas[b+], J. Delhalle[+], and J.G. Fripiat[+]

*IBM Corporation, D55, Bldg 701-2
Poughkeepsie, New York 12603 (USA).
+Laboratoire de Chimie Théorique Appliquée
Facultés Universitaires Notre-Dame de la Paix
61, rue de Bruxelles B-5000 Namur (Belgium).

1. INTRODUCTION

Since the first theoretical works of the sixties (1) on LCAO techniques in polymer quantum chemistry, the field has known a rapid development.

Periodic systems are computationally easier to treat than aperiodic ones since the translational symmetry can be fully exploited in order to reduce to manageable dimensions the formidable task of computing electronic states of an extended system.

Ab initio programs for polymers are available and are currently applied in several groups (2) like Erlangen, Vienna, Budapest, Poughkeepsie and Namur. However, due to different computational strategies those programs do not produce rigorously the same numbers even for polymers studied with the same basis sets and within the exactly same geometrical conformations. This is a result of different cut-off of integrals and originates mainly from an internal structure which is, in all cases, a logical extension of molecular programs. It is illuminating to note that all polymer packages use standard "molecular" packages taken from the IBMOL, GAUSSIAN or other series.

Furthermore, it has to be realized that sophisticated "ab initio" methods, already time consuming for middle-and large sized molecules become very onerous when applied to polymers of

J. Ladik et al. (eds.), Quantum Chemistry of Polymers - Solid State Aspects, 1–21.

chemical or biological interest. In this sense, the use of double-zeta or polarization LCAO basis is still exceptional.

The conclusion is that additional work should be made in order to implement efficient ab initio programs for polymers which would request much smaller computer times than the existing ones.

In our opinion, this will be achieved by implementing into a general system the new fast techniques for evaluating integrals over gaussian basis, the explicit use of helical symmetry and efficient methods for computing long-range electrostatic effects.

This requests a general methodology, more "polymer-minded" than "molecule-minded". It is precisely the purpose of this paper to present some current trends in such directions.

The paper is organized as follows :

i) For sake of completeness, we remind basic equations of LCAO Bloch-like procedures ;

ii) We discuss the important question of cut-off procedures in existing programs and point out the differences in various programs ;

iii) We introduce a compact formula for the calculation of electron densities and related quantities. This formula takes full advantage of polymer translational periodicity and, as far as we know, has not, as such, been previously exploited ;

iv) In the last part we shortly investigate possible ways of reducing computing times by use of symmetry elements other than the translational ones like screw axes.

The discussed topics are presently implemented into an original computer program.

2. GENERAL FORMULAS FOR THE SOLUTION OF LCAO-SCF EQUATIONS IN THE CASE OF PERIODIC 1D SYSTEMS

The crucial point in Bloch's theory of one-dimensional (1D) systems is that, at point ($\vec{r}$) related by direct space lattice translation (ja ; j = integer, a = length of the 1D unit cell) electron densities (ρ) are invariant and one-electron wave functions or "orbitals"(ϕ) only differ by an exponential of unity modulus and of imaginary argument (ikja) :

$$\rho(\vec{r} + ja) = \rho(\vec{r}) \tag{1}$$

$$\phi(\vec{r} + ja) = e^{ikja}\,\phi(\vec{r}) \tag{2}$$

Those last equations call for the obvious comments :

i) Even if dealing with 1D periodic system, the orbitals are truly 3D ;

ii) If ja is a direct lattice vector, k must be defined in a reciprocal space (not necessarily a vector of the reciprocal lattice) ;

iii) A k point and its translations k + jg ($g = 2\pi/a$) in reciprocal space have the same effect on a given orbital ϕ.

This is the basis of the well-known band theory as applied to periodic polymers, i.e. orbital energies (ε_k) are labelled with respect to k and are periodic in reciprocal space so that their knowledge is only required in a unit cell or in the first Brillouin zone (BZ). For the usual Hamiltonians, orbital energies are symmetrical with respect to k = o (Γ point), $\varepsilon_k = \varepsilon_{-k}$.

In a LCAO expansion of orbitals, eq. (2) implies equivalent relations between coefficients (C_p^o, C_p^k) of atomic orbitals (χ_p^o, χ_q^h) respectively centered in the origin unit cell (0) and in another unit cell (h) distant from the previous one by a (ha) translation :

$$C_p^h = e^{ikha} C_p^o \tag{3}$$

Hereinafter, lower indices (p,q,r or s) will refer to the labelling of a given orbital (χ_p, χ_q, χ_r or χ_s) within a unit cell while upper indices (j,h,l) refer to the position of a given unit cell. When applying Hartree-Fock or Self-Consistent-Field (SCF) concepts, we consider the motion of a single electron in the field of fixed nuclei and in averaged Coulomb and Pauli (exchange) fields of all electrons. The SCF-LCAO-polymeric equations of a polymer consisting of (2N + 1) unit cells are given in Table 1, using notations previously introduced, and compared to traditional LCAO-MO equations. In the Table, we drop the o upper indices and do not detail the extent of the summations, $[-N \leqslant j,h,l \leqslant +N \;;\; 1 \leqslant p,q,r,s \leqslant \omega]$. A further comment is that, in the polymer case, the matrices are complex Hermitian ones while real symmetric in the molecular case.

Table 1 : Comparison of SCF-LCAO equations for molecules (left) and polymers (right).

Orbital

Molecules	Polymers	
$\phi_j(\underline{r}) = \sum_p C_{jp}\chi_p(\underline{r})$	$\phi_n(k,\underline{r}) = \sum_j \sum_p (2N+1)^{-1/2} . C_{np}(k)e^{ikja}\chi_p^j(\underline{r})$	(4)

Secular system

Molecules	Polymers	
$\sum_p C_{jp}(h_{pq}-\varepsilon_j S_{pq}) = 0$	$\sum_p C_{np}(k)\{\sum_j e^{ikja}[h_{pq}^{\ j} - \varepsilon_n(k)S_{pq}^{\ j}]\} = 0$	(5)

Determinant

Molecules	Polymers	
$\lvert h_{pq}-\varepsilon S_{pq}\rvert = 0$	$\lvert\sum_j e^{ikja} h_{pq}^{\ j} - \varepsilon(k)\sum_j e^{ikja} . S_{pq}^{\ j}\rvert = 0$	(6)
	$\lvert h_{pq}(k)-\varepsilon(k)\ S_{pq}(k)\rvert = 0$	(7)

Integrals

Molecules	Polymers	
$S_{pq} = \int \chi_p(\underline{r})\chi_q(\underline{r})dv$	$S_{pq}^{\ j} = \int \chi_p(\underline{r})\chi_q^j(\underline{r})dv$	(8)
$h_{pq} = \int \chi_p(\underline{r})h(\underline{r})\chi_q(\underline{r})dv$	$h_{pq}^{\ j} = \int \chi_p(\underline{r})h(\underline{r})\chi_q^j(r)$	(9)
$= T_{pq} - \sum_A Z_A V_{pq}\rvert_A + \sum_r \sum_s D_{rs}\{2(pq\rvert rs)-(pr\rvert qs)\}$	$= T_{pq}^{\ j} - \sum_h \sum_A Z_A V_{pq}^{\ j}\rvert_A^h + \sum_{hlrs}\sum\sum\sum D_{rs}^{hl}\{2(p\overset{j}{q}\rvert \overset{hl}{rs})-(p\overset{h}{r}\rvert \overset{jl}{qs})\}$	(10)
$T_{pq} = \int \chi_p(\underline{r})\{-\frac{1}{2}\nabla_r^2\}\chi_q(\underline{r})dv$	$T_{pq}^{\ j} = \int \chi_p(\underline{r})\{-\frac{1}{2}\nabla_r^2\}\chi_q^j(\underline{r})dv$	(11)
$V_{pq}\rvert_A = \int \chi_p(\underline{r})\lvert\underline{r}-\underline{R}_A\rvert^{-1}\chi_q(r)dv$	$V_{pq}^{\ j}\rvert_A^h = \int \chi_p(\underline{r})\lvert\underline{r}-\underline{R}_A-ha\rvert^{-1}\chi_q^j(\underline{r})dv$	(12)
$(pq\rvert rs) = \iint \chi_p(\underline{r})\chi_q(\underline{r})\lvert\underline{r}-\underline{r}'\rvert^{-1} . \chi_r(\underline{r}')\chi_s(\underline{r}')dv\,dv'$	$(p\overset{j}{q}\rvert \overset{hl}{rs}) = \iint \chi_p(\underline{r})\chi_q^j(\underline{r})\lvert\underline{r}-\underline{r}'\rvert^{-1} . \chi_r^h(\underline{r}')\chi_s^l(\underline{r}')dv\,dv'$	(13)

3. CUT-OFF PROCEDURES IN EXISTING PROGRAMS

The crucial point in an ab initio polymer program is the way the lattice summations appearing in equations such as eq. (10) are truncated.

Different strategies are applied in most existing programs ; they constitute the basic differences between them and are responsible for the varying numerical outputs. This topic has already been covered by Karpfen (3) and more recently by Kertesz (4). We briefly discuss it here in order to stress the originality a specific polymer program must have. Note also that the cut-off procedure directly determines the number of two-electron integrals to be computed and thus directly influences the computer time required by a given application. We discuss truncation problems for the Coulomb and exchange part of the Hartree-Fock matrix elements (eq. 12) :

$$\sum_h \sum_l \sum_{r,s} D^{hl}_{rs} \{2(p^{0}q^{j}|r^{h}s^{l}) - (p^{0}r^{h}|q^{j}s^{l})\} \tag{14}$$

but the same arguments can be developed for the nuclei-electron attraction part.

3.1. Cell-Wise Summation, Symmetric Cut-Off ; Namur Group (5).

In the framework of a Nth-neighbor approximation, the summation over j runs from -N to +N and that over l from h-N to h + N. That way looks reasonable to take into account symmetrically both charge distributions :

$$P^{j}_{pq}(\underline{r}_1) = \chi^{o}_{p}(\underline{r}_1)\chi^{j}_{q}(\underline{r}_1)$$

$$P^{hl}_{rs}(\underline{r}_2) = \chi^{h}_{r}(\underline{r}_2)\chi^{l}_{s}(\underline{r}_2) \tag{15}$$

A correct handling of the summation over h is more difficult to find, the Namur group cuts it into three parts :

i. from -N to +N ; this determines a short-range domain where all two-electron integrals are exactly calculated (if larger than a given threshold, actually 10^{-8} a.u.). The largest separation between two atomic orbitals involved in a two-electron integral occurs when j=N, h=-N and l=h-N=-2N. This "short-range" domain covers thus 3N+1 cells.

ii. from -2N to -(N+1) and from (N+1) to 2N. This corresponds to the intermediate-range region where computation of the electrostatic terms can be simplified by using asymptotic approximations of the F_ν error functions, involved in the calculations of two-electron integrals between gaussian functions.

iii. From $-\infty$ to $-(2N+1)$ and from $(2N+1)$ to $+\infty$, the Namur group makes use of a multipole expansion to exactly compute up to infinity long-range electrostatic terms (6) between non-overlapping charge distributions P_{pq}^{o1} and P_{rs}^{hl}. This constitutes the most important originality of the Namur ab initio program and affords properly stabilized and, where required, symmetric matrix elements within a Nth-neighbor approximation. That point will be discussed in detail later on (§ 4 and 5).

3.2. Cell-Wise Summation ; Vienna Group (7).

In this program, it is considered that, within the Nth-neighbor approximation, the interaction domain only covers N+1 unit cells (instead of 3N+1 in Namur's program). It is then possible to build a molecular cluster containing N+1 unit cells and to use a molecular program to perform the computation of the two-electron part. All integrals are kept which involve at least one orbital centered in the reference unit cell. Following Karpfen, a single FORTRAN statement is sufficient to convert the integral part of a conventional MO program to the corresponding crystal orbital integral part. One advantage is that the simple structure involved allows to make use of performing and worldwide fully tested ab initio molecular packages. A disadvantage lies in the creation of artificial asymmetries. The reason for this is that summations over h and l explicitely depend on the value of j. Typically, e.g. in the case of the 2nd neighbor approximation, since the interaction domain is limited to N+1=3 cells :

i. when $j = 0$, h runs from -2 to $+2$,

ii. when $j = 1$, h runs from -1 to $+2$,

iii. when $j = 2$, h runs from 0 to $+2$;

obviously, these asymmetries decay as the value of N is increased.

In order to compute long-range terms, the Vienna group has recently introduced a multipole expansion technique on a per cell basis (8).

3.3. Symmetric Cut-Off ; Budapest Group (2b)

The program is very similar to the previous one, the difference lies on the fact that, within a N+1 unit cell cluster, more integrals than in the Vienna group's case are neglected on symmetry grounds. A symmetric cut-off is achieved by means of a distance parameter ρ. As soon as the distance between any of two atomic orbitals involved in a two-electron integral is larger than ρ, this integral is neglected. This procedure provides symmetrized matrix elements. However, no long-range interactions are taken into account.

3.4. Overlap Determined Cell-Wise Summation ; Erlangen Group (9)

Suhai has recently introduced a cut-off procedure rather similar to that of method 3.1. but where the summation over 1 is extended from h-N' to h+N'. The value of N' is determined on the basis of the exponential decay of the distribution P^{hl}_{rs} (eq. 15). When that decay is very fast, N' can be chosen smaller than N resulting in significantly faster computations.

4. AN EFFICIENT FORMULA FOR COMPUTING ELECTRON DENSITIES AND RELATED PROPERTIES

The influence of a "molecular-like reasoning" is evident in the way electron densities and related electron-electron repulsion have been computed until now.

Trivially extending the well-known formula for electron densities in molecules, where D_{pq} is an element of the density matrix between orbitals p and q,

$$\rho(\underline{r}) = \sum_p \sum_q D_{pq} \chi_p(\underline{r}) \chi_q(\underline{r}), \tag{16}$$

we find in the polymer case :

$$\begin{aligned}\rho(\underline{r}) &= \sum_h \sum_l \sum_{r,s} D^{hl}_{rs} \chi^h_r(\underline{r}) \chi^l_s(\underline{r}) \\ &= \sum_h \sum_l \sum_{r,s} D^{l-h}_{rs} \chi^h_r(\underline{r}) \chi^l_s(\underline{r})\end{aligned} \tag{17}$$

This formula does not explicitly take into account the important property of periodicity we discussed before. It is to be realized that such an important property is only <u>implicitly</u> introduced by the identities between translationally equivalent elements of the density matrix

$$\begin{aligned}&D^{01}_{rs} = D^{12}_{rs} = D^{\bar{1}0}_{rs} = D^{23}_{rs} = D^{\bar{2}\bar{1}}_{rs} = \ldots \\ &D^{h}_{rs} = D^{1h+1}_{rs} = D^{\bar{1}h-1}_{rs} = D^{2h+2}_{rs} = D^{\bar{2}h-2}_{rs} = \ldots\end{aligned} \tag{18}$$

An explicit use of the translational periodicity can be made by the following straightforward manipulation when using a new index m = l-h (or l = m+h)

$$\rho(\underline{r}) = \sum_h \sum_l \sum_{r,s} D_{rs}^{l-h} \chi_r^h(\underline{r}) \chi_s^l(\underline{r})$$

$$= \sum_h \sum_m \sum_{r,s} D_{rs}^{m} \chi_r^h(\underline{r}) \chi_s^{h+m}(\underline{r})$$

$$= \sum_m \sum_{r,s} D_{rs}^{m} \sum_h \chi_r^h(\underline{r}) \chi_s^{h+m}(\underline{r})$$

$$= \sum_m \sum_{r,s} D_{rs}^{m} \rho_{rs}^{m}(\underline{r}) \qquad (19)$$

with

$$\rho_{rs}^{m}(\underline{r}) = \sum_h \chi_r^h(\underline{r}) \chi_s^{h+m}(\underline{r}). \qquad (20)$$

It is easily shown that these partial cell electron densities, $\rho_{rs}^m(\underline{r})$ are periodic in direct space and are, at a given point $\underline{r}$, the contributions of two overlapping atomic basis function distant from m translations :

$$\rho_{rs}^{0}(\underline{r}) = \sum_h \chi_r^h(\underline{r}) \chi_s^{h}(\underline{r})$$

$$\rho_{rs}^{1}(\underline{r}) = \sum_h \chi_r^h(\underline{r}) \chi_s^{h+1}(\underline{r}) \qquad (21)$$

... ...

Usual relations of polymer quantum chemistry can indeed be deduced from that interesting formula (eq. 19). For example, we can easily deduce averaged values of one-electron periodic operator $\mathcal{O}(\underline{r})$ within such an electron density $\rho(\underline{r})$. It is appealing to make use of the common trick (10) for evaluating such quantities $\langle\mathcal{O}\rangle_{av}$ by the formula :

$$\langle\mathcal{O}\rangle_{av} = \int dv\, \mathcal{O}(\underline{r}) \rho(\underline{r}', \underline{r}) \qquad (22)$$

where we introduce the convention that in treating average values, the operator $\mathcal{O}(\underline{r})$ in the integrands acts only on the unprimed coordinate $\underline{r}$ and that, after these operations have been carried out, we put $\underline{r}'=\underline{r}$. Obviously, that trick is not required if the operator $\mathcal{O}(\underline{r})$ is not a differential operator. Note that eq. 19 is trivially extended to

$$\rho(\underline{r}',\underline{r}) = \sum_h \sum_l D_{rs}^{l-h} \chi_r^h(\underline{r}')\chi_s^l(\underline{r})$$

$$= \sum_m D_{rs}^m \sum_h \chi_r^h(\underline{r}')\chi_s^{h+m}(\underline{r}) \qquad (23)$$

Thus we are now able to deduce easily usual formulas :

$$\langle O\rangle_{av} = \int dv\, O(\underline{r})\rho(\underline{r}',r)$$

$$= \sum_m \sum_{r,s} D_{rs}^m \sum_h \langle\chi_r^h|O|\chi_s^{h+m}\rangle \qquad (24)$$

If the operator $O(\underline{r})$ is periodic, then

$$O(\underline{r}) = O(\underline{r} + ja) \qquad (25)$$

and

$$\langle\chi^0|O|\chi^m\rangle = \langle\chi^1|O|\chi^{1+m}\rangle = \ldots =$$

$$= \langle\chi^l|O|\chi^{l+m}\rangle$$

$$= \langle\chi^h|O|\chi^{h+m}\rangle \qquad (26)$$

so that

$$\sum_{h=-N}^{N} \langle\chi_r^h|O|\chi_s^{h+m}\rangle = (2N+1)\ \langle\chi_r^0|O|\chi_s^m\rangle$$

$$= (2N+1)O_{rs}^m \qquad (27)$$

and we obtain

$$\langle O\rangle_{av} = (2N+1)\sum_m \sum_{r,s} D_{rs}^m O_{rs}^m \qquad (28)$$

A first example is the averaged value of the electron density (which is the total number of electrons). In that case :

$$O = 1\ ,$$

$$O_{rs}^m = S_{rs}^m\ ,$$

and

$$\int \rho(\underline{r})dv = (2N+1)\sum_m \sum_{r,s} D_{rs}^m S_{rs}^m = (2N+1).n$$

where n is the total number of electrons per unit cell.

Another example is that of the averaged kinetic energy, when we have

$$\mathcal{O} = -\frac{\hbar^2}{2m} \nabla^2(\underline{r})$$

$$\mathcal{O}_{rs}^{\ m} = \langle \chi_r^0 | -\frac{\hbar^2}{2m}\nabla^2(\underline{r}) | \chi_s^m \rangle = T_{rs}^{\ m}$$

and

$$\int dv \{-\frac{\hbar^2}{2m}\nabla^2(\underline{r})\}\rho(\underline{r}',\underline{r}) = (2N+1) \sum_m \sum_{r,s} D_{rs}^{\ m} T_{rs}^{\ m} \tag{30}$$

At first sight, the previous discussion could look very formal. It has however extremely important consequences which can be easily seen, for example, when computing the Coulomb part of a Hartree-Fock matrix element

$$C_{pq}^{\ j} = \iint dv\, dv'\, \chi_p(\underline{r})\chi_q^j(\underline{r})\,|\underline{r}-\underline{r}'|^{-1}\rho(\underline{r}',\underline{r}') \tag{31}$$

When using the standard formula for electron density (eq. 17) we end up with the traditional contribution :

$$C_{pq}^{\ j} = \sum_h \sum_l \sum_{r,s} D_{rs}^{\ 1-h} (pq^j|rs^{hl}) \tag{32}$$

Since the elements of the density matrix D_{rs}^{1-h} evolve iteratively in a SCF calculation , it is thus unavoidable to store roughly (when neglecting integral redundancy) $(2N+1)^3\omega^4$ integrals if the number of unit cells explicitely considered is 2N+1 and the size of the AO basis set in the unit cell is ω.

In the standard case of polyethylene described with a minimal basis set (ω= 14) and in a next-nearest neighbour approximation (N=2), the number of two-electron integrals to compute and to store is around 2.6 10^6. In the calculation of DNA recently reported by Otto, Clementi and Ladik (2d) (ω = 125, N=2), the number would be around 3.1 10^{10}. These figures must be corrected for taking into account integral redundancy (divide by a factor of $\simeq$ 8) and the fact that only non-zero (non-negligible) integrals are stored. However, it means a tremendous number of storing and retrieval operations in the SCF process.

This is precisely where use of our compact electron density formula (19) brings great improvement into actual computer programs. Without any loss of generality, the Coulomb part of a matrix element becomes :

$$C_{pq}^{j} = \sum_{m}\sum_{r,s} D_{rs}^{m}\{\sum_{h}(pq|\overset{j}{}\,\overset{h}{r}\,\overset{h+m}{s})\}$$

$$= \sum_{m}\sum_{r,s} D_{rs}^{m}\{\overset{j}{pq}|\overset{\star}{r}\,\overset{\star+m}{s}\}$$

It is striking to note that the terms between curly brackets $\{\overset{j}{pq}|\overset{\star}{r}\,\overset{\star+m}{s}\} = \sum_{h}(\overset{j}{pq}|\overset{h}{r}\,\overset{h+m}{s})$ do not evolve with iterations and can be computed and stored at once, reducing the number of stored terms by a factor of (2N+1) and minimizing I/O operations. Furthermore, the computation of $\{\overset{j}{pq}|\overset{\star}{r}\,\overset{\star+m}{s}\}$ is algorithmically easy due to the presence of many common terms since the charge distribution $P_{r\,s}^{h\,h+m}$ is invariant upon translation. Let us note however that such summations are divergent. Those divergence difficulties are easily removed by proper combinations of electron-nuclei and electron-electron terms as will be discussed in § 4 and 5.

The same methodology can be applied to the exchange interactions. The exchange part of a matrix element contains a permutation operator $\mathcal{P}rr'$ and is expressed in occurrence of the density $\rho(\underline{r},\underline{r}')$ previously defined. It can be shown to be :

$$E_{x_{pq}}^{j} = -\frac{1}{2}\iint dvdv'\chi_p(\underline{r})|\underline{r}-\underline{r}'|^{-1}\rho(\underline{r},\underline{r}')\mathcal{P}rr'\chi_q^{j}(\underline{r})$$

$$= -\frac{1}{2}\iint dvdv'\chi_p(\underline{r})|\underline{r}-\underline{r}'|^{-1}\rho(\underline{r},\underline{r}')\chi_q^{j}(\underline{r}')$$

$$= -\frac{1}{2}\sum_{m}\sum_{r,s} D_{rs}^{m}\sum_{h}(\overset{h}{p}r|\overset{j}{q}\,\overset{h+m}{s})$$

$$= -\frac{1}{2}\sum_{m}\sum_{r,s} D_{rs}^{m}\{\overset{\star}{p}r|\overset{j}{q}\,\overset{\star+m}{s}\} \qquad (34)$$

The same conclusions as for the Coulomb part hold for the exchange contributions, i.e. the storage must not be achieved at the level of individual $(\overset{h}{p}r|\overset{j}{q}\,\overset{h+m}{s})$ integrals but at the level of the translationally combined terms $\{\overset{\star}{p}r|\overset{j}{q}\,\overset{\star+m}{s}\}$ resulting in faster I/O operations both for the integral and SCF parts of the computer program and opening a way for fast optimized algorithms.

It is unclear why, up to now, little attention has been paid to such a formulation which, however, was already implicit in eq. 7 in a 1980 paper by Pisani and Dovesi (11).

5. A BRIEF ANALYSIS OF THE LATTICE SUM PROBLEMS

In order to discuss the long-range behaviour of electrostatic interactions, we rearrange eq. 12 into :

$$F_{pq}^{j} = T_{pq}^{j} \qquad \text{kinetic}$$

$$+ \sum_h \sum_l \sum_r \sum_s D_{rs}^{hl} \{ (\overset{j}{pq}|\overset{hl}{rs}) - n^{-1} S_{rs}^{hl} \sum_A Z_A V_{pq}^{j}|_A^h \}$$

Coulomb electron-electron + electron-nuclei

$$- \frac{1}{2} \sum_h \sum_l \sum_r \sum_s D_{rs}^{hl} (\overset{h}{pr}|\overset{jl}{qs}) \qquad \text{exchange} \tag{35}$$

In eq. 35, we have rescaled the nuclei-electron attraction term to introduce the same double summation over h and l as in electron-electron Coulomb terms by making use of the normalization-like condition

$$\sum_h \sum_r \sum_s D_{rs}^{hl} S_{rs}^{hl} = \sum_h \sum_r \sum_s D_{rs}^{Oh} S_{rs}^{Oh} = n \tag{36}$$

where n is the number of electron pairs per unit cell.

In the first implemented programs, much attention has not been given to the convergency properties of eq. 10 (or of eq. 35). Nobody anticipated for serious difficulties other than due to the nature and the length of atomic basis such as encountered in molecular quantum chemistry. From straight chemical intuition, it was usually believed that the range of effectively interacting cells would be very small though no consistent rule for the truncation of j, h and l summations could actually be supplied. These teneous arguments have much delayed a formal but necessary mathematical analysis on the convergence of these series to define better the numerical applicability of eqs. 10 or 35.

However, it must be pointed out that as early as 1956, Löwdin (10) has already stressed the formal difficulties which would arise in the numerical solutions of eq. 35. Furthermore, starting with the work by O'Shea and Santry in 1974 (12a) and Ukrainski in 1975 (12b), it became gradually apparent that model chains embody also size related difficulties mathematically expressed as conditionally (Coulombic interactions) and sometimes slowly (exchange contributions) convergent series (lattice sums). The basic implications were also fully appreciated since terms of the series involve multicenter integrals and straightforward summations are prohibitive (12c).

In the forthcoming discussion, we want to present a brief analysis of the problems raised by lattice summations, we limit our attention to Fock matrix elements since they incorporate simultaneously Coulomb and exchange lattice sums related difficulties ; the conclusion can be easily extended to total energy expressions.

To ease the analysis it is convenient to rearrange eq. 33 following the recipe given in § 3 eq. 35 :

$$F_{pq}^{\,j} = T_{pq}^{\,j}$$

$$+ \sum_r \sum_s \sum_l D_{rs}^{\,l} \sum_h \{(p\overset{j}{q}|\overset{h}{r}\overset{h+1}{s}) - (n)^{-1} S_{rs}^{\,l} \sum_A Z_A V_{pq}^{\,j}|_A^h\}$$

$$- \frac{1}{2} \sum_r \sum_s \sum_h \sum_l D_{rs}^{hl} (p\overset{h}{r}|\overset{j}{q}\overset{l}{s}) \qquad (37)$$

In order to characterize the behavior of the series over j, h and l, we need first to analyze the decay with respect to the lattice indices of $(p\overset{j}{q}|\overset{h}{r}\overset{h+1}{s})$, $(n^{-1})S_{rs}^{l}Z_A V_{pq}^{j}|_A^h$ and $(p\overset{h}{r}|\overset{j}{q}\overset{l}{s})$ integrals. They all correspond to electrostatic interactions between charge distribution represented by pair-products of atomic functions, respectively :

$$P_{pq}^{\,j} = \chi_p(\underline{r})\chi_q^j(\underline{r}) \quad \text{with} \quad P_{rs}^{h\,h+1} = \chi_r^h(\underline{r})\chi_s^{h+1}(\underline{r})$$

$$P_{pq}^{\,j} = \chi_p(\underline{r})\chi_q^j(\underline{r}) \quad \text{with rescaled nuclear attraction} \quad (2n)^{-1} S_{rs}^{\,l} Z_A$$

$$P_{pr}^{\,h} = \chi_p(\underline{r})\chi_r^h(\underline{r}) \quad \text{with} \quad P_{qs}^{jl} = \chi_q^j(\underline{r})\chi_s^l(\underline{r})$$

Since the majority of calculations is made using gaussian atomic basis sets, we will refer hereinafter to their nice mathematical features without loss of generality in the argumentation.

The value of the above integrals is proportional to overlap integrals

$$S_{pq}^{\,j}\, S_{rs}^{\,l} \quad \text{for} \quad (p\overset{j}{q}|\overset{h}{r}\overset{h+1}{s})$$

$$S_{pq}^{\,j} \quad \text{for} \quad V_{pq}^{\,j}|_A^h$$

and

$$S_{pr}^{\,l}\, S_{qs}^{\,l-h} \quad \text{for} \quad (p\overset{h}{r}|\overset{j}{q}\overset{l}{s})$$

The overlap integrals, in turn, fall off like exponentials of the distance from the center of the participating atomic functions, the actual rate of decay being fixed by the exponent and nature (s,p,d,...) of the functions. The superscripts j,h,l on these integrals measure the interdistance (in integer numbers of the 1D lattice parameter a) between the atomic functions which are reasonably localized in the direct space. Thus, one can anticipate a fast decay of these integrals and a small interaction range for cells. This is indeed the case <u>except</u> when the same lattice index occurs on both terms of a product of atomic functions, e.g. $\chi_r^h \chi_s^{h+j}$ (or $\chi_r^{h+j} \chi_q^j$). In such an event, the asymptotic decay with respect to h (and j) is not exponential but like $a^{-1} S_{pq}^j S_{rs}^l |h + \gamma_{pqrs}^{jl}|^{-1}$ (and $a^{-1} S_{ps}^l S_{rq}^h |j+\gamma_{psrq}^{lh}|^{-1}$). A similar behaviour is found for the nuclear attraction term $n^{-1} S_{rs}^l Z_A V_{pq}^j|_A^h$ whose asymptotic decay is then $a^{-1} n^{-1} S_{pq}^j S_{rq}^h Z_A |h+\gamma_{pq}^j(A)|^{-1}$. From eq.(37) it is easily observed that electron-electron repulsion, electron-nuclear attraction and exchange integrals all embody such a behaviour, the first two with respect to h and the last one with respect to j. In the following, we analyze separately Coulombic terms and exchange terms since the origin and the mathematical characterization are different.

5.1. Coulombic Contributions.

From the asymptotic regime already identified, namely $a^{-1} S_{pq}^j S_{rs}^l |h+\gamma_{pqrs}^{jl}|^{-1}$ for $(pq|{}_{r\,s}^{h\,h+l})$ and $a^{-1} n^{-1} S_{pq}^j S_{rs}^l Z_A |h+\gamma_{pq}^j(A)|^{-1}$, individual series over such terms behave like the logarithmically divergent harmonic series of exponent one. A finite result can be obtained when the individual contributions are grouped, as suggested by the curly brackets in eq.(37), under the electroneutrality conditions, $n = \sum_A Z_A$. Then content of the curly brackets is smaller than a convergent majorant with positive terms,

$$a^{-1} S_{pq}^{j} S_{rs}^{l} \left| |h+\gamma_{pqrs}^{jl}|^{-1} - n^{-1} \sum_A Z_A |h+\gamma_{pq}^{j}(A)|^{-1} \right|$$

$$< a^{-1} S_{pq}^{j} S_{rs}^{l} \left| |h+\gamma_{pqrs}^{jl}|^{-1} - |h+\gamma_{pq}^{*j}|^{-1} \right| < K h^{-2}, \; h \neq 0 \qquad (38)$$

where $n^{-1} \sum_A Z_A = 1$ and γ_{pq}^{*j} is such that the above inequality is satisfied. Such series are also known as the Madelung conditionally convergent lattice sums. As observed from eq. (37) the final rate of convergence with respect to h is very low, $\sim |h|^{-2}$. The magnitude of the contributions is roughly proportional to the difference between γ_{pqrs}^{jl} and γ_{pq}^{*j} which, in a way, is related to the "separation of charges" or the polarity of the system.

5.2. Exchange Contributions

An important and recent analysis of exchange terms is given independently by Delhalle and Harris (13). In this paper we summarize the part of this paper which is relevant to our discussion.

The exchange contribution

$$-\frac{1}{2}\sum_{h}\sum_{l}\sum_{r}\sum_{s} D_{rs}^{l-h}(\overset{h}{pr}|\overset{jl}{qs}) \tag{39}$$

corrects for the self-electron repulsion in the Coulomb part of the SCF one-electron operator and also includes the effect of the Pauli principle on the independent electron model. Clearly, the magnitude of the integrals $(\overset{h}{pr}|\overset{jl}{qs})$ is determined by the distribution P_{pr}^{h} and P_{qs}^{jl} and the distance between their barycenters. The terms in h and l summations are very important for h near 0 and for l near j

$$\simeq -\frac{1}{2}\sum_{r}\sum_{s}\{D_{rs}^{j}(\overset{0}{pr}|\overset{jj}{qs})$$

$$+ D_{rs}^{j-1}(\overset{1}{pr}|\overset{jj}{qs}) + D_{rs}^{j+1}(\overset{\bar{1}}{pr}|\overset{jj}{qs}) + \ldots$$

$$+ D_{rs}^{j+1}(\overset{0}{pr}|\overset{jj+1}{qs}) + D_{rs}^{j-1}(\overset{0}{pr}|\overset{jj-1}{qs}) + \ldots$$

$$\simeq \frac{1}{2} D_{pq}^{j}(\overset{0}{pp}|\overset{jj}{qq}) \tag{40}$$

It is thus to point out that exchange contributions can be important for those Fock matrix elements between widely separated orbitals since the decrease of both two-center repulsion integrals $(\overset{0}{pp}|\overset{jj}{qq})$ and elements of density matrices D_{pq}^{j} is very slow whith distance and is certainly not an exponential one.

Up to rather recently, little had been undertaken to characterize the convergence of D_{pq}^{j} with respect to j. However, troublesome numerical results have boosted such an analysis.

From few existing studies, mainly based on simplified methods such as the Hückel one, it was known that D_{pq}^{j} can decay as slowly as $(-1)^{j}|j|^{-1}$ leading to a $(-1)^{j}|j|^{-2}$ asymptotic decay of the exchange terms as first reported by Ukrainski in a semi-empirical work on a metallic polyacetylene model (12b). In a detailed analysis of ab initio results on a hydrogen chain model, the Namur group gained insight into the exchange behavior (14).

They reached the conclusions that poor convergence of the exchange energy is connected to the HOMO-LUMO separation (energy gap). In cases where the Hartree-Fock scheme produces good quality results (saturated systems, insulators, ...), they observe that the exchange potential is of short range. By contrast, in metallic situations, essentially characterized by degeneracy or near degeneracy of HOMO and LUMO levels the exchange potential is of long range. However, let us note incidentally that this situation precisely corresponds to the case where a single-determinant approximation to the ground-state wave function does not even provide a qualitatively correct description of the system.

Those results were further developed by a study due to Monkhorst and Kertesz (15) ; using a two-band model system within Hückel methodology they recovered the $(-1)|j|^{-2}$ asymptotic decay of exchange in a half-filled band (metallic) situation of this model and established a simple relationship between band gap, band width and an (exponential) large-j behavior of the density matrix elements in the filled band (insulating) situation.

More recently, as quoted previously, Delhalle and Harris (13) have provided a general analysis of the convergence of D_{pq}^{j} in terms of basic theorems on the convergence of Fourier series coefficients. They show that the convergence of D_{pq}^{j} is essentially determined by the analytic properties of $D_{pq}(k)$

$$D_{pq}(k) = \sum_{j} D_{pq}^{j} e^{ikja} \tag{41}$$

throughout the first Brillouin zone. In general, it can be stated that to typical insulators with large band gaps will correspond quickly decaying density matrix elements, D_{pq}^{j}, and, at the other extreme, metallic systems will lead to poorly converging D_{pq}^{j}.

These findings have an immediate return on the computational performances of the direct space approach. As we have previously shown, the decay of F_{pq}^{j} is not only affected by the localized nature of the involved basis set but also, through its exchange part, by the convergence of D_{pq}^{j}.

6. MULTIPOLE EXPANSION FOR LONG-RANGE COULOMB INTERACTIONS

As noted previously, in actual polymer calculations, it is obviously impossible to deal with very large values of the number of unit cells since the two electron part of an LCAO-CO calculation involves an enormous number of integrals. There is a need for limiting that number of unit cells to some amenable value N (empirically ranging from 1 -nearest neighbour approximation- to 5). Long-range interactions behave like conditionally and slowly convergent series and actually are significantly contributing far beyond this number N. An analysis to this problem has solved this question by multipole expansion (6). F_{pq}^{j} (eq. 37) is cast in the form (12c) :

$$F_{pq}^{j} = T_{pq}^{j} + \sum_{l=-N}^{N} \sum_{r,s} D_{rs}^{l} \sum_{h=-N}^{+N} [(pq|rs)^{j\,h\,h+l} - (n)^{-1} S_{rs}^{l} \sum_{A} Z_A V_{pq}^{j}|_A^h]$$

$$+ L_{pq}^{j}(N) - \frac{1}{2} \sum_{h}\sum_{l}\sum_{r}\sum_{s} D_{rs}^{l-h} (pr|qs)^{1\,jl} \qquad (42)$$

with

$$L_{pq}^{j}(N) = \Big(\sum_{h=-\infty}^{\infty} - \sum_{h=-N}^{+N} \Big)\Big[-\sum_{A} Z_A V_{pq}^{j}|_A^h + \sum_{l=h-N}^{h+N} \sum_{r,s} D_{rs}^{hl} (pq|rs)^{j\,h\,l} \Big] \qquad (43)$$

The detailed deduction of the multipole expansion gives

$$L_{pq}^{j}(N) = \sum_{k=0}^{\infty} \sum_{l=0}^{\infty} U_{pq}^{j(k,l)} a^{-(k+l+1)} \Delta_N^{(k+l+1)} \qquad (44)$$

where

$$\Delta_N^{(n)} = \sum_{h=1}^{\infty} h^{-n} - \sum_{h=1}^{N} h^{-n} = \zeta(n) - \sum_{h=1}^{N} h^{-n}, \qquad (45)$$

$\zeta(n)$ is the Riemann zeta function, $U_{pq}^{j(k,l)}$ is the interaction of the 2^k-pole and 2^l-pole moments,

$$U_{pq}^{j(k,l)} = \sum_{m=-s(k,l)}^{s(k,l)} (k+l)!(-1)^m[(-1)^k+(-1)^l].$$

$$\cdot[(k+|m|)!(l+|m|)!]^{-1} M_{pq}^{j(k,m)} M^{(l,m)^{*}}. \qquad (46)$$

In the above expressions, s is equal to the smaller of numbers k and l. Capital letter $M^{(l,m)}$ refers to the mth component of the 2^lth electric moment expressed in spherical coordinates and related to the charge distributions associated with the orbital product $P_{pq}^{j}(\underline{r})$:

$$M_{pq}^{j(k,m)} = \langle \chi_p(\underline{r}) | r^k P_k^{|m|} (\cos\theta) e^{im\phi} | \chi_q^j(\underline{r}) \rangle$$

or to the total charge (electrons + nuclei) associated with each translational unit :

$$M^{(k,m)} = - \sum_A Z_A r_A^k P_k^{|m|}(\cos\theta_A) e^{im\phi_A}$$

$$+ \sum_{j=-N}^{N} \sum_{p,q} D_{pq}^{j} \langle \chi_p(\underline{r}) | r^k P_k^{|m|}(\cos\theta) e^{im\phi} | \chi_q^j(\underline{r}) \rangle \qquad (47)$$

Because of electroneutrality constraints, $U_{pq}^{j(k,o)} = 0$ and to avoid coordinate dependence of the results, truncation is performed to sum only $U_{pq}^{j(k,1)}$ terms of identical k+1+1 values. We obtain the working formula :

$$LR_{pq}^{j}(N) = \sum_{\substack{k=3,5,.. \\ k=odd}} \Delta_N^{(k)} a^{-k} \sum_{1=1}^{k-1} U_{pq}^{j(k-1-1,1)} \qquad (48)$$

where a selective control can be made by the N and k parameters.

7. USE OF SYMMETRY IN POLYMER CALCULATIONS

In most theoretical investigations a stereoregular polymer is described by an infinite, extended and isolated chain constructed from a periodic sequence of monomer units. In addition to translational symmetry, stereoregular polymers possess some other symmetry elements like screw axes, mirror, or glide planes. The related operations combine into groups, the line groups (16a-c).

In spite of the usefulness of group theory in reducing the computational labor and providing a better understanding of the physical reality, polymer quantum calculations rarely have made use of more than translational periodicity. One should mention precursors who have gone beyond translational symmetry: McCubbin (16d), Blumen and Merkel (16e). The latters have succeeded in implanting cyclic screw axis symmetry into an ab initio program.

At about the same time, Bozovic and his colleagues from Belgrade have undertaken a systematic and rather extensive work on the line groups for which they gave a compact derivation (16c). By now all the necessary irreducible representations are obtained (16f). In the next future a posteriori analyses (16g) and uses of symmetry adapted functions (16h) based on line groups are expected to develop. Computational effort will be reduced by minor adaptations since the integrals to be calculated occur to be combinations of the same ones found in programs using translational symmetry alone (16e,h). In addition the data output will be substantially improved since one-electron energies will come out symmetry assigned,thus allowing for use of connectivity properties (16g) and important conclusions will be drawn on the states and processes.

8. CONCLUSIONS

This work demonstrates that specific algorithms are now available for efficient ab initio computations of electronic properties of stereoregular polymers.

Attention has been paid to :

i) Specific formulas for electron densities, Coulomb and exchange matrix elements,fully taking into account the polymer translational symmetry ;

ii) Well-balanced evaluation of long-range electron-electron and electron-nuclei electrostatic interactions ;

iii) Importance of developing programs with line group symmetry adapted functions.

Those aspects have been implemented into a specific modular program. Versions will be made available for IBM and Digital DEC 20 computers.

ACKNOWLEDGEMENTS

It is a pleasure to thank the National Foundation for Cancer Research and the NATO Scientific Affairs Division for their partial support of this work. We wish in addition to thank IBM-Poughkeepsie and IBM-Belgium for making possible the stay of J.M. André and the Belgian National Science Foundation (FNRS) for a fellowship to J.L. Brédas. J.M. André is very indebted for fruitful and illuminating discussions with Dr. E. Clementi.

REFERENCES

[a]Permanent address : Laboratoire de Chimie Théorique Appliquée Facultés Universitaires N.D. de la Paix, Namur (Belgium).
[b]Chercheur qualifié du Fonds National Belge de la Recherche Scientifique (FNRS).

1.a. André,J.-M., Gouverneur, L. and Leroy, G. 1967, Int. J. Quantum Chem. 1, 427.
b. André, J.-M., Gouverneur, L. and Leroy, G. 1967, Int. J. Quantum Chem. 1, 451.

c. Del Re, G., Ladik, J. and Biczó, G. 1967, Phys. Rev. 155, 997.
d. André, J.-M. 1969, J. Chem. Phys. 50, 1536.
2. see, for instance :
a. André, J.-M. 1970, Comput. Phys. Commun. 1, 391.
b. Kertesz, M., 1976, Acta Phys. Acad. Sci. Hung. 41, 127.
c. Suhai, S. and Ladik, J. 1977, Solid State Commun. 22, 227.
d. Otto, P., Clementi, E. and Ladik, J. 1982, IBM Technical Report POK-13..
3. Karpfen, A. 1981, Int. J. Quantum Chem. 19, 1207.
4. Kertesz, M. 1982, Advan. Quantum Chem. 15, 161.
5.a. Brédas, J.L. 1979, Ph.D. Thesis, Namur.
b. Brédas, J.L., André, J.M., Fripiat, J.G. and Delhalle, J. 1978, Gazz. Chim. Ital. 108, 307.
c. Brédas, J.L., André, J.M. and Delhalle, J. 1980, Chem. Phys. 45, 109.
6.a. Piela, L. and Delhalle, J. 1978, Int. J. Quantum Chem. 13, 605.
b. Piela, L. and Delhalle, J. 1978, Ann. Soc. Sci. Bruxelles 92, 42.
c. Piela, L. in "Recent Advances in the Quantum Theory of Polymers", André, J.M. et al., Eds., Lecture Notes in Physics, vol. 113, (Springer Verlag, Berlin, 1980), pp 104 et sq.
d. Delhalle, J., Piela, L., Brédas, J.L. and André, J.M. 1980, Phys. Rev. B22, 6254.
7.a. Karpfen, A. and Schuster, P. 1976, Chem. Phys. Lett. 44, 459.
b. Karpfen, A. 1978, Theoret. Chim. Acta 50, 49.
8. Beyer, A. and Karpfen, A., 1982, Chem. Phys. 64, 343.
9. Suhai, S. 1980, J. Chem. Phys. 73, 3843.
10. Löwdin, P.O. 1956, Advan. Phys. 5, 1.
11.a. Pisani, C. and Dovesi, R. 1980, Int. J. Quantum Chem. 17, 501.
b. Dovesi, R., Pisani, C. and Roetti, C. 1980, Int. J. Quantum Chem. 17, 517.
12.a. O'Shea, F. and Santry, D.P. 1974, Chem. Phys. Lett. 25, 164.
b. Ukrainski, I.I. 1975, Theoret. Chim. Acta. 38, 139.
c. Delhalle, J., André, J.M., Demanet, Ch. and Brédas, J.L. 1978, Chem. Phys. Lett. 54, 186.
13. Delhalle, J. and Harris, F.E., to be published.
14.a. Piela, L., André, J.M., Fripiat, J.G. and Delhalle, J. 1981, Chem. Phys. Lett. 77, 143.
b. André, J.M., Brédas, J.L., Thémans, B. and Piela, L. 1983, Int. J. Quantum Chem. 23, 1065.
15. Monkhorst, H.J. and Kertesz, M. 1981, Phys. Rev. B24, 3015.
16.a. Hermann, C. 1928, Z. Kristallogr. 69, 250.
b. Alexander, E. 1929, Z. Kristallogr. 70, 367.
c. Vujicić, M., Bozović, I.B. and Herbut, F. 1977, J. Phys. A10, 1271.
d. McCubbin, W.L. 1966, Phys. Status Solidi 16, 289.
e. Blumen, A. and Merkel, C. 1977, Phys. Status Solidi B83, 425.

f. Bozović, I.B., Vujicić, M. and Herbut, F. 1978, J. Phys. A 11, 2133.
g. Bozović, I.B., Delhalle, J. and Damnjanović, M. 1981, Int. J. Quantum Chem. 20, 1143.
h. Bozović, I.B. and Delhalle, J. to be published.

THEORETICAL INTERPRETATION OF VALENCE XPS SPECTRA OF STEREO-REGULAR ORGANIC POLYMERS

J. Delhalle and J.-M. André

Laboratoire de Chimie Théorique Appliquée
Facultés Universitaires Notre-Dame de la Paix
Rue de Bruxelles, 61 B-5000 Namur (Belgium).

1. INTRODUCTION

The basic theoretical framework for quantum mechanical calculations of chain-like systems was largely established by the end of the sixties [1-3], yet the first calculations did not attract much interest from the concerned scientists. Attention of solid state physicists on electronic properties was almost exclusively centered on elemental inorganic materials, and polymer chemists, providers of the materials, expressed strong reservations as to the value of the model used to calculate electronic properties of structurally so complex systems.

To some extent this reluctance was prejudiced; the isolated and regular model chain used in electronic structure calculations is inherently the same as in vibrational spectroscopy calculations which have proved so worthy to assign most, if not all, spectral features of polymers[4]. It became essential to find points of comparison between polymer quantum calculations and experiment to assess the value of the formers.

At about the same period, X-ray photoelectron spectroscopy (XPS) was emerging as a major experimental technique by which to assess the electronic structure of various systems, including polymers. Successful comparisons would have not only served as checks on the calculated data with respect to X-ray photoelectron spectra, but also provided some confidence in the future development of the nascent polymer quantum chemistry.

The purpose of this lecture is to describe a specific

J. Ladik et al. (eds.), Quantum Chemistry of Polymers - Solid State Aspects, 23–31.

interplay, chronologically the first of that kind, between quantum mechanical calculations on model organic polymers and measurements of valence X-ray photoelectron spectra of their existing counterparts[5]. We hope thereby to show the relevance of polymer quantum chemistry in interpreting and predicting shapes of XPS spectra in terms of polymer architecture. This contribution is by no means a review of the field, and moreover only those systems on which a personal experience has been gained will serve as illustration in the sequel.

2. X-RAY PHOTOELECTRON SPECTROSCOPY

In X-ray photoelectric experiments on polymers, a monochromatic radiation of frequency ν is used to ionize a solid sample held in a vacuum chamber. Typical indicent photon energy is of the order of 1.5 keV and the vacuum has to be better than 10^{-8} Torr. In the process an outgoing electron carries away a positive kinetic energy, E_{kin}, which is the balance between the initial, (i), and the final, (f), state energies, E_i and E_f, of the system

$$E_i + h\nu = E_f + E_{kin}.$$

The measured kinetic energy can then be related to the binding energy, E_b, of the electrons inside the sample,

$$E_{kin} = h\nu - (E_f - E_i) = h\nu - E_b ,$$

and therefore XPS provides an experimentally based description of the electronic structure of the sample.

The 1.5 keV photon sources have sufficient energy to excite both core and valence electrons of most organic polymers. By far most important applications of XPS are related to the recording and study of core levels. On a wide scan XPS spectrum, the unscattered electrons result in characteristic peaks whose energy and intensity ratios serve to identify the elements in the material, to characterize the molecular environment, and, in some cases, to provide estimates of the relative abundance of species. A good example of a wide scan XPS spectrum of an alternating ethylene-tetrafluroethylene copolymer can be found in Fig. 2 of reference 6.

Hereafter we concentrate on valence bands. Most responsive to bonding, the valence levels are expected to disclose interesting information. Unfortunately , up until recently, they did not receive much attention; two closely connected difficulties are at the basis of such a delay (or prudence) :

a. on the one hand, the structure of a polymer is usually

complex (a detailed characterization remains one of the most important challenges of today's polymer science) which makes problematic the choice of a suitable geometrical model to be used in the theoretical calculations.

b. on the other hand, the valence band is more difficult to record and analyze than core levels, and, when it is available its interpretation requires theoretical support.

The spectra used in the illustrations have been obtained by the Namur group of electron spectroscopy, reference 7 is a convenient source of illustrative spectra. Few remarks should be added about the measurements :

a. the spectra have been obtained on a Hewlett Packard spectrometer (HP 5950 A) using a monochromatized AlK_{α} radiation (1486.6 eV).

b. valence bands represent a small portion of a photoelectron spectrum, and, due to low photoelectric cross-sections of valence electrons, the recording of one valence spectrum typically requires 12 hours.

c. charging effects are controlled by the use of an electron flood gun.

d. most polymers are studied as thin films.

e. calibrations of the recorded spectra are usually made by mixing polymers with a reference compound of similar electric properties. Precision is usually of the order of 0.1 to 0.2 eV.

3. XPS AND BAND STRUCTURES

As already stated in the previous section, XPS studies of polymers are made on solid samples. On the basis of the high energy of incident photons the outgoing electrons can tentatively be viewed as coming from the bulk, i.e. from deeper than the first five top layers. Accordingly XPS spectra can be rationalized in terms of the three-step photoemission model

a. excitation from an initial state (i) to a final state (f)

b. transfer of the electron to the surface

c. escape from the surface

Hereafter we will restrict our attention to step 1 and consequently assume that the shape of the spectra is not essentially dependent on steps 2 and 3. This is obviously a simplification that we have chosen to live with in our present studies mainly because there is no easy way, at least to our knowledge, to include in a practical way steps 2 and 3 in the calculations.

The theoretical interpretation will be based on band theory results where electronic states "extend" over the length of the polymer molecule. In such a case the removal of an electron should have little effect on the energy level structure and Koopmans' theorem should be a valid approximation to account for the photoemission process involved in step 1. More precisely step 1 of the photoionization corresponds to an excitation between two states with the same k-point in the reduced Brillouin zone of respective energies $\varepsilon_i(k)$ and $\varepsilon_f(k)$. In a photoelectric experiment the number of photoelectrons with a particular energy is measured, this is represented by the energy distribution of the joint density of states which is expected to contain much of the same physical information as experiment does.

In the case of chains it amounts to a one-dimensional integration over the Brillouin-zone (BZ) of a rather complicated integrand such as found in the theoretical evaluation of optical spectra, namely

$$J(E) \sim \sum_i^{occ} \int_{BZ} \delta[\varepsilon_f(k) - \varepsilon_i(k) - h\nu]\,\delta[E - \varepsilon_i(k)]\,P_{fi}^{\nu}(k)\,dk$$

where the summation extends over all occupied states, $\varepsilon_f(k)-\varepsilon_i(k)-h\nu = 0$ defines the range of allowed excitations and $P_{fi}^{\nu}(k)$ is the probability for the ejection of an electron from state $|k_i\rangle$ to state $|k_f\rangle$ by a photon of energy $h\nu$. Notice that J(E) differs from $\mathcal{D}(E)$, the density of occupied states, by the factors $P_{fi}^{\nu}(k)$ and $\delta[\varepsilon_f(k) - \varepsilon_i(k) - h\nu]$,

$$\mathcal{D}(E) \sim \sum_i^{occ} \int_{BZ} \delta[E - \varepsilon_i(k)]\,dk.$$

$\mathcal{D}(E)$ corresponds to the number of occupied states per unit energy range. Only under conditions of an unstructured continuum, constant probability and constant transport and escape, would a photoelectron spectrum be a direct measure of the density of states, $\mathcal{D}(E)$.

Even though there are evidences that in some systems high photon energies photoelectron spectra direcly reflect the main features of $\mathcal{D}(E)$[8], we have found that transition probability,

$P^{\nu}_{fi}(k)$, usually cannot be eliminated in polymers [9,10]. $P^{\nu}_{fi}(k)$ is a rather difficult quantity to evaluate rigorously, and our approach will be to rely on the intensity model proposed by Gelius[11]. This model involves Mulliken populations and relative photoionization cross-sections, σ_p's, to approximate $P^{\nu}_{fi}(k)$:

$$P^{\nu}_{fi}(k) \sim \sum_{p} C^{*}_{np}(k)\left[\sum_{q} S_{pq}(k)\; C_{np}(k)\right]\sigma_p .$$

The σ_p's are the relative photoionization cross-sections of a particular atomic subshell, they have been obtained from a fitting procedure on reference systems.

In practice, once the technical but difficult problem of connecting discrete one-electron energies, $\varepsilon_n(k_i)$ into continuous bands has been solved[12], J(E) is approximated by a histogram[13], $\bar{I}(E_n)$,

$$\bar{I}(E_n) \sim \Delta E^{-1} \sum_{i}^{occ} \int_{E_n - \Delta E/2}^{E_n + \Delta E/2} \delta[E - \varepsilon_i(k)]\, P^{\nu}_{fi}(k)\,dk$$

It is generally accepted that for solid state measurements, the apparent width of the photoelectron peaks ranges from 0.6 to 0.9 eV. Also the experimental resolution function is reasonably well reproduced by a Gaussian of full width, Γ, at half-maximum of 0.7 eV (Γ = 2.345σ). Consequently the final "theoretical spectrum" based on a strict one-electron picture, with no account of transport to and escape from the surface, is obtained according to[13]

$$I(E_n) \sim \int_{-\infty}^{\infty} \bar{I}(E)\, \exp\left[-(E-E_n)^2/2\sigma^2\right] dE$$

In spite of all the simplifications which have led to the above theoretical simulation of XPS spectrum, encouraging results have been obtained as briefly discussed in the next section.

4. COMPARISON BETWEEN THEORY AND EXPERIMENT

In this section we review some of our joint works with the XPS group of Namur. We report (a) on the first confrontations made on polyethylene between band structure calculations and XPS measurements, then (b) consider some effects due to changes in the primary structure of the chain (substitution, structural isomerism), and (c) present attempts at detecting more subtle effects on the shape of XPS spectra (conformation).

4.1. Early comparisons

Among classical polymers, polyethylene is not only one of the simplest from the viewpoint of its monomeric unit but has a reasonably well characterized structure[14,15]. Therefore it serves as a reference in many theoretical calculations. The first comparison of an XPS spectrum of a large paraffin molecule, C_nH_{2n+2}, with a band structure goes back to 1972 with the work of Wood et al.[16]. These authors compared a CNDO/2 band structure of an all-trans model polyethylene to the valence spectrum of $C_{36}H_{74}$ relating the observed peaks to those energy regions where the first derivative of energy bands are small. This task became more convenient with density of states calculations[17] but inclusion of photoionization cross-sections together with convolution by a suitable experimental resolution function was necessary to make appropriate comparisons[9]. The calculated spectrum based on a Floating Spherical Gaussian Orbital band structure of an all-trans model polyethylene (Fig. 3 of reference 18) is a typical example of the kind of agreement that can be reached between theory and experiment. Due to the difficult problem of assigning an absolute energy to XPS spectra of insulators (calibration) the theoretical and experimental spectra are usually superimposed by bringing into coincidence a well identified peak. In the case of the FSGO calculation reported above, the most intense C-C peak was chosen for that purpose. The overall agreement is satisfactory, especially at large binding energies. The low intensity region is more problematic from the experimental side since it comprises those states where the C-H orbitals are dominant and characterized by low probabilities of photoionization.

Such successful comparisons between theory and experiment in the case of polyethylene raised questions as to how sensitive a valence XPS spectrum could be on the structure of polymers.

4.2. Changes in primary structure

A first attempt toward this aim came with a theoretical work, based on extended Hückel calculations[19], considering the response of the electronic levels to a progressive substitution of hydrogens in an all-trans polyethylene. The series of linear fluoropolymers ($-CH_2-CH_2-$, $-CHF-CH_2-$, $-CF_2-CH_2-$, $-CHF-CHF-$, $-CHF-CFH-$, $-CF_2-CHF-$, $-CF_2-CF_2-$) so obtained was artificially considered in a zig-zag planar conformation. Theoretical conclusions were that calculated changes should be observable in XPS measurements as it was confirmed soon after[20,10]. A more systematic study of polymer primary structures by XPS and theoretical calculations have then been undertaken[21] confirming both the observability of substitutional effects by milder substituents and the usefulness of theoretical calculations in this connection.

In 1976, Clark et al. came out with interesting XPS measurements on a series of polybutylacrylates, $-CH_2-CH(CO-O-R)-$, with $R = C_4H_9(t)$, $C_4H_9(i)$ and $C_4H_9(n)$ emphasizing the usefulness of valence lines in distinguishing structurally isomeric polymer systems for which core level spectra appear identical. Going deeper into this matter and using once again the simple extended Hückel technique we could predict important differences in the energy level distribution of three model compounds, $-CH_2-CH(OCH_3)$, $-(CH_2)_3-O-$ and $-CH_2-CH(CH_3)-O-$. The differences were such that each compound could be unambiguously identified from their valence XPS spectrum[5,22]. Again this was confirmed experimentally[23,7]. These results added to the potential of XPS as a tool for relating geometrical characteristics of polymers to their electronic structure.

4.3. Conformational changes

Among these successful results one compound, isotactic polypropylene, did not compare well with experiment[21]. In fact we had naively assumed a zig-zag planar backbone for this compound. In 1975 theoretical calculations[24] on four conformations of polyethylene, T,G, TG and TGTG' revealed differences in the shape of their density of states that should be observable through XPS measurements provided these conformations could actually be isolated, which is not the case. However it was an incentive to reconsider this previous calculation on polypropylene erroneously assumed in an all-trans conformation. As a result of this simple modification in the model geometry, $2/3_1$-helix, the theoretical spectrum came in close agreement with experiment[25]. This suggested also that very mild structural changes can influence the shape of the valence XPS spectra and thereby become observable.

5. CONCLUSIONS

It should be stressed that these investigations, the first of their kind, are at an early stage. Undoubtedly they will be superseded by more definitive works. A tentative project along these lines was already outlined in reference 5 and only few attempts have been made so far[7]. Investigations are now resuming in our group to assess in a more definite manner the potential of photoemission experiments to relate polymer structure to the distribution of electronic levels.

ACKNOWLEDGMENT

The authors express their thanks to their friends and collaborators, J.J. Pireaux, J. Riga and J. Verbist from the Laboratoire de Spectroscopie Electronique.

REFERENCES

1. S. Yomosa, J. Phys. Soc. Japan, 19, 1718 (1964)
2. G. Del Re, J. Ladik, G. Biczo, Phys. Rev., 155, 997 (1967)
3. J.M. André, L. Gouverneur, G. Leroy, Internat. J. Quantum Chem., 1, 427 (1967)
4. B. Fanconi, Ann. Rev. Phys. Chem., 31, 265 (1980)
5. J. Delhalle, in "Recent Advances in the Quantum Theory of Polymers" J.M. André et al. eds, Lecture Notes in Physics, 113 (Springer-Verlag, Berlin, 1980) p. 255
6. J.J. Pireaux, J. Riga, R. Caudano, J. Verbist, Y. Gobillon, J. Delhalle, S. Delhalle, J.M. André, J. Polym. Sci. : Polym. Chem. Ed., 17, 1175 (1979)
7. J.J. Pireaux, J. Riga, R. Caudano, J.J. Verbist, in "Photon, Electron and Ion Probes of Polymer Structure and Properties" D.W. Dwight, T.J. Fabish, H.R. Thomas eds., (ACS Symposium Series, Washington D.C., 1981), 162, 169-202
8. D.A. Shirley, Phys. Rev., B5, 4709 (1972)
9. J. Delhalle, S. Delhalle, J.M. André, Chem. Phys. Lett., 34, 430 (1975)
10. J. Delhalle, S. Delhalle, J.M. André, J.J. Pireaux, J. Riga, R. Caudano, J.J. Verbist, J. Electron Spectr., 12, 293 (1977)
11. U. Gelius, in "Electron Spectroscopy", D.A. Shirley ed. (North-Holland, Amsterdam, 1972) p. 311
12. J. Delhalle, D. Thelen, J.M. André, Computers and Chem., 3, 1 (1979)
13. J. Delhalle, S. Delhalle, Internat. J. Quantum Chem., 11 , 349 (1977)
14. S. Kavesh, J.M. Schultz, J. Polym. Sci. A28, 243 (1970)
15. G. Avitabile, R. Napolitano, B. Pirozzi, K.D. Rouse, M.W. Thomas, B.T.M. Willis, J. Polym. Sci. Polym. Letters Ed., 13, 351 (1975)
16. M.H. Wood, M. Barber, I.H. Hillier, J.M. Thomas, J. Chem. Phys., 56, 1788 (1972)
17. J. Delhalle, J.M. André, S. Delhalle, J.J. Pireaux, R. Caudano, J.J. Verbist, J. Chem. Phys., 60, 595 (1974)
18. J.L. Brédas, J.M. André, J. Delhalle, Chem. Physics, 45, 109 (1980)
19. J. Delhalle, Chem. Phys., 5, 306 (1974)
20. J.J. Pireaux, J. Riga, R. Caudano, J.J. Verbist, J.M. André, J. Delhalle, S. Delhalle, J. Electron Spectr., 5, 531 (1974)
21. J.J. Pireaux, J. Riga, R. Caudano, J.J. Verbist, J. Delhalle, S. Delhalle, J.M. André, Y. Gobillon, Physica Scripta, 16, 329 (1977)

22. C. Puissant, "Etude théorique de polymères contenant un atome d'oxygène", Mémoire de licence, Facultés Notre-Dame de la Paix, (Namur, Belgium), 1978
23. C. Sevrin, "Etude par ESCA de polyéthers et de polycarbonates", Mémoire de licence, Facultés Notre-Dame de la Paix, (Namur, Belgium), 1979
24. S. Delhalle, J. Delhalle, C. Demanet, J.M. André, Bull. Soc. Chim. Belges, 84, 1071 (1975)
25. J. Delhalle, R. Montigny, C. Demanet, J.M. André, Theoret. Chim. Acta, 50, 343 (1979)

43. G. [illegible], "Étude [illegible] de [illegible]", Mémoire de licence, Facultés Notre-Dame de la Paix, Namur, Belgique, [illegible].

44. [illegible], Étude par ESR de polyéthers et de polycarbonates, Mémoire de licence, Facultés Notre-Dame de la Paix, Namur, Belgique, [illegible].

45. [illegible]

46. [illegible]

AB INITIO STUDIES ON THE STRUCTURE AND PHONON SPECTRA OF SIMPLE POLYMERS

Alfred Karpfen

Institut für Theoretische Chemie und
Strahlenchemie der Universität Wien
A-1090 Wien, Währingerstr.17, Austria

1. INTRODUCTION

The determination of the structure of molecules and solids is one of the most important topics of both theoretical chemistry and physics and of experimental research in the various fields of spectroscopy. Knowledge of the specific arrangement of atoms in stable molecular species and in condensed matter and the development of concepts which explain why certain structures are preferred and others disfavored is the basis for the interpretation of many experimental results. An accurate description of energy surfaces of molecules, clusters, polymers and solids is therefore of central importance for the understanding of their static and dynamic properties.

In the case of isolated, small molecules the topological characterization of energy surfaces, the classification of stationary points such as stable minima or saddle points, the evaluation of detailed equilibrium geometries and the computation of harmonic and anharmonic force constants are nowadays quite routinely performed at different levels of sophistication, ranging from empirical force fields to refined ab initio methods including a substantial amount of electron correlation contributions. Energy surfaces of realistic polymers, on the other hand, have almost exclusively been obtained via empirical potential functions or the concept of an energy surface has even been discarded in favor of a purely statistical approach to the conformational problems in polymers |1|. While the latter method is the only practical alternative to describe the thermodynamic properties of extended chain macromolecules in solution more advanced techniques may be applied for an investigation of the structure of crystalline polymers.

J. Ladik et al. (eds.), Quantum Chemistry of Polymers - Solid State Aspects, 33–55.

Provided the repetition unit is small enough and interchain forces are sufficiently weak semiempirical or ab initio studies on isolated polymer chains may be performed with the aim to obtain equilibrium structures of different periodic arrangements and to compare their relative stabilities.

In this contribution a survey of ab initio crystal orbital calculations on diverse polymeric systems performed within the framework of the restricted Hartree Fock approximation is given emphasizing the computation of energy surfaces and properties derived from them. Computed equilibrium geometries of simple, covalent polymers and of hydrogen bonded chains are compared with the geometries of small, related molecules and clusters and, if available, with experimentally observed structures. As representative examples for conjugated polymers we present results on polyyne and on polymethineimine, and on the structure and relative stabilities of conceivable isomeric modifications of polyacetylene and polydiacetylene. Applications to saturated, covalent polymers are provided by a discussion of different periodic, helical conformations in polyethylene and polyoxymethylene. Particular attention is paid to the case of hydrogen bonded polymers which may be viewed as parts of anisotropic molecular crystals. For three simple hydrogen bonded model polymers built from hydrogen fluoride, hydrogen cyanide and formic acid molecules, respectively, the important role of non-additive contributions to intermolecular forces is demonstrated. Due to restrictions in the space available only a few of the obtained results are given explicitely in the text. Electronic band structures of most of the polymers considered in this survey will be discussed by others in this volume.

2. METHODICAL ASPECTS

The ab initio crystal orbital method |2-4| is by now an established tool for the evaluation of ground state properties of polymers. The explicit use of translational symmetry allows a more efficient evaluation of polymer properties than the alternative strategy consisting of a computation of clusters of increasing size and subsequent extrapolation of their properties to the limiting case of infinite chain length.

Methodical advances recently achieved in this approach and collections of more recent applications to a great variety of polymeric systems may be found in several conference reports |5-7| and in review papers |8-13|. Since the basic equations of the crystal orbital formalism will be covered by André in this volume we restrict the discussion to a few practical aspects concerning the various numerical convergence problems arising in actual crystal orbital calculations.

Compared to conventional restricted Hartree Fock calculations on molecules two additional problems have to be dealt with, namely numerical integrations over the Brillouin zone and a proper treatment of lattice summations. Numerical experience by all research groups active in the field of crystal orbital calculations has shown that the one-dimensional Brillouin zone integrations pose no particular problem and computed total energies are rather insensitive to the density of points in k-space provided a certain critical number is exceeded. Substantial progress in the treatment of lattice summations has been achieved via the introduction of electrostatic approximations to the calculation of long range interactions |14-16|. With slight modifications implementation of this formalism is possible in any crystal orbital program irrespective of the particular way in which the exact calculation of electron repulsion integrals in the short range region is truncated |17| and allows a numerically satisfactory evaluation of total energies in non-metallic systems. Caution is, however, in order in choosing the appropriate onset for electrostatic approximations in terms of neighbor interactions. Unreasonable structures may apparently be stabilized if this onset is chosen too early.

One must, however, always keep in mind that errors in computed properties due to the use of too small basis sets are in general considerably larger than errors which arise as a consequence of an early truncation of lattice sums. Experience in the design of atomic basis sets accumulated in numerous ab initio calculations on molecules and clusters may therefore be transferred to the case of polymers without any modifications. Errors in the description of molecular properties or in the calculation of intermolecular interactions which occur because of the application of poor basis sets certainly show up in polymer calculations as well. Of course the same is valid for the intrinsic failures of the Hartree Fock approximation.

In all applications discussed in the following chapters the many-dimensional energy surfaces have been scanned pointwise as a function of some appropriately chosen set of internal coordinates. Energy values thus obtained have been subjected to polynomial fits in order to find the equilibrium geometry and internal, harmonic force constants. In a few cases (polyyne, hydrogen fluoride, hydrogen cyanide) these force constants have been used for an evaluation of vibrational frequencies and phonon dispersion curves within the framework of the harmonic approximation using standard methods of polymer vibrational spectroscopy (see e.g. refs. |18,19|).

3. CONJUGATED POLYMERS

In figure 1 the structures of those conjugated polymers are displayed which have been the subject of extensive ab initio in-

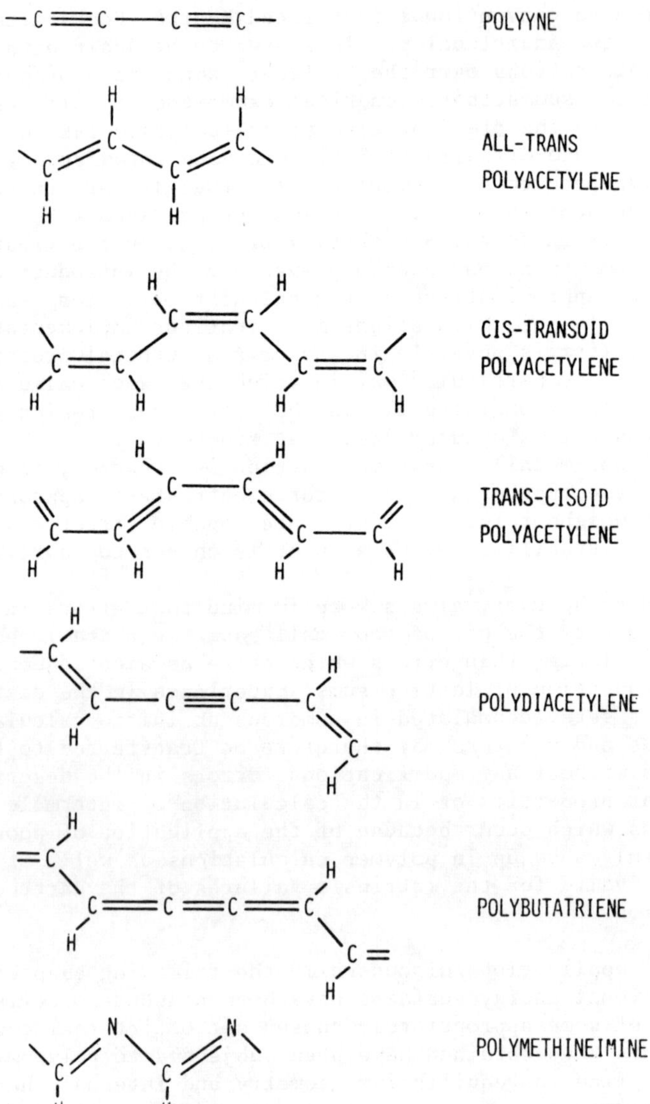

Figure 1: Structures of conjugated polymers considered in this survey.

vestigations by various groups in the last years. Since most of the obtained results have recently been reviewed in a very detailed manner |12| we confine the discussion to a few items only.

In all cases where the question concerning the relative stabilities of equidistant versus bond alternating structures arises (polyyne |20,21|, polyacetylene |22-27|, polymethineimine |28,29|) the latter are more stable within the framework of the restricted Hartree Fock approximation. For polyyne and polyacetylene this issue is in accord with the well known concept of a Peierls distortion |30|. The occurence of Hartree Fock instabilities (see e.g. refs. |31,32|) in the case of the equidistant, metallic structures of polyyne (cumulene) and all-trans polyacetylene points, however, to the need for improved methods going beyond the independent particle model. First efforts in this direction |27| show that at the level of second order Moller-Plesset perturbation theory the alternant configuration of polyacetylene is still preferred energetically although as expected the energy difference to the equidistant structures becomes smaller.

Computed cis-trans energy differences should be less influenced by the intrinsic failures of the Hartree Fock approximation. In table 1 we compare energies of all-trans, cis-transoid, and trans-cisoid polyacetylenes. In agreement with the experimental results

Table 1: Destabilization of cis-transoid and trans-cisoid polyacetylene with respect to the all-trans isomer. Energy values are given in kcal/mol per C_2H_2.

	STO-3G |24|	7,3/3 |13|	FSGO |25|	CNDO/2 |33|
cis-transoid	1.9	2.4	19.8	1.9
trans-cisoid	2.1	4.0	20.7	4.3

on solid polyacetylene |34| and the cis-trans energy difference in butadiene which is about 2.5 kcal/mol |35| the all-trans structure is the most stable isomer of polyacetylene. Nearly complete structure optimization has been performed in refs. |13,24| for all three isomers assuming only full planarity. Resulting equilibrium geometries as obtained with a 7s3p/3s basis are collected in table 2. As in the case of ab initio studies on molecules geometry optimization is a necessary prerequisite for the reliable prediction of energy differences.

As a further example the case of polydiacetylene is presented. For a survey of recent experimental data see the contribution of Bloor in this volume. From the two possible backbone structures, acetylenic or butatrienic, the former is considerably more stable

Table 2: Computed equilibrium geometries of polyacetylene isomers |13|.

	all-trans	cis-transoid	trans-cisoid
$r_{C=C}$ |bohr|	2.513	2.519	2.515
r_{C-C} |bohr|	2.738	2.755	2.764
r_{C-H} |bohr|	2.041	2.028	2.033
∢ CCC |°|	124.4	126.9	127.3
∢ C=C-H |°|	119.1	116.5	118.8
lattice period |bohr|	4.65	8.35	8.58

|36-38|. An energy difference of about 12 kcal/mol per C_4H_2 unit has been obtained with a 7s3p/3s basis |37|. Practically all polydiacetylenes for which reliable X-ray structures are available do indeed correspond to the acetylenic structure. A more complete list of computed bond lengths and harmonic stretching force constants for conjugated hydrocarbon polymers has recently been compiled |13|.

4. SATURATED, COVALENT POLYMERS

In this chapter we will first discuss equilibrium structures of polyethylene and polyoxymethylene (see figure 2) and compare

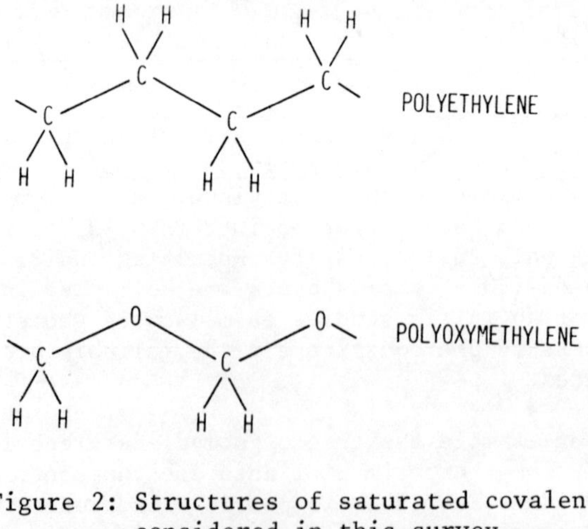

Figure 2: Structures of saturated covalent polymers considered in this survey.

them with experimental data. Subsequently a short description of torsional potentials in these polymers is given.

4.1 Polyethylene

The first pioneering ab initio crystal orbital studies on polyethylene have been presented by André et al. |39-41| and were devoted to an analysis of the valence band structure. We have recently performed similar calculations with the aim to obtain the equilibrium structure of the all-trans conformation of polyethylene |13,42,43| . Results obtained with a 7s3p/3s basis are shown in table 3 together with those of X-ray and neutron investigations.

Table 3: Equilibrium structure of all-trans polyethylene.

	calc.	X-ray	neutron diffraction
	|42,43|	|44-47|	|48|
r_{C-C} |bohr|	2.939	2.89	2.91
r_{C-H} |bohr|	2.082	2.02	2.04
∢CCC |°|	112.3	112.0	111.7
∢HCH |°|	107.4	107.0	109.4

Moreover the longitudinal elastic modulus or Young's modulus of polyethylene has also been investigated. We compare the computed value with experimental results in table 4.

Table 4: Young's modulus of polyethylene. All values in GPa.

calculated |42|	345
neutron scattering |49|	329
Raman |50,51|	358,290

4.2 Polyoxymethylene

Contrary to the case of polyethylene an all-gauche conformation is energetically preferred in polyoxymethylene. The first initio study of polyoxymethylene has quite recently been carried out |52| applying again a 7,3/3 basis. A few of the obtained structural details are compared with X-ray structures in table 5.

Table 5: Equilibrium structure of polyoxymethylene.

		calculated |52|	X-ray |53|
r_{C-O}	|bohr|	2.693	2.700
∢OCO	|°|	112.7	112.9
τ_{OCOC}	|°|	70.8	77.0

4.3 Conformational Stability

An accurate evaluation of the energetics of different conformations in polymers is of basic importance for an understanding of their physical properties. We therefore investigated also other periodic conformations of polyethylene and polyoxymethylene including extensive geometry optimization. Results on the energies of $(t)_\infty$, $(g)_\infty$, and $(tg)_\infty$ helices are collected in table 6.

Table 6: Relative stabilities of different regular,helical structures of polyethylene and polyoxymethylene. Energy values in kcal/mol.

	polyethylene $\Delta E/C_2H_4$ |13,43|	polyoxymethylene $\Delta E/CH_2O$ |52|
$(t)_\infty$	0.0	12.8
$(g)_\infty$	1.74, 1.85[a)]	0.0
$(tg)_\infty$	1.08[a)]	4-5

a) STO-3G values

The preference for the $(t)_\infty$ conformation in polyethylene and of the $(g)_\infty$ conformation in polyoxymethylene is satisfactorily reproduced. Whereas in the case of polyethylene periodic $(g)_\infty$ and $(tg)_\infty$ helices do correspond to minima on the energy surface the $(t)_\infty$ and possibly also the $(tg)_\infty$ helices do not correspond to minima. The energy difference between $(t)_\infty$ and $(g)_\infty$ conformations in polyoxymethylene is probably overestimated by our calculations.

5. HYDROGEN BONDED CHAINS

Hydrogen bonding is one of the most important mechanisms of intermolecular interaction. The structure of many biopolymers and molecular crystals is dominated by the occurence of hydrogen bonds. An extensive survey of experimental and theoretical studies on the phenomenon of hydrogen bonding may be found in ref. |54|. More recent results of quantum mechanical calculations on hydrogen bonded systems are collected in a few review articles |55-60|. Most of the theoretical treatments are devoted to an elucidation of dimer and cluster properties. Systematic ab initio studies on periodic hydrogen bonded chains are rare although many spectroscopically observable features usually ascribed to hydrogen bonding are present in dimers to a small extent only. The reasons for this lack are probably the high demand on the quality of the basis sets that must be applied and the large number of internal degrees of freedom that have to be optimized. Consequently calculations of this type become rather tedious and time consuming. In the following the most important results of detailed ab initio studies on three different hydrogen bonded chains - hydrogen fluoride, hydrogen cyanide, and formic acid - are presented. In the first two cases extended basis sets including polarization functions have been applied whereas for formic acid a minimal basis has been applied only. Finally a more general discussion of the trends in the changes of different ground state properties upon polymer formation is provided.

5.1 Hydrogen Fluoride Chains

The crystal structure of hydrogen fluoride is strongly anisotropic |61,62|. Bent hydrogen bonded chains are formed with interchain distances slightly outside the range to be expected from a consideration of van der Waals radii. The description of structural and other properties of an isolated hydrogen fluoride chain seems therefore to be a reasonable starting point for an evaluation of crystal properties. Details of the geometry of the vapor phase dimer $(HF)_2$ have been obtained by accurate microwave studies |63|. Vibrational spectra of HF, $(HF)_2$ |64| and of solid HF |65-68| are also available. In a recent thermodynamic study hydrogen bond enthalpies and entropies of open chain and cyclic HF oligomers have been fitted to a great number of vapor phase data |69|. Because of the large amount of experimental data available hydrogen fluoride is a good test case. In the last years we carried out several studies of the infinite chain of hydrogen fluoride molecules |70-72|. In figure 3 we compare computed equilibrium structures and hydrogen bond energies of HF, $(HF)_2$ |73| and $(HF)_\infty$ |72| with experimental data. All results have been obtained with extended basis sets (10s6p2d/6s1p) and a complete optimization of all internal degrees of freedom. Both the dimer and the infinite chain have fully planar equilibrium structures.

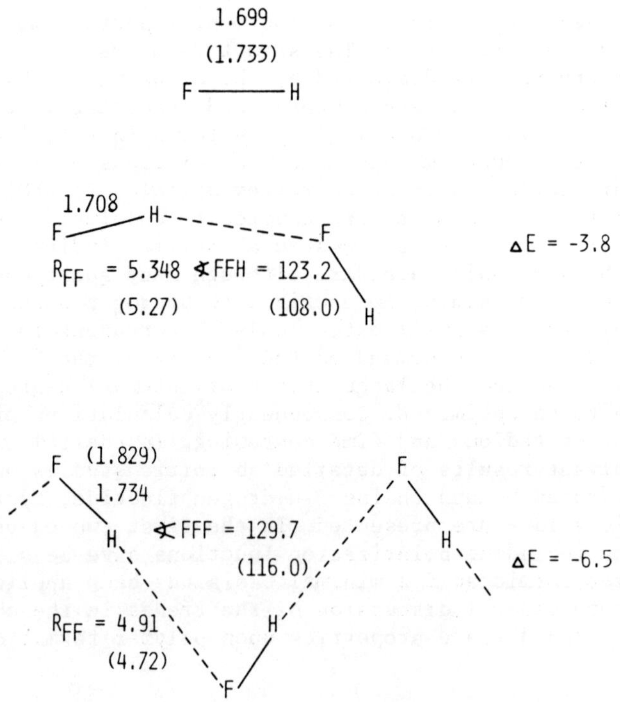

Figure 3: Equilibrium geometries and stabilization energies of HF, $(HF)_2$, and $(HF)_\infty$. Distances in bohr, energy values in kcal/mol. Experimental values in brackets.

The intermolecular distance is significantly reduced in the crystal whereas the intramolecular HF distance is widened. Both these features are reproduced at the Hartree Fock level although quantitative agreement could not be achieved at this stage of methodical sophistication. The inherent deficiencies of the Hartree Fock model together with an analysis of the computed potential surfaces within the framework of the harmonic approximation allows currently to give a semiquantitative picture only. Hydrogen bond energies in the polymer are significantly enhanced over their value in the dimer.

The change in HF stretching frequencies upon crystal formation is even more drastic than the observed structural modifications. In table 7 we compiled experimental and computed HF stretching frequencies of HF, $(HF)_2$, and $(HF)_\infty$. A slightly smaller (9,5,1/5,1) basis has been used for HF and $(HF)_\infty$.

Table 7: HF stretching frequencies. All values in cm^{-1}.

	ν_{HF} exper.		ν_{HF} calc.	
HF	3962	|64|	4494	|72|
$(HF)_2$	3895 3857	|64|	4440 4396	|73|
$(HF)_\infty$	3404 3065	|65|	4170 3967	|72|

Frequency shifts are modest in the dimer whereas a large reduction is observed in the crystal. Again the trend is qualitatively reproduced at the Hartree Fock level. The symmetric HF vibration has a lower frequency than the antisymmetric one in agreement with the experimental order. Because of the lack of single crystals of hydrogen fluoride difficulties arise in the interpretation and assignment of vibrational frequencies in the region of intermolecular modes. With the aid of crystal orbital calculations these difficulties can partly be removed. A more complete discussion of vibrational spectra of $(HF)_\infty$ including a calculation of phonon dispersion curves is given in ref. |72|.

5.2 Hydrogen Cyanide

The crystal structure of hydrogen cyanide |74| is even more anisotropic than the the crystal structure of hydrogen fluoride. Perfectly linear chains are formed. Again interchain distances are sufficiently large to allow a meaningful comparison with theoretical predictions for isolated chains. In figure 4 we compare results on the equilibrium geometries and hydrogen bond energies of $(HCN)_2$ and $(HCN)_\infty$ as obtained with a (10,6,1/6,1) basis |75| with corresponding experimental data. Computed intramolecular vibration frequencies are shown in table 8.

We observe again a reduction of the intermolecular distance,

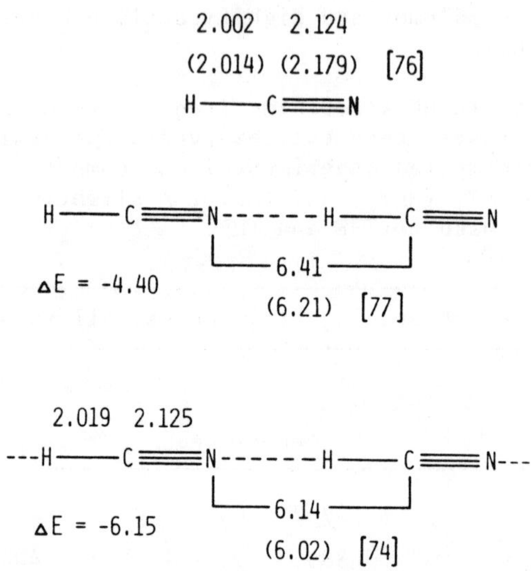

Figure 4: Equilibrium geometries and hydrogen bond energies of HCN, $(HCN)_2$, and $(HCN)_\infty$. Distances in bohr, energy values in kcal/mole. Experimental values in brackets.

Table 8: Stretching frequencies of HCN and $(HCN)_\infty$. All values in cm^{-1}.

			ν_{C-H}	$\nu_{C\equiv N}$
HCN	calc.	[75]	3645	2428
	exper.	[78]	3312	2089
$(HCN)_\infty$	calc.	[75]	3507	2427
	exper.	[79]	3130	2099

an increase in the hydrogen bond energy, an increase in the C-H bond distance and a reduction of the C-H frequency in the infinite chain. It should, however, be noted that these features are less pronounced in HCN than in the case of HF. The C≡N triple bond distance and hence also the C≡N frequency remain practically unchanged upon polymer formation.

5.3 Formic Acid

Depending on the substituent carboxylic acids RCOOH may form rather complicated crystal structures |80,81|. The simplest carboxylic acids, formic and acetic acid, form catemeric chains in the solid state |82-88| whereas a cyclic structure is preferred in the vapor phase dimer |89-91|. Formic acid being the first member of the carboxylic acid series seems to be best suited for a theoretical analysis. The isolated formic acid molecule may occur in the syn- and anti- conformation (see figure 5). From extensive micro-

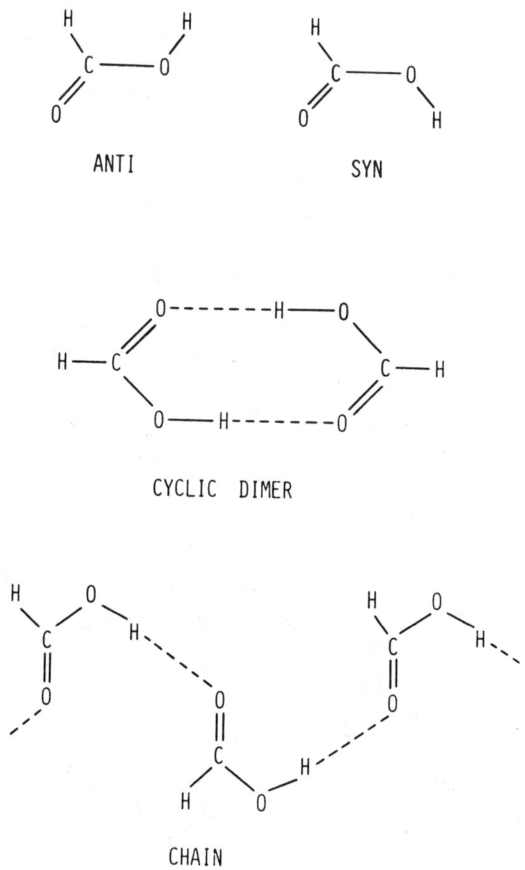

Figure 5: Syn- and anti- conformations of formic acid, cyclic vapor phase dimer and infinite chain in solid β-formic acid.

wave spectroscopical work |92-94| structural details of syn- and anti- conformations are well established. From recent CEPA calculations |95| the energy difference between these two isomers is found to be about 5 kcal/mol with the syn-conformer as the more stable structure. The cyclic dimer of formic acid is an appropriate system for the study of different proton transfer mechanisms and has therefore been investigated intensively with the aid of different quantum mechanical methods |96-106|. Various crystal packing arrangements of formic acid have been compared using frozen monomer geometries and ab initio wavefunctions in combination with electrostatic summation techniques or with interatomic potential functions fitted to SCF calculations |102,103|. The stability of various catemer motifs of formic acid has been studied with the aid of empirical force fields |80,81|. Only a single crystal orbital study is available in which two possible modifications of formic acid chains are compared |107|. No structure optimization has, however, been performed. Here we present results on six different periodic formic acid chains (see firgure 6). Structures I to III correspond to periodic arrays of syn-conformations, whereas structures IV to VI originate from anti-conformations. These structures have been

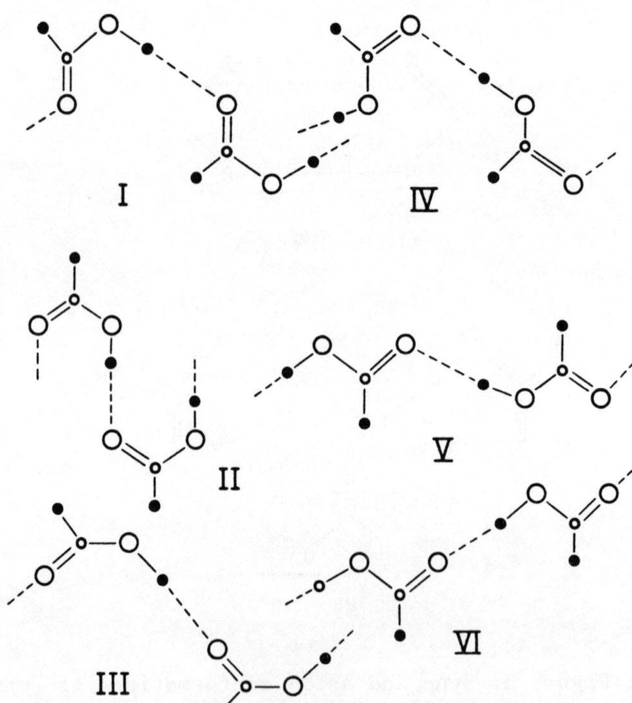

Figure 6: Periodic formic acid chains considered in this study.

selected because their stability has been discussed in refs. |80,81|. Clearly other, more complicated periodic patterns, are also conceivable. The size of this system allowed us merely to use a minimal STO-3G basis. The calculations are hence of an explorative nature only. We nevertheless give a more detailed discussion of our results on formic acid here than for the two previous cases because formic acid is an interesting system, representative of the structural variety observed in carboxylic acids and in amides. For all six structures considered intra- and intermolecular degrees of freedom have been optimized with the only restriction of an assumed planarity of all chains and the existence of an S_2 screw axis in chain direction for structures I to V.

In table 9 a few selected features of the computed equilibrium geometries of structures I to VI are compared with the corresponding results for the syn- and anti-conformer of the formic acid molecule |108,109| and of the cyclic dimer |98|. Relative stabilities with respect to the syn-conformation are shown in table 10.

Table 9: Selected optimized bond distances of formic acid monomers, dimer, and chains as obtained with an STO-3G basis. All values in bohr.

Structure	$r_{C=O}$	r_{C-O}	r_{O-H}	r_{O---O}
syn-monomer |108|	2.294	2.604	1.873	-
anti-monomer |109|	2.289	2.612	1.867	-
cyclic dimer |98|	2.324	2.551	1.908	4.80
I	2.341	2.522	1.930	4.694
II	2.318	2.547	1.896	4.819
III	2.306	2.567	1.878	5.019
IV	2.336	2.530	1.931	4.652
V	2.338	2.524	1.932	4.641
VI	2.325	2.565	1.872	4.822

The most stable structure turns out to be pattern I which corresponds to the β-form of solid formic acid in agreement with the actual low temperature structure found by X-ray and neutron diffraction studies |82-84|. Considering the stabilization energy per hydrogen bond structure I is also more stable than the cyclic dimer in line with the preference for a chain structure in the crystal versus the occurence of predominantly cyclic dimers in the vapor

Table 10: STO-3G stabilization energies per molecule of formic acid monomers, dimer, and chains relative to the syn-conformer. All values in kcal/mol.

Structure		ΔE
anti-monomer	|109|	+4.46
cyclic dimer	|98|	-7.55
I		-9.41
II		-6.28
III		-5.97
IV		-8.79
V		-7.10
VI		-5.03

phase. Next in stability is pattern IV which originates from structure I through a simultaneous shift of all protons accompanied by a structural relaxation interchanging C-O single and double bonds resulting in a chain of anti-conformers. Whereas the STO-3G energy difference between syn- and anti-conformers is about 4.5 kcal/mol for the case of isolated formic acid molecules the corresponding energy difference between structures I and IV is only about 0.6 kcal/mol per molecule. Although this trend goes in the expected direction it might be overestimated at the STO-3G level. It is well established that the main weakness of STO-3G calculations on intermolecular interaction is the prediction of too small intermolecular distances in comparison to experimental data or Hartree-Fock limit results. This applies also to the case of formic acid. Trends in the structural relaxations are nevertheless plausibly described.

Compared to the cyclic dimer the intermolecular O---O distance in formic acid (structure I) is reduced by 0.1 bohr. The corresponding experimental shortening is also 0.1 to 0.15 bohr |83,84,89, 90|. As in the previously discussed cases the O-H bond involved in hydrogen bonding is lengthened markedly in the chain, even stronger than in the cyclic dimer. Relaxation of C-O distances is substantial tending to equalize them. In all catemeric structures the difference in C-O single and double bond distances is, however, still distinctly visible contrary to the results of early X-ray investigations |82| where almost equal C-O distances have been found. These measurements have been performed at higher temperatures. More recent low-temperature results |83,84| reveal that the C-O distances are significantly different in line with our computed structures.

There are, however, indications for a phase transition in formic acid around 208 K |110,111|. Which kind of structural change actually occurs, a transition to structure IV, an order-disorder transition involving structures I and IV, or a significant change in interchain orientation or -distances is still an open question |111-113|. An ab initio calculation of vibrational spectra of formic acid chains is currently out of reach. We compiled therefore available Raman results on O-H frequencies in isolated formic acid, in the cyclic dimer and in the crystal in table 11 in order to enable a comparison with the previously discussed cases of hydrogen fluoride and hydrogen cyanide.

Table 11: O-H stretching frequencies in formic acid monomer, dimer and crystal. All values in cm^{-1}.

		ν_{O-H}
syn-monomer	|114|	3569
cyclic dimer	|115|	3110
crystal	|116|	2752

Experimental frequency shifts are large and similar in magnitude to the case of hydrogen fluoride. A more detailed discussion of our results on formic acid chains including the application of larger basis sets will be given elsewhere |117|.

5.4. Some General Features of Hydrogen Bonding in Infinite Chains

Among other properties hydrogen bonds X-H----Y are usually characterized by the following features |55|:

i) a smaller H----Y distance than the sum of their van der Waals radii.
ii) increase of the X-H bond length.
iii) reduction of the X-H stretching frequency.
iv) increase of the polarity of the X-H bond accompanied by an increase in dipole moments and dipole moment derivatives.
v) smaller NMR chemical shifts of protons in hydrogen bonds.

Although all these features do occur in open chain hydrogen bonded dimers formed between neutral molecules in the vapor phase, they are mostly present to a minor extent only. Exceptions are complexes between strong acids and strong bases. If one assumes, as it is often done, an additive behavior of intermolecular interaction in sequential, chain-like hydrogen bonded polymers only small changes of dimer properties will result. The fact that these changes are substantial in many hydrogen bonded crystals is a clear indication for non-additive contributions to intermolecular interactions and demonstrates the inappropriateness of additive models in this case.

One may either discuss these non-additivities in terms of a perturbation theory of intermolecular forces where it turns out that the main contributions to non-additivity stems from polarization forces |59,118-122| or alternatively one may consider changes in the potential surfaces for proton motion on going from the monomer to the polymer.

As an indication supporting the former point of view we compare in table 12 computed dipole moments of isolated molecules and of dipole moments of molecules embedded in a chain for the three cases considered previously.

Table 12: Dipole moments of isolated molecules and dipole moments per molecule in infinite, hydrogen bonded chains. All values in Debye.

	molecule	chain
HF |72|	2.30	2.65
HCN |75|	3.25	3.73
HCOOH |117|	0.67	1.13

Dipole moments per molecule are substantially enhanced in the chain compared to the dipole moments of the isolated molecules for all three model systems. This enhancement arises as a consequence of mutual polarization of interacting molecules. Unless effective pair potentials are used where this change of molecular properties is incorporated additive models of intermolecular interaction fail in the case of hydrogen bonded systems.

The second point of view, namely an investigation of potential surface for proton motion in these hydrogen bonded chains, is perhaps even more instructive. Proton motions in open chain dimers between neutral molecules follow in most cases a single minimum potential in case the intermolecular distance is fixed in the vicinity of the equilibrium geometry. The minimum corresponds to the structure X-H----Y, whereas the ionic structure X^-----$H-Y^+$ is energetically disfavored and gives at most rise to a small shoulder in the potential surface. In hydrogen bonded chains, however, a collective in phase movement of all protons corresponds to one of the spectroscopically active X-H vibrations and is therefore important for the dynamics. For symmetry reasons |70| this type of motion leads to a symmetric double minimum potential in the case of the hydrogen fluoride chain and therefore to a drastic change in the F-H force constant and probably also to a more pronounced role of anharmonic contributions.

We have observed that in the case of HCN all property changes connected with chain formation, particularly the reduction of C-H frequencies have been much weaker than for HF. The same simultaneous proton motion would lead to a chain of HNC molecules, an isomer which is considerably less stable than HCN. Consequently, an an asymmetric double minimum potential is formed at best.

Formic acid lies between these two cases. The energy difference between syn- and anti- conformation is sufficiently reduced in the chain to give rise to a nearly symmetrical double minimum potential for simultaneous in phase proton motions. Observed frequency shifts are therefore comparable in magnitude to the case of hydrogen fluoride.

6. CONCLUSIONS

Hartree Fock crystal orbital calculations on the ground state properties of polymers lead to results which are directly comparable to analogous calculations on molecules. The transition from molecules to polymers or from clusters to molecular crystals may therefore be investigated at the same level of sophistication. Analysis of the changes in geometry, vibrational spectra and other properties allows an assessment of the importance or insignificance of long range interactions or of cooperative phenomena. Further technical improvements of the currently available crystal orbital programs, the use of more efficient optimization techniques, and explicit inclusion of electron correlation will polymer calculations bring soon to a level where quantitative predictions are possible and will therefore provide ample material for the interpretation of experimental results in polymer spectroscopy.

Acknowledgement: The author wishes to thank Prof. P. Schuster for his interest and continuous support. Assistance in performing some of the calculations and in computer programming by Drs. A. Beyer, R.Höller, and J.Petkov and ample supply with computing resources by the Interfakultäres EDV-Zentrum, Wien are gratefully acknowledged. This work has been supported by the Austrian "Fonds zur Förderung der wissenschaftlichen Forschung", Project No. 3669.

REFERENCES

1 P.J.Flory, "Statistical Mechanics of Chain Molecules", Interscience, New York (1969)
2 T.E.Peacock and R.McWeeny, Proc. Phys. Soc. London, 74, 385 (1959)
3 G.Del Re, J.Ladik, and G.Biczo, Phys.Rev. 155, 997 (1967)
4 J.M.André, L.Gouverneur, and G.Leroi, Intern. J. Quantum Chem. 1, 451 (1967)
5 "Electronic Structure of Polymers and Molecular Crystals", J.M.André and J.Ladik, eds., NATO-ASI Series Vol. B9, Plenum, New York (1975)
6 "Quantum Theory of Polymers", J.M.André, J.Delhalle, and J. Ladik, eds., NATO-ASI Series Vol. C39, Reidel Publ., Dordrecht, Netherlands (1978)
7 "Recent Advances in the Quantum Theory of Polymers", J.M. André, J.L.Brédas, J.Delhalle, J.Ladik, G.Leroi, and C.Moser, eds., Lecture Notes in Physics Series, No. 113, Springer, Berlin (1980)
8 M.Kertesz, Acta Phys. Hung. 41, 107 (1976)
9 J.M.André, Adv. Quantum Chem. 12, 65 (1980)
10 J.Ladik and S.Suhai in "Molecular Interactions", H.Ratajczak and W.J.Orville-Thomas, eds., Wiley, New York (1980) p.151
11 J.Ladik and S.Suhai in "Theoretical Chemistry", Royal Society of Chemistry, London (1981) Vol. 4, p.49
12 M.Kertesz, Adv. Quantum Chem. 15, 161 (1982)
13 A.Karpfen, Phys. Scripta T1, 79 (1982)
14 L.Piela and J.Delhalle, Intern. J. Quantum Chem. 13, 605 (1978)
15 J.Delhalle, L.Piela, J.L.Brédas, and J.M.André, Phys. Rev. B22, 6254 (1980)
16 L.Piela, J.M.André, J.L.Brédas, and J.Delhalle, Intern. J. Quantum Chem. S14, 405 (1980)
17 A.Karpfen, Intern. J. Quantum Chem. 19, 1207 (1981)
18 L.Piseri and G.Zerbi, J.Mol. Spectry. 26, 254 (1968)
19 H.Tadokoro, "Structure of Crystalline Polymers", Wiley, New York (1979) pp. 179-322
20 M.Kertesz, J.Koller, and A.Azman, J.Chem.Phys. 68, 2779 (1978)
21 A.Karpfen, J.Phys. C12, 3227 (1979)
22 M.Kertesz, J.Koller, and A.Azman, J.Chem.Phys. 67, 1180 (1977), Chem.Comm. 575 (1978)
23 A.Karpfen and J.Petkov, Solid State Comm. 29, 251 (1979), Theor. Chim. Acta 53, 65 (1979)
24 A.Karpfen and R.Höller, Solid State Comm. 37, 179 (1981)
25 J.L.Brédas, J.M.André, and J.Delhalle, J.Mol.Struct. 87, 237 (1982)
26 S.Suhai, J.Chem.Phys. 73, 3843 (1980)
27 S.Suhai, Phys.Rev. B27, 3506 (1983), Chem. Phys. Lett. 96, 619 (1983)

28 A.Karpfen, Chem.Phys.Lett. 64, 299 (1979), in ref. |7|, p 137
29 M.Kertesz, J.Koller, and A.Azman, Chem.Phys. Lett. 69, 225 (1980)
30 R.Peierls, "Quantum Theory of Solids", Oxford University, New York (1955) p 108
31 J.Cizek and J.Paldus, J.Chem.Phys. 47, 3976 (1967), J.Paldus and J.Cizek, Phys. Rev. A6, 2268 (1970)
32 H.Fukutome, Prog. Theor. Phys. 40, 998, 1227 (1968)
33 T.Yamabe, K.Tanaka, H.Terame-e, H.Fukui, A.Imamura, H.Shirakawa, and S.Ikeda, J.Phys. C12, L247 (1979), Solid State Comm. 29, 329 (1979)
34 C.K.Chiang, C.R.Fincher,jr., Y.W.Park, A.J.Heeger, H.Shirakawa, E.J.Louis, S.C.Gau, and A.G.McDiarmid, Phys. Rev. Lett. 39, 1098 (1977)
35 L.A.Carreira, J.Chem.Phys. 62, 3851 (1974)
36 M.Kertesz, J.Koller, and A.Azman, Chem.Phys. 27, 273 (1978)
37 A.Karpfen, J.Phys. C13, 5673 (1980)
38 S.Suhai, Chem. Phys. 54, 91 (1980)
39 J.M.André and G.Leroy, Chem.Phys.Lett. 5, 71 (1970)
40 J.M.André, J.Delhalle, S.Delhalle, R.Caudano, J.J.Pireaux, and J.J.Verbist, Chem. Phys. Lett. 23, 206 (1973)
41 J.Delhalle, J.M.André, S.Delhalle, J.J.Pireaux, R.Caudano, and J.J.Verbist, J.Chem.Phys. 60, 595 (1974)
42 A.Karpfen, J.Chem.Phys. 75, 238 (1981)
43 A.Karpfen and A.Beyer, submitted
44 C.W.Bunn, Trans. Faraday Soc. 35, 482 (1939)
45 H.M.M.Shearer and V.Vand, Acta Cryst. 9, 379 (1956)
46 P.W.Teare, Acta Cryst. 12, 294 (1959)
47 S.Kavesh and J.M.Schultz, J.Polym. Sci. Part A 2, 243 (1970)
48 M.Stamm, E.W.Fischer, M.Dettenmaier, and P.Convert, Disc. Faraday Soc. 68, 263 (1979)
49 R.A.Feldkamp, G.Venkaterman, and J.S.King, in "Neutron Inelastic Scattering (IAEA, Vienna, 1968), Vol. II, p 159
50 R.G.Schaufele and T.Shimanouchi, J.Chem.Phys. 47, 3065 (1967)
51 G.Strobl and E.Eckel, J.Polym. Sci. Polym.Phys.Ed. 59, 93 (1962)
52 A.Karpfen and A.Beyer, submitted
53 Y.Takahashi and H.Tadokoro, J.Polym.Sci.Polym.Phys.Ed. 17, 123 (1979)
54 "The Hydrogen Bond- Recent Developments in Theory and Experiments", Vols. I-III, P.Schuster, G.Zundel, and C.Sandorfy, eds., North-Holland, Amsterdam (1976)
55 P.Schuster, in "Intermolecular Interactions from Diatomics to Biopolymers", B.Pullman, ed. Wiley, Chichester (1978) p 363
56 P.Kollman, J.Amer.Chem.Soc. 99, 4875 (1977)
57 P.Kollman, in "Applications of Electronic Structure Theory" (Modern Theoretical Chemistry, Vol. 4) H.F.Schaefer,III, ed. Plenum, New York (1977) p 109
58 P.Schuster, Angew. Chem. Int. Ed. 20, 546 (1981)

59 P.Schuster, A.Karpfen, and A.Beyer, in "Molecular Interactions", H.Ratajczak and W.J.Orville-Thomas, eds., Wiley, New York (1980) p 117
60 A.Beyer, A.Karpfen,and P.Schuster, in press
61 M.Atoji and W.N.Lipscomb, Acta Cryst. 7, 173 (1954)
62 M.W.Johnson, E.Sandor, and E.Arzi, Acta Cryst. B31, 1998 (1975)
63 Th.R.Dyke, B.J.Howard, and W.Klemperer, J.Chem.Phys. 56, 2442 (1976)
64 D.F.Smith, J.Mol. Spectry. 3, 473 (1959)
65 J.S.Kittelberger and D.F.Hornig, J.Chem.Phys. 46, 3099 (1967)
66 A.Axmann, W.Biem, P.Borsch, F.Hoszfeld, and H.Stiller, Disc. Faraday Soc. 7, 69 (1969)
67 R.Tubino and G.Zerbi, J.Chem.Phys. 51, 4509 (1969)
68 A.Anderson, B.H.Torrie, and W.S.Tse, Chem.Phys. Lett. 70, 300 (1980), J.Raman Spectry. 10, 148 (1981)
69 R.L.Redington, J.Chem.Phys. 75, 4417 (1981), J.Phys. Chem. 86, 552, 561 (1982)
70 A.Karpfen and P.Schuster, Chem.Phys. Lett. 44, 459 (1976)
71 A.Karpfen, Chem.Phys. 47, 401 (1980)
72 A.Beyer and A.Karpfen, Chem.Phys. 64, 343 (1982)
73 H.Lischka, Chem.Phys. Lett. 66, 108 (1979)
74 W.J.Dulmage and W.N.Lipscomb, Acta Cryst. 4, 330 (1951)
75 A.Karpfen, Chem.Phys. in press
76 I.Suzuki, M.A.Pariseau, and J.Overend, J.Chem.Phys. 44, 3561 (1966)
77 L.W.Buxton, E.J.Campbell, and W.J.Flygare, Chem.Phys. 56, 399 (1981)
78 G.E.Hyde and D.F.Hornig, J.Chem.Phys. 20, 647 (1952)
79 H.B.Friedrich and P.F.Krause, J.Chem.Phys. 59, 4942 (1973)
80 L.Leiserowitz, Acta Cryst. B32, 775 (1976)
81 Z.Berkovitch-Yellin and L.Leiserowitz, J.Amer.Chem.Soc. 104, 4052 (1982)
82 F.Holtzberg, B.Post, and I.Fankuchen, Acta Cryst. 6, 127 (1953)
83 I.Nahringbauer, B34, 315 (1978)
84 A.Albinati, K.D.Rouse, and M.W.Thomas, Acta Cryst. B34, 2188 (1978)
85 R.E.Jones and D.H.Templeton, Acta Cryst. 11, 484 (1958)
86 P.G.Jönsson, Acta Cryst. B27, 893 (1971)
87 I.Nahringbauer, Acta Chem.Scand. 24, 453 (1970)
88 A.Albinati, K.D.Rouse, and M.W.Thomas, Acta Cryst. B34, 2184 (1978)
89 A.Almenningen, O.Bastiansen, and T.Motzfeld, Acta Chem. Scand. 23, 2848 (1969), ibid. 24, 747 (1970)
90 J.Karle and L.O.Brockway, J.Amer. Chem. Soc. 66, 574 (1944)
91 J.L.Derissen, J.Mol.Struct. 7, 67 (1971)
92 W.H.Hocking, Z.Naturf. 31a, 1113 (1976)
93 E.Bjarnov and W.H.Hocking, Z.Naturf. 33a, 610 (1978)
94 R.Wellington-Davis, A.G.Robiette, M.C.L.Gerry, E.Bjarnov, and G.Winnewisser, J.Mol. Spectry. 81, 93 (1980)

95 C.Zirz and R.Ahlrichs, Theor. Chim.Acta 60, 355 (1981)
96 E.Clementi, J.Mehl, and W.von Niessen, J.Chem.Phys. 54, 508 (1971)
97 E.Ady and J.Brickmann, Chem.Phys. Lett. 11, 302 (1971)
98 J.E.Del Bene and W.L.Kouchenour, J.Amer. Chem. Soc. 98, 2041 (1976)
99 S.Iwata and K.Morokuma, Theor.Chim.Acta, 44, 323 (1977)
100 P.Bosi, G.Zerbi, and E.Clementi, J.Chem.Phys. 66, 3376 (1977)
101 A.Sokalski, H.Romanowski, and A.Jaworski, Adv. Mol. Rel. Int. Proc. 11, 29 (1977)
102 P.H.Smit, J.L.Derissen, and F.B. van Duijneveldt, J.Chem. Phys. 67, 274 (1977), ibid. 69, 4241 (1978), Mol. Phys. 37, 501,521 (1979)
103 J.L.Derissen and P.H.Smit, Acta Cryst. A34, 842 (1978)
104 S.Scheiner, and C.W.Kern, J.Amer.Chem.Soc. 101, 4081 (1979)
105 J.Lipinsky and W.A.Sokalsky, Chem.Phys. Lett. 76, 88 (1980)
106 F.Graf,R.Meyer,T.-K.Ha, and R.R.Ernst, J.Chem.Phys. 75, 2914 (1981)
107 M.Kertesz,J.Koller,E.Zakrajzek, and A.Azman, Z.Naturf. 31a, 637 (1976)
108 J.E.Del Bene, G.T.Worth, F.T.Marchese, and M.E.Conrad, Theor. Chim.Acta 36, 195 (1975)
109 M.R.Peterson and I.G.Csizmadia, J.Amer.Chem.Soc. 101, 1076 (1979)
110 Y.Mikawa, R.J.Jakobsen, and J.W.Brasch, J.Chem.Phys. 45, 4750 (1966)
111 H.R.Zelsmann, F.Bellon, Y.Marechal, and B.Bullemer, Chem. Phys. Lett. 6, 513 (1970)
112 G.N.Robertson and M.C.Lawrence, Chem.Phys. 62, 131 (1981)
113 J.Grip and E.J.Samuelson, Phys. Scripta, 24, 52 (1981)
114 J.E.Bertie and K.H.Michaelian, J.Chem.Phys. 76, 886 (1982)
115 R.C.Millikan and K.S.Pitzer, J.Amer.Chem.Soc. 80, 3515 (1958)
116 H.R.Zelsmann, Y.Marechal, A.Chasson, and P.Faure, J.Mol. Struct. 29, 357 (1975)
117 A.Karpfen, in preparation
118 B.Pullman, P.Claverie, and J.Caillet, Proc.Natl.Acad.Sci. U.S. 57, 1663 (1967)
119 P.Claverie, in "Intermolecular Interactions: From Diatomics to Biopolymers, B.Pullman, ed., Wiley, New York (1978) p 69
120 A.Beyer, A.Karpfen and P.Schuster, Chem.Phys.Lett. 67, 369 (1979)
121 E.Clementi, W.Kolos, G.C.Lie, and G.C.Ranghino, Intern. J. Quantum Chem. 17, 377 (1980)
122 W.Kolos, in "New Horizons of Quantum Chemistry", P.O.Löwdin and B.Pullman, eds., Reidel Publ., Dordrecht, (1983) p 243

THEORY OF EXCITONS IN ORGANIC SOLIDS

T.C. Collins

Associate Vice President for Academic Affairs, University of Missouri, 309 University Hall, Columbia, Missouri 65211, USA

ABSTRACT. This set of lectures is designed to establish the general equations as well as the approximations that are used in formulating the equations which are used to describe excitons. As example of the use of exciton theory, the visible optical spectrum of two typical polydiacetylene crystals calculated by S. Suhai (PTS with an acetylene-like structure and TCDU with a butatriene-like structure) are presented using a first principle Green's function formalism of charge transfer exciton theory. Also the formalism of a model of a high-temperature superconductor based on the exciton mechanism is presented.

1. INTRODUCTION

Excitons are quasiparticle-like excitations to the N-body system whereas most first principle calculations of crystals, namely electron energy band calculations, obtain results by taken an electron away or adding an electron to the system, the (N $\pm$ 1)-body system. The (N $\pm$ 1)-body approximation for the low-lying excitations of the N-body systems is in general not a very good one for organic solids. Thus, we will focus attention on the exciton in this set of lectures.

The introduction of excitons was made by Frenkel /1/ in his attempts to gain insight into the transformation of light into heat in solids. He was able to explain the transformation by first-order perturbation of a system of N atoms with one electron per atom. If one represents the ground state atomic wave function by Ψ_I and the excited atomic wave function by Ψ_{II}, the ground state of the crystal will be (neglecting that the wave function

J. Ladik et al. (eds.), Quantum Chemistry of Polymers - Solid State Aspects, 57–82.

should be antisymmetric)

$$\Phi_o = \prod_{i=1}^{N} \Psi_I(i). \tag{1.1}$$

The state of the crystal in which one electron is excited and the rest are in the normal state are formed from linear combinations of

$$\Phi_i = \Psi_{II}(i) \prod_{j \neq i} \Psi_I(j). \tag{1.2}$$

There are N sets of coefficients $\{C_i\}_N$ corresponding to the splitting of the undisturbed states of the N isolated atoms with the total energy equal to $\epsilon_{II} + (N-1)\epsilon_I$ into N different states denoted as the "excitation multiplet" of the crystal. The corresponding value of the coefficients are determined by the equations

$$\sum_{l=1}^{N} U_{pl} C_l = \epsilon' C_p \tag{1.3}$$

where the U_{pl}'s are matrix elements of the mutual potential energy of all of the atoms U with respect to the functions of Eq. (1.2).

One finds U is of the form

$$U = \sum_{\alpha < \beta} U(\alpha, \beta, R_{\alpha\beta}), \tag{1.4}$$

where $U(\alpha, \beta, R_{\alpha\beta})$ is the mutual potential energy of the atoms α and β and $R_{\alpha\beta}$ is the distance between the respective nuclei. Assuming the atomic wave functions are orthogonal and normal, the matrix element for $p \neq l$ becomes

$$U_{pl} = \int \Phi_p^* U \Phi_l \, d\tau = U_{pl}(R_{pl}), \tag{1.5}$$

where $U_{pl}(R_{pl})$ is a function of the distance R_{pl} alone. When $p = l$, one obtains

$$U_{pp} = \sum_{l \neq p} U_{pl},$$

where

$$U_{pl} = \iint d\tau_p d\tau_l \, U(p,l;R_{pl}) |\Psi_{II}(p)|^2 |\Psi_I(l)|^2. \tag{1.6}$$

Eq. (1.6) is interpreted to be the average value of the mutual energy of atoms p and l when one of them is in the normal state and the other in the excited state.

Frenkel then noted that Eqs. (1.3) are similar to those which determine the normal modes of vibration of a system of coupled classical oscillators with one degree of freedom each. If the coefficients C_k are considered as the amplitudes of these oscillators, and the energies ϵ' are replaced by the square of the frequencies, the matrix elements U_{pl} with $p \neq l$ can then be interpreted as the coupling coefficients and the U_{pp} as the coefficients of the quasi-elastic force for the uncoupled oscillators.

By fixing the nuclei at the lattice points of the crystal that has rectangular form, the normal modes of vibration do not depend on the shape of the function $U_{pl}\,(R_{pl})$ nor on the value of U_{pp}. These vibrations can be described as standing waves with the wave components

$$g_i = n_i/2A_i \quad , \tag{1.7}$$

where A_i is an edge of the crystal and the set $\{n_i\}$, $i = 1,2,3$, specifies the mode of normal vibrations and may take all integral values between zero and $N_i - 1$. The solutions of Eqs. (1.3) have the form

$$C_{\{n_i\}\,\{p_i\}} = Q_{\{n_i\}} \prod_{i=1}^{3} \cos \frac{\pi a_i}{A_i} p_i n_i \, . \tag{1.8}$$

In Eq. (1.8) the set $\{p_i\}$ designates the lattice site; a_i is the lattice constant, and Q_{n_i} is the normalization coefficient. Thus the exciton eigenfunction for this example has the form

$$\chi_{\{n_i\}} = Q_{\{n_i\}} \sum_{p_i} \prod_{i=1}^{3} \left(\cos \frac{\pi a_i}{A_i} p_i n_i\right) \Phi_{\{p_i\}} \quad . \tag{1.9}$$

Let us turn now to the case where the excitation is not in a single atom or molecule but is much more delocalized. Considering a crystal which has the valence band filled and the conduction band empty in the ground state, Wannier /2/ has developed exciton energy levels that have the form of a hydrogen atomic series in this type of system.

The Hamiltonian of the above crystal is

$$H_o = -\sum_i \frac{\hbar^2}{2m}\nabla_i^2 + \sum_i U(\vec{r}_i) + \sum_{i<j} \frac{e^2}{|\vec{r}_i - \vec{r}_j|} \quad , \tag{1.10}$$

where the interaction of the nuclei with themselves and the

electrons are again represented by U, and the spin-dependent interactions are neglected. The energy for the full band using antisymmetric product of one-electron functions becomes

$$E_o = \langle \Phi_o | H | \Phi_o \rangle = \sum_{i,\sigma} \int \psi^*_{i\sigma}(x) \left[\frac{-\hbar^2}{2m} \nabla^2 + U(\vec{r}) \right] \psi_{i\sigma}(x) d\tau$$

$$+ (1/2) \sum_{\substack{i,j \\ \sigma,\sigma'}} \iint d\tau_1 d\tau_2 \psi^*_{i\sigma}(x_1) \psi^*_{j\sigma'}(x_2) \left[\frac{e^2}{|\vec{r}_i - \vec{r}_j|} (1 - \hat{P}_{12}) \right] \psi_{j\sigma'}(x_2) \psi_{i\sigma}(x_1). \tag{1.11}$$

The operator $\hat{P}_{12}$ exchanges x_1 and x_2 and σ represents the spin state of the wave function. If the one-electron functions are Bloch functions, they satisfy the equation

$$\left[\frac{-\hbar^2}{2m} \nabla^2 + V(\vec{r}) \right] \Phi_{n,\vec{k},\sigma} = \epsilon_{n\vec{k}} \Phi_{n,\vec{k},\sigma} , \tag{1.12}$$

where $V(\vec{r})$ is the average potential, n refers to the band and $\vec{k}$ is the wave vector. Using Eq. (1.12), one reduces Eq. (1.11) to

$$E_o = 2 \sum_{\vec{k},v} \left\{ \epsilon_{v,\vec{k}} + \int d\vec{r} \Phi_{v,\vec{k}}(\vec{r}) \left[U(\vec{r}) - V(\vec{r}) \right] \Phi_{v,\vec{k}}(\vec{r}) \right\}$$

$$+ (1/2) \sum_{\substack{\vec{k},\vec{k}' \\ \sigma,\sigma'}} \int d\tau_1 d\tau_2 \Phi^*_{v\vec{k},\sigma}(x_1) \Phi^*_{v\vec{k}',\sigma'}(x_2) \left[\frac{e^2}{|\vec{r}_1 - \vec{r}_2|} (1 - \hat{P}_{12}) \right] \times$$

$$\Phi_{v\vec{k}',\sigma'}(x_2) \Phi_{v\vec{k},\sigma}(x_1), \tag{1.13}$$

where the factor 2 arises from the spin summation, and v refers to the valence band.

Writing the Hamiltonian matrix elements in the Bloch representation of one electron in the conduction band and the remaining electrons in the valence band, one has

$$\langle v\vec{k}_n \, c\vec{k}_e | H_o | v\vec{k}'\, c\vec{k}'_e \rangle = \delta_{\vec{k}_n \vec{k}'_n} \delta_{\vec{k}_e, \vec{k}'_e} \left[E_o + W_e(\vec{k}_e) \, W_v(\vec{k}_n) \right.$$

$$+ \delta_{\vec{k}_e + \vec{k}_n - \vec{k}'_e - \vec{k}'_n, o} \langle v, \vec{k}_n \sigma_n ; c, \vec{k}_e \sigma_e | \frac{e^2}{|\vec{r}_1 - \vec{r}_2|} (\hat{P}_{12} - 1) | v\vec{k}'_{n,} \sigma'_{n'}, c, \vec{k}'_e \sigma'_e \rangle \tag{1.14}$$

where

$$W_c(\vec{k}_e) = \mathcal{E}_c(\vec{k}_e) + \langle c\vec{k}_e | (U - V) \, c\vec{k}_e \rangle + \sum_{\vec{k}_n} \langle v\vec{k}_n c\vec{k}_e | \frac{e^2}{|\vec{r}_1 - \vec{r}_2|} (2 - \hat{P}_{12}) | \times$$

$$| v\vec{k}_n c\vec{k}_e \rangle , \tag{1.15}$$

and

$$W_v(\vec{k}_n)=\varepsilon_v(\vec{k}_n)+\langle v\vec{k}_n|(U-V)|v\vec{k}_n\rangle+\sum_{\vec{k}'_n}\langle v\vec{k}_n v\vec{k}'_n|\frac{e^2}{|\vec{r}_1-\vec{r}_2|}(2-\hat{P}_{12})|v\vec{k}_n v\vec{k}'_n\rangle. \tag{1.16}$$

The wave vector $\vec{K} = \vec{k}_e - \vec{k}_n$ is significant in that it characterizes the way in which the total wave function transforms under translational symmetry operations of the crystal. Therefore one has mixing among only those zero-order functions having the same reduced value of $\vec{k}_e - \vec{k}_n$ plus a reciprocal lattice vector. Further, as was seen in the first example by Frenkel, the Hamiltonian matrix elements formed from localized states (such as Wannier functions)

$$\langle v\vec{R}_i,\ c\vec{R}_j|H_o|v\vec{R}'_i,c\vec{R}'_j\rangle \tag{1.17}$$

depend only on the differences $\vec{R}'_i - \vec{R}_i$, $\vec{\beta} = \vec{R}_j - \vec{R}_i$ and $\vec{\beta}' = \vec{R}'_j - \vec{R}'_i$. Thus Wannier introduced a third representation called "exciton waves" which depended on $\vec{K}$ and $\vec{\beta}$. These representations are related by the unitary tranformations

$$\Phi_{vc}(\vec{K},\vec{\beta}) = N^{-1/2}\sum_{\vec{k}} e^{-i\vec{\beta}\ \vec{k}} B_{vc}(\vec{k} - \vec{K},\ \vec{k}) \tag{1.18}$$

$$= N^{-1/2}\sum_{\vec{R}} e^{i\vec{K}\ \vec{R}}\ A_{vc}(\vec{R},\vec{R} + \vec{\beta}), \tag{1.19}$$

where B_{vc} and A_{vc} are the antisymmetric products of Bloch and and Wannier functions having one electron excited, respectively and $\vec{k} = (1/2)(\vec{k}_e + \vec{k}_n)$.

The matrix elements of H_0 in the exciton representation have the form

$$\langle vc\vec{K}\vec{\beta}|H_o|vc\vec{K}\vec{\beta}'\rangle = \delta_{\vec{\beta}\vec{\beta}'}E_o + e^{i\vec{K}(\vec{\beta}-\vec{\beta}')/2}N^{-1}\sum_{\vec{k}} e^{i\vec{k}\ (\vec{\beta}-\vec{\beta}')}\left[W_c(\vec{k}+(1/2)\vec{K})\right.$$
$$\left. - W(\vec{k}-(1/2)\vec{K})\right] + \sum_{\vec{R}} e^{i\vec{K}\cdot\vec{R}}\langle v o\sigma_n, c\vec{\beta}\sigma_e|\frac{e^2}{|\vec{r}_1-\vec{r}_2|}(\hat{P}_{12}-1)|v\vec{R}\sigma_n, c(\vec{R}-\vec{\beta}')\sigma_e\rangle \tag{1.20}$$

Note the first two terms are in the Bloch functions and the last term is expressed in Wannier functions. Also, there are no finite matrix elements of different $\vec{K}$'s because of translation symmetry. Thus one looks for the linear combination of exciton functions with respect to the index $\vec{\beta}$ which diagonalizes the Hamiltonian matrix

$$\chi_{vc}(\vec{K}) = \sum_{\vec{\beta}} U_{vc\vec{K}}(\vec{\beta})\ \Phi_{vc}(\vec{K},\vec{\beta}) \tag{1.21}$$

where the coefficients $U(\vec{\beta})$ have to satisfy

$$\sum_{\vec{\beta}'}\{\langle vc\vec{K}\vec{\beta}|H_o|vc\vec{K}\vec{\beta}'\rangle - \delta_{\vec{\beta}\vec{\beta}'}E\}U_{vc\vec{K}}(\vec{\beta}') = 0. \tag{1.22}$$

To find the solution to this equation, Wannier neglected the off-diagonal part and the exchange part of the last term in Eq. (1.20). Further, he noted that the Coulomb term approached $-e^2/\beta$ for large values of β. So the problem now becomes one of finding the correct linear combination of $U(\vec{\beta})$ for the effective Hamiltonian

$$H^o_{\vec{\beta}\vec{\beta}'}(K) = \delta_{\vec{\beta}\vec{\beta}'}E_o + e^{i\vec{K}(\vec{\beta}-\vec{\beta}')/2}N^{-1}\sum_{\vec{k}}e^{i\vec{k}(\vec{\beta}-\vec{\beta}')}\,W_c(\vec{k}+(1/2)\vec{K})-W_v(\vec{k}-(1/2)\vec{K}) - \delta_{\vec{\beta}\vec{\beta}'}e^2/\beta\ . \tag{1.23}$$

The remaining term

$$H'_{\vec{\beta}\vec{\beta}'} = \sum_{\vec{R}} e^{i\vec{K}\cdot\vec{R}}[\langle v o\sigma_n, c\vec{\beta}\sigma_e|\frac{e^2}{|\vec{r}_1-\vec{r}_2|}(\hat{P}_{12}-1)|v\vec{R}\sigma'_n, c(\vec{R}+\vec{\beta}')\sigma'_e\rangle + \delta_{\vec{\beta}\vec{\beta}'}e^2/\beta \tag{1.24}$$

can be treated as a first-order perturbation correction.

Substituting Eq. (1.23) into Eq. (1.22) one obtains

$$E_oU'_K(\vec{\beta}) + N^{-1}\sum_{\vec{\beta}',\vec{k}} e^{i\vec{k}(\vec{\beta}-\vec{\beta}')}[W_c(\vec{k}+(1/2)\vec{K})-W_v(\vec{k}-(1/2)\vec{K})\,U'_{\vec{K}}(\vec{\beta}') - e^2/\beta\,U'_{\vec{K}}(\vec{\beta}) = E\,U'_{\vec{K}}(\vec{\beta}), \tag{1.25}$$

where the transformation

$$U'_{\vec{K}}(\vec{\beta}) = e^{-i\vec{K}\,\vec{\beta}'/2}U_{\vec{K}}(\vec{\beta})\ . \tag{1.26}$$

The next step is to transform from a difference equation to a differential equation. To do this use the Fourier transformations. If one has

$$N^{-1}\sum_{\vec{\beta}',\vec{k}} e^{i\vec{k}\,(\vec{\beta}-\vec{\beta}')}\,W(\vec{k})F(\vec{\beta}')\ , \tag{1.27}$$

use

$$F(\vec{\beta}) = N^{-1/2}\sum_{\vec{k}} e^{-i\vec{k}\,\vec{\beta}}\,G(\vec{k}). \tag{1.28}$$

This gives

$$N^{-1/2}\sum_{\vec{k}} e^{-\vec{k}\,\vec{\beta}}\, W(\vec{k})\, G(\vec{k}). \tag{1.29}$$

For functions W which can be expanded into powers of the components of $\vec{k}$, one can write Eq. (1.27) as

$$W(-i\nabla_{\vec{\beta}})\, F(\vec{\beta}). \tag{1.30}$$

Using Eq. (1.30), Eq. (1.25) becomes

$$\left[W_c(-i\nabla_{\vec{\beta}}+(1/2)\vec{K})-W_v(-i\nabla_{\vec{\beta}}-(1/2)\vec{K})-e^2/\beta\right]U'_{\vec{K}}(\vec{\beta}) = (E-E_o)\, U'_{\vec{K}}(\vec{\beta}). \tag{1.31}$$

Assume that the crystal is simple enough to allow for spherically symmetric development of the form

$$W_v = E_v - \hbar^2k^2/2m_h^* \tag{1.32}$$

and

$$W_c = E_c + \hbar^2k^2/2m_e^* \quad , \tag{1.33}$$

where m_e^* and m_h^* are the effective masses of the electron and hole, respectively. Further, transforming to the center of mass coordinates plus the use of

$$U'(\vec{\beta}) = e^{i\vec{K}\,\vec{\beta}}U(\vec{\beta}) \tag{1.34}$$

to eliminate the cross term of $\vec{K}\,\nabla_{\vec{\beta}}$, $U(\vec{\beta})$ must satisfy the equation

$$(-\hbar^2/2\mu\,\nabla^2 - e^2/\beta)U(\vec{\beta}) = (E-E_o-E_G- \hbar^2k^2/2(m_e^*+m_h^*))U(\vec{\beta}) \tag{1.35}$$

where $E_G = E_c - E_v$ and μ is the reduced mass of the exciton. Since the operator on the left of Eq. (1.35) is the hydrogenic operator, the eigenvalues have the form

$$E_{v\vec{K}} = E_o + E_G - \mu e^4/2\hbar^2n^2 + \hbar^2k^2/2(m_e^*+m_h^*) \quad , \tag{1.36}$$

where v represents the quantum numbers associated with the hydrogenic problem. n is the principle quantum number, and the last term of Eq. (1.36) is the kinetic energy of the exciton. The exciton wave functions thus must have the form

$$\chi_{cv}(v,\vec{K}) = \sum_{\vec{\beta}} \exp\left[i\,\frac{m_e^*}{m_e^*+m_h^*}\,\vec{K}\cdot\vec{\beta}\right]U_v(\vec{\beta})\,\Phi_{vc}(\vec{\beta},\vec{K}) \; . \tag{1.37}$$

For a more complete description of the effects of Eq. (1.24), and the effects of magnetic fields consult reference /3/, Chapter 2.

In the next section we will outline a Green's function formalism to obtain the exciton energies. The calculations of S. Suhai on two typical polydiacetylene crystals (PTS with an acetylene-like structure and TCDU with a butatriene-like structure) are presented in section III. In section IV the formalism of a model of a high-temperature superconductor based on the exciton mechanism is presented, and in the final section a conclusion is given.

II. A FIRST PRINCIPLE GREEN'S FUNCTION FORMALISM FOR EXCITONS

In order to calculate the excitations to the N-body system, one must consider the second-order Green's function. In particular the derivation of the polarization propagator of the particle-hole (PH) excitation is the term that needs to be outlined. This term describes the response of the system to a perturbation of the form

$$H^{ex}(t) = \int \hat{\psi}^{+}(\vec{x})\, V^{ex}(\vec{x},t)\, \hat{\psi}(\vec{x})\, d\vec{x} \tag{2.1}$$

where $V^{ex}(\vec{x},t)$ is an external perturbing potential and $\hat{\psi}(x)$ is a field operator of the form

$$\hat{\psi}(\vec{x}, t = o) = \sum_n U_n(\vec{x})\, \hat{C}_n(t = o)\ . \tag{2.2}$$

Here $\{U_n(\vec{x})\}$ is a complete set of orthonormal spin orbitals, and $\{\hat{C}_n(t)\}$ are Heisenberg operators obeying Fermi statistics.

The PH polarization propagator $\Pi(x,x')$ is given by

$$i\Pi(x,x') = \langle\psi_o|\hat{T}[\hat{\psi}^{+}(x)\hat{\psi}(x)\hat{\psi}^{+}(x')\hat{\psi}(x')]|\psi_o\rangle\ ,\quad x = \vec{x},t \tag{2.3}$$

where $|\psi_o\rangle$ is the exact Heisenberg ground state of the system, and $\hat{T}$ is the time-ordering operator. On expanding in the complete, orthonormal set $\{U_n\}$,

$$i\Pi(x,x') = \sum_{\alpha\beta\lambda\mu} U^{*}_{\mu}(\vec{x})U_{\lambda}(\vec{x})U^{*}_{\alpha}(x')U_{\beta}(x')\ i\Pi_{\lambda\mu;\alpha\beta}(t-t'). \tag{2.4}$$

The relation between the PH polarization propagator and the inverse dielectric function is

$$\varepsilon^{-1}(x,x') = \delta^{4}(x-x') + \int dx''\ V(x-x'')\Pi(x'',x')\ , \tag{2.5}$$

or

$$\varepsilon^{-1}_{\lambda\mu;\alpha\beta}(t-t') = \delta(t-t')\delta_{\lambda\alpha}\delta_{\mu\beta} + \sum_{\gamma\delta} V_{\mu\lambda;\gamma\delta}\Pi_{\delta\gamma;\alpha\beta}(t-t'). \tag{2.6}$$

The frequency transform of $\Pi_{\lambda\mu;\alpha\beta}(t-t')$ is defined as

$$\Pi_{\lambda\mu;\alpha\beta}(t-t') = (1/2\pi) \int d\omega \, \Pi_{\lambda\mu;\alpha\beta}(\omega) \, e^{-i\omega(t-t')}, \quad (2.7)$$

and using the Lehmann representation $\Pi_{\lambda\mu;\alpha\beta}(\omega)$ becomes

$$\Pi_{\lambda\mu;\alpha\beta}(\omega) = \sum_n \left[\frac{\langle\Psi_0|\hat{c}_\mu^+\hat{c}_\lambda|\Psi_n\rangle\langle\Psi_n|\hat{c}_\alpha^+\hat{c}_\beta|\Psi_0\rangle}{\omega - (E_n^N - E_0^N) + i\eta} - \frac{\langle\Psi_0|\hat{c}_\alpha^+\hat{c}_\beta|\Psi_n\rangle\langle\Psi_n|\hat{c}_\mu^+\hat{c}_\lambda|\Psi_0\rangle}{\omega + (E_n^N - E_0^N) - i\eta} \right],$$

$$\eta \to 0^+ , \quad (2.8)$$

where the intermediate states $|\Psi_n\rangle$ refer to excited states of the N-particle system. The poles of $\Pi_{\lambda\mu;\alpha\beta}$ give the excited state of the system which can be reached by a density perturbation.

By using the interaction representation, the structure of $\Pi_{\lambda\mu;\alpha\beta}$ is depicted as

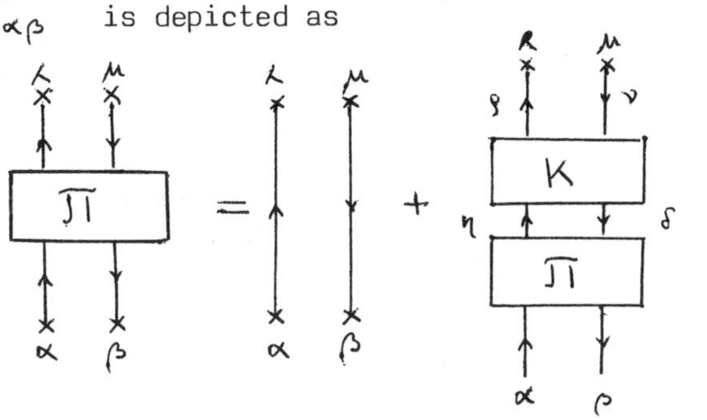

(2.9)

The directed lines are single-particle propagators which in general have self-energy renormalization. $\Pi_{\lambda\mu;\alpha\beta}(\omega)$ also contains disjoint graphs, however these terms are independent of t-t' and contribute only at $\omega = 0$. Since the main interest is for $\omega \neq 0$, we omit these terms. Using Eq. (2.9), the Bethe-Salpeter equations are

$$\Pi_{\lambda\mu;\alpha\beta}(\omega) = \Pi^0_{\lambda\mu;\alpha\beta}(\omega) + \sum_{\rho\nu;\eta\delta} \int \frac{d\omega_1 d\omega_2}{2\pi^2} \Pi_{\lambda\mu;\rho\nu}(\omega;\omega_1) K_{\rho\nu;\eta\delta}(\omega;\omega_1-\omega_2) \Pi_{\eta\delta;\alpha\beta}(\omega,\omega_2) \quad (2.10)$$

where

$$\Pi^0_{\lambda\mu;\alpha\beta}(\omega;\omega_1) \equiv -i\, G_{\lambda\alpha}(\omega_1)\, G_{\beta\mu}(\omega_1+\omega) \quad (2.11)$$

and

$$\Pi_{\lambda\mu;\alpha\beta}(\omega) = \int \frac{d\omega_1}{2\pi} \tilde{\Pi}_{\lambda\mu;\alpha\beta}(\omega;\omega_1) \; . \tag{2.12}$$

Note that $G_{\lambda\delta}$ refers to the full one-particle Green's function. Eq. (2.10) is an integral equation, but if the kernel $K(\omega;\omega_1-\omega_2)$ depends only on ω, the Bethe-Salpeter equation would reduce to a Dyson's equation:

$$\Pi_{\lambda\mu;\alpha\beta}(\omega) = \Pi^{o}_{\lambda\mu;\alpha\beta}(\omega) + \sum_{\rho\nu;\eta\delta} \Pi^{o}_{\lambda\mu;\alpha\beta}(\omega)\, K_{\rho\nu;\eta\sigma}(\omega)\, \Pi_{\eta\delta;\alpha\beta}(\omega) \; , \tag{2.13}$$

It is this equation that will be investigated.

In matrix notation

$$\underline{\underline{\Pi}}(\omega) = \underline{\underline{\Pi}}^{o}(\omega) + \underline{\underline{\Pi}}^{o}(\omega)\, \underline{\underline{K}}(\omega)\, \underline{\underline{\Pi}}(\omega) \; . \tag{2.14}$$

This equation is factorable, and its inverse is

$$\underline{\underline{\Pi}}^{-1}(\omega) = \underline{\underline{\Pi}}^{o-1}(\omega) - \underline{\underline{K}}(\omega) \tag{2.15}$$

The matrix elements of $\underline{\underline{\Pi}}(\omega)$ have a pole at $\omega = E^N - E^N_0$ unless the numerator in Eq. (2.8) vanish; however, the structure of $\underline{\underline{\Pi}}(\omega)$ in the vicinity of the pole is analytic, and for real ω ,

$$\underline{\underline{\Pi}}^{+}(\omega) = \underline{\underline{\Pi}}(\omega) \; . \tag{2.16}$$

Thus, for real ω , $\underline{\underline{\Pi}}(\omega)$ can be diagonalized with a unitary fransformation

$$\underline{\underline{U}}(\omega)\, \underline{\underline{\Pi}}(\omega)\, \underline{\underline{U}}^{-1}(\omega) = \underline{\underline{\Pi}}^{D}(\omega) \; . \tag{2.17}$$

From the inverse

$$\underline{\underline{U}}(\omega)\, \underline{\underline{\Pi}}^{-1}(\omega)\, \underline{\underline{U}}^{-1}(\omega) = \underline{\underline{\Pi}}^{D-1}(\omega) \; , \tag{2.18}$$

one sees that $\underline{\underline{U}}(\omega)$ also diagonalizes $\underline{\underline{\Pi}}^{-1}(\omega)$. Since some of the diagonal matrix elements $\underline{\underline{\Pi}}^{D}(\omega)$ have poles at the exact excitation energies of the system, the corresponding diagonal matrix elements of $\underline{\underline{\Pi}}^{D-1}(\omega)$ have zeros at the same points. Thus, the zero eigenvalues of $\underline{\underline{\Pi}}^{-1}(\omega)$ correspond to the collective energy levels of the system, and one needs to solve the frequency dependent eigenvalue problem

$$\underline{\underline{\Pi}}^{-1}(\omega)\, \vec{C}(\omega) = \Lambda(\omega)\, \vec{C}(\omega) \; , \tag{2.19}$$

with $\Lambda(\omega) = 0$. Combining Eq. (2.19) with Eq. (2.15), one obtains

$$[\underline{\underline{\Pi}}^{0-1}(\omega) - \underline{\underline{K}}(\omega)]\,\vec{C}(\omega) = 0, \tag{2.20}$$

from which the excitation energies ω and the coresponding eigenvector $\vec{C}(\omega)$ can be obtained.

Let us look at the structure of $\underline{\underline{K}}(\omega)$. The first-order expression for $\underline{\underline{K}}(\omega)$ is

$$K^{(1)}_{\rho\nu;\eta\sigma} \equiv V_{\sigma\eta;\rho\nu} - V_{\rho\eta;\sigma\nu}, \tag{2.21}$$

where

$$V_{\sigma\eta;\rho\nu} = \int d\vec{r}_1 d\vec{r}_2 U^*_\sigma(\vec{r}_1)U^*_\rho(\vec{r}_2)\frac{1}{|\vec{r}_1-\vec{r}_2|}U_\nu(\vec{r}_2)U_\eta(\vec{r}_1). \tag{2.22}$$

Note that $K^{(1)}$ is frequency independent, so that Eq. (2.13) is correct to the first order in the electron-electron interaction. The one-particle Green's function may be written as

$$G_{\lambda\alpha}(\omega) = \sum_n \frac{A^n_\lambda(\omega)A^{n*}_\alpha(\omega)}{\omega - \omega_n(\omega)} \tag{2.23}$$

The usual pole structure is implied with ω_n referring to the ionization energy of the (N+1)-body system. The $A^n_\lambda(\omega)$ is a one-particle amplitude between the N- and (N$\pm$1)-body system. For free particles, Eq. (2.23) becomes

$$G^0_{\lambda\alpha}(\omega) = \delta_{\lambda\alpha}\left[\frac{\theta(\alpha - F)}{\omega - \omega^0_\alpha + i\eta} + \frac{\theta(F-\alpha)}{\omega - \omega^0_\alpha - i\eta}\right]. \tag{2.24}$$

Here F refers to the Fermi energy and ω^0_α to the eigenvalue of the free-particle Hamiltonian.

To get $\underline{\underline{\Pi}}^0(\omega)$, use Eqs. (2.11),(2.12) and (2.23). After performing the frequency integration $\underline{\underline{\Pi}}^0(\omega)$ becomes

$$\underline{\underline{\Pi}}^0_{\lambda\mu;\alpha\beta}(\omega) = \sum_{m,n}\left\{\frac{g_m A^m_\beta(\omega_m)A^{m*}_\mu(\omega_m)A^n_\lambda(\omega_m-\omega)A^{n*}_\alpha(\omega_m-\omega)}{\omega - [\omega_m - \omega_n(\omega_m-\omega)]} + \frac{g_m A^n_\beta(\omega_m+\omega)A^{n*}_\mu(\omega_m+\omega)A^m_\lambda(\omega_m)A^{m*}_\alpha(\omega_m)}{\omega + [\omega_m - \omega_n(\omega_m+\omega)]}\right\}. \tag{2.25}$$

Here and through this section, the index m refers to ionization of the (N+1)-particle system, while n refers to ionizations of the (N-1) system; all other indices are general. The factor g_m is

$$g_m = \left[1 - \partial \omega_m(\omega)/\partial\omega \right]_{\omega=\omega_m} . \tag{2.26}$$

If the free-particle Green's function in the expression for $\underline{\underline{\Pi}}^{o}(\omega)$ is used, one obtains

$$\Pi^{oo}_{\lambda\mu;\alpha\beta}(\omega) = \delta_{\lambda\alpha}\delta_{\beta\mu}\left[\frac{\theta(\alpha-F)\theta(F-\beta)}{\omega-(\omega^o_\alpha-\omega^o_\beta)+i\eta} - \frac{\theta(F-\alpha)\theta(\beta-F)}{\omega+(\omega^o_\beta-\omega^o_\alpha)-i\eta}\right] . \tag{2.27}$$

Inserting $\underline{\underline{K}}^{(1)}(\omega)$ into Eq. (2.14) for $\underline{\underline{K}}(\omega)$, one obtains the structure depicted as

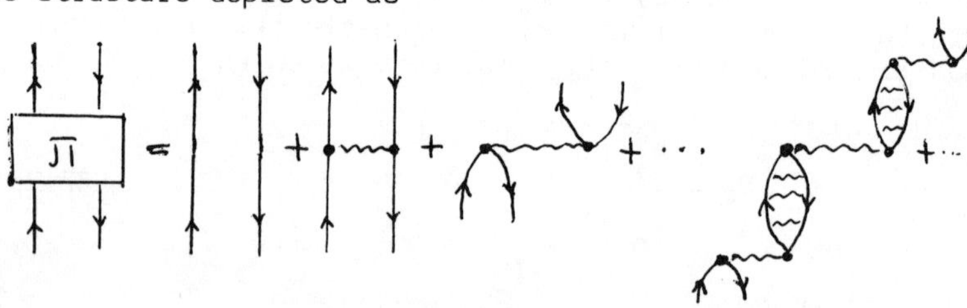

(2.28)

Using $\underline{\underline{\Pi}}^{oo}(\omega)$ instead of $\underline{\underline{\Pi}}^{o}(\omega)$ in Eq. (2.14), and $\underline{\underline{K}}^{(1)}(\omega)$ in place of $\underline{\underline{K}}(\omega)$, one obtains the random-phase approximation (RPA) equations for $\underline{\underline{\Pi}}(\omega)$. In the RPA, Eq. (2.20) becomes

$$(\omega - \omega^o_{ml})C^n_{ml} - \sum_{m'l'}\left[(V_{ml,l'm'} - V_{mm',ll'})C^n_{m'l'}\right.$$

$$\left. + (V_{ml,m'l'} - V_{ml',m'l})C^n_{l'm'}\right] = 0 \tag{2.29}$$

and

$$-(\omega + \omega^o_{ml})C^n_{lm} - \sum_{m'l'}\left[(V_{lm,l'm'} - V_{lm',l'm})C^n_{m'l'}\right.$$

$$\left. + (V_{lm,m'l'} - V_{ll',m'm})C^n_{l'm'}\right] = 0, \quad \omega^o_{ml} \equiv \omega^o_m - \omega^o_l . \tag{2.30}$$

In the RPA, one may make the identification

$$C^n_{\beta\alpha} = \langle \Psi_o | \hat{c}^+_\alpha \hat{c}_\beta | \Psi_n \rangle . \tag{2.31}$$

By using this equation, the correct spin structure for excited state $|\Psi_n\rangle$ for closed-shell systems becomes

$$C^n_{\beta\alpha,S} = (1/\sqrt{2})(C^n_{\beta\uparrow\alpha\downarrow} - C^n_{\beta\downarrow\alpha\uparrow}); \quad C^n_{\beta\alpha;T} = \begin{cases} C^n_{\beta\uparrow\alpha\uparrow} \\ C^n_{\beta\downarrow\alpha\downarrow} \\ (1/\sqrt{2})(C^n_{\beta\uparrow\alpha\downarrow} + C_{\beta\downarrow\alpha\uparrow}). \end{cases} \tag{2.32}$$

The indices S and T refer to singlet and triplet, and the spin is indicated explicitly. Rearranging Eqs. (2.29) and (2.30) with the use of Eq. (2.32) and assuming the basis set is real, one obtains

$$\sum_{m'l'} (\Delta E_{ml;m'l'} C^n_{m'l'} + M_{ml;m'l'} C^n_{l'm'}) = \omega C^n_{ml}$$

and (2.33)

$$\sum_{m'l'} (\Delta E_{ml;m'l'} C^n_{l'm'} + M_{ml;m'l'} C^n_{m'l'}) = -\omega C^n_{lm}$$

where

$$\Delta E_{ml;m'l'} = \delta_{mm'}\delta_{ll'}(\omega^0_m - \omega^0_l) + 2\gamma V_{ml;m'l'} - V_{mm';ll'},$$

$$M_{ml;m'l'} = 2\gamma V_{ml;m'l'} - V_{ml';m'l},$$

$$\gamma = 1 \text{ (singlet)}; \qquad \gamma = 0 \text{ (triplet)}.$$

In addition, using restricted Hartree-Fock orbitals, $V_{ml;m'l'}$ is the matrix element over spatial orbitals assumed to the independent of spin. The matrix M describes multipair excitations. In M = 0, Eqs. (2.33) are separable. Then their diagonal elements yield

$$\omega_T \approx \omega^0_{ml} - V_{mm;ll}, \quad \omega_S \approx \omega^0_{ml} - V_{mm;ll} + 2V_{ml;ml}. \tag{2.34}$$

The structures of Eq. (2.20) would be similar to that of Eq. (2.33) if $\underline{\underline{\Pi}}^0(\omega)$ in Eq. (2.20) is retained. This is approximately true in most cases. The major difference would be that ω^0_n would be replaced by ω_n in Eqs. (2.33). Since ω_n differs from ω^0_n by the renormalization effects coming from the particle or hole self-energies, an investigation of the self-energies would yield these effects. Also, it is clear from the structure of Eqs. (2.33) and (2.34) that $\underline{\underline{K}}(\omega)$ contains the particle-hole interactions. Thus one has a clear separation between structures that contribute to the relaxation and correlation of the hole or particle and those which contribute to particle-hole interactions.

To outline the foregoing, make the replacements

$$A_\lambda^n(\omega) = A_\lambda^n \quad , \quad g_m = 1 \quad , \quad \omega_m(\omega) = \omega_m .$$

Eq. (2.25) becomes

$$\Pi_{\lambda\mu;\alpha\beta}(\omega) = \sum_{ml} \left[\frac{A_\beta^m A_\mu^{m*} A_\lambda^l A_\alpha^{l*}}{\omega-(\omega_m-\omega_l)+i\eta} - \frac{A_\beta^l A_\mu^{l*} A_\lambda^m A_\alpha^{m*}}{\omega+(\omega_m-\omega_l)-i\eta} \right] . \tag{2.35}$$

The poles $\Pi(\omega)$ occur at

$$\omega = \pm (\omega_m - \omega_n) .$$

If the poles $\Pi(\omega)$ are not far from this value, then the preceeding replacement is valid. On using the definition

$$\Pi^{o'}_{ml;m'l'} = \delta_{mm'} \delta_{ll'} \sum_{\lambda\mu\alpha\beta} (A_\beta^m)^{-1} (A_\mu^{m*})^{-1} \Pi^{o}_{\lambda\mu;\alpha\beta} (A_\alpha^{l*})^{-1} (A_\lambda^l)^{-1} , \tag{2.36}$$

Eq. (2.14) transforms to

$$\underline{\underline{\Pi}}'(\omega) = \underline{\underline{\Pi}}^{o'}(\omega) + \underline{\underline{\Pi}}^{o'}(\omega) \underline{\underline{K}}'(\omega) \underline{\underline{\Pi}}'(\omega) . \tag{2.37}$$

Here, $\underline{\underline{K}}'(\omega)$ has the same structure as $\underline{\underline{K}}(\omega)$ except that it contains

$$V'_{no;pq} = \sum_{\delta\eta;\rho\nu} A_\delta^n A_\eta^{o*} V_{\delta\eta;\rho\nu} A_\rho^{p*} A_\nu^q . \tag{2.38}$$

Thus one can solve equations for the renormalized excitations of the same form as the RPA equations,

$$\left[\underline{\underline{\Pi}}^{o'-1}(\omega) - \underline{\underline{K}}^{(1)'}(\omega) \right] \vec{C}(\omega) = 0 ,$$

except that in Eqs. (2.33), ω_n would replace ω_n^o and $V'_{no,pq}$ would replace $V_{no,pq}$.

Finally, the construction of $\underline{\underline{\Pi}}(\omega)$ using the Bethe-Salpeter amplitude is

$$\underset{\lambda\mu;\alpha\beta}{\Pi}(\omega) = \sum_n \left[\frac{C^n_{\lambda\mu}(\omega)C^{n*}_{\alpha\beta}(\omega)}{\omega - \omega^{ex}_n(\omega) + i\eta} - \frac{C^n_{\beta\alpha}(\omega)C^{n*}_{\mu\lambda}(\omega)}{\omega + \omega^{ex}_n(\omega) - i\eta} \right] . \qquad (2.39)$$

To perform these calculations one needs the self-energy structure of the one-particle Green's function. The general structure of the one-particle Green's function presented in Eq. (2.23) is given by the Dyson's equation

$$G_{\lambda\alpha}(\omega) = G^0_{\lambda\alpha}(\omega) + \sum_{\eta\sigma} G^0_{\lambda\eta}(\omega) \sum_{\eta\sigma}(\omega)\, G_{\sigma\alpha}(\omega). \qquad (2.40)$$

The self-energy $\Sigma_{\eta\sigma}(\omega)$ contains all many-body corrections to the one-particle Green's function. Similar to Eq. (2.20), one seeks solutions to the equation

$$\sum_\alpha \left[G^{0-1}_{\lambda\alpha}(\omega) - \Sigma_{\lambda\alpha}(\omega) \right] A^n_\alpha(\omega) = 0, \qquad (2.41)$$

which was used to construct Eq. (2.23). The general structure of $G_{\lambda\alpha}(\omega)$ in Lehmann representation is

$$G_{\lambda\alpha}(\omega) = \sum \left[\frac{\langle \Psi^N_0 | \hat{c}_\lambda | \Psi^{N+1}_n \rangle \langle \Psi^{N+1}_n | \hat{c}^+_\alpha | \Psi_0 \rangle}{\omega - (E^{N+1}_n - E^N_0) + i\eta} + \frac{\langle \Psi^N_0 | \hat{c}^+_\alpha | \Psi^{N-1}_n \rangle \langle \Psi^{N-1}_n | \hat{c}_\lambda | \Psi^N_0 \rangle}{- (E^N_0 - E^{N-1}_n) - i\eta} \right] . \qquad (2.42)$$

Thus the $\{\omega_n\}$ refer to ionizations of the (N+1)-particle system. One interpretation of ω_1 (or ω_m) is to think of it as the energy necessary to create a hole among the occupied orbitals (or to fill a hole in the virtuals), and $\Sigma_{\lambda\alpha}(\omega)$ describes the relaxation and rearrangement of electrons around the hole.

The structure of $\Sigma_{\lambda\alpha}(\omega)$ may be determined using many-body perturbation theory. If one chooses self-consistence field-restricted Hartree-Fock orbitals, all diagrams contributing to canonical Hartree-Fock are omitted. The second-order expression for $\Sigma_{\lambda\alpha}(\omega)$ is

$$\Sigma^{(2)}_{\lambda\alpha}(\omega) = \sum_{mlm'} \frac{V_{\lambda m';ml}(2V_{ml;\alpha m'} - V_{m'l;\alpha m})}{\omega + \omega^o_l - \omega^o_{m'} - \omega^o_m + i\eta} + \sum_{mll'} \frac{V_{\lambda l';ml}(2V_{ml;\alpha l'} - V_{ml',\alpha l})}{\omega + \omega^o_m - \omega^o_l - \omega^o_{l'} - i\eta} \quad (2.43)$$

It is interesting to note that $\Sigma_{\lambda\alpha}$ involve ionizations of the (N$\pm$2)-body system as well as excitations of the N-body system. We again refer to Ref. /3/ for more discussions.

III. EXCITON SPECTRUM OF EXAMPLE POLYDIACETYLENES

In this section we demonstrate the contributions of excitons to the visible optical spectrum of two typical polydiacetylene (PDA) crystals as calculated by S. Suhai /4/. The calculations predict the first absorption maximum of PTS at 2.1 - 2.2 eV and of TCDU at 1.7 - 1.8 eV, with an exciton band width of 2.9 and 3.5 eV, respectively. In this investigation the electron-hole interaction was treated by first order perturbation theory in the framework of the Lax-Slater-Koster resolvent method.

Using atomic units, the Hamiltonian of the crystal is

$$\hat{H} = \int \hat{\psi}^+(x)\{-(1/2)\Delta + V_p(x)\}\hat{\psi}(x)dx + (1/2)\int\int \hat{\psi}^+(x)\hat{\psi}^+(x')\frac{1}{|\vec{r}-\vec{r}'|}\hat{\psi}(x')\hat{\psi}(x)\,dxdx' \quad (3.1)$$

The field operators $\hat{\psi}^+$ and $\hat{\psi}$ obey the anticommutation fermion rules, and we expand them in complete orthonormal set of Wannier spinorbitals belonging to the valence (v) and conduction (c) bands.

$$\hat{\psi}(x) = \sum_l w_v(x-l)\,\hat{a}_{lv} + \sum_l w_c(x-l)\,\hat{a}_{lc} \quad , \quad (3.2)$$

and

$$\hat{\psi}^+(x) = \sum_l w^*_v(x-l)\,\hat{a}^+_{lc} + \sum_l w^*_c(x-l)\,\hat{a}^+_{lc} \quad . \quad (3.3)$$

Substituting Eqs. (3.2) and (3.3) into Eq. (3.1), one obtains

$$\hat{H} = \hat{h} + \hat{g} \quad , \quad (3.4)$$

where

$$\hat{h} = \sum_{l,m} h^{c}_{lm}\, \hat{a}^{+}_{lc}\, \hat{a}_{mc} + \sum_{l,m} h^{v}_{lm}\, \hat{a}^{+}_{lv}\, \hat{a}_{mv} \;, \tag{3.5}$$

and

$$\hat{g} = \sum_{\substack{l_1,l_2,l_3,l_4 \\ j_1,j_2,j_3,j_4}} g\left(\begin{smallmatrix} l_1 l_2 \\ j_1 j_2 \end{smallmatrix}\middle|\middle|\begin{smallmatrix} l_3 l_4 \\ j_3 j_4 \end{smallmatrix}\right) \hat{a}^{+}_{l_1 j_1}\, \hat{a}^{+}_{l_2 j_2}\, \hat{a}_{l_3 j_3}\, \hat{a}_{l_4 j_4} \;,$$

$$j_i = v, c \;. \tag{3.6}$$

The mono- and bielectronic integrals are defined as

$$h^{j}_{lm} = \int w^{*}_{j}(x-l) \left\{ -(1/2)\Delta + V_p(x) \right\} w_{j}(x-m)\,dx \tag{3.7}$$

and

$$g\left(\begin{smallmatrix} l_1 l_2 \\ j_1 j_2 \end{smallmatrix}\middle|\middle|\begin{smallmatrix} l_3 l_4 \\ j_3 j_4 \end{smallmatrix}\right) = \iint w^{*}_{j_1}(x-l_1)\, w^{*}_{j_2}(x'-l_2)\, \frac{1}{|\vec{r}-\vec{r}\,'|}\, w_{j_3}(x'-l_3) \times$$

$$w_{j_4}(x-l_4)\,dx\,dx' \;. \tag{3.8}$$

Introducing the electron-hole picture, we define the band operators as $\hat{a}^{+}_{lc} = \hat{a}^{+}_{l}$, $\hat{a}_{lc} = \hat{a}_{l}$ and $\hat{d}^{+}_{l} = \hat{a}_{lv}$, $\hat{d}_{l} = \hat{a}^{+}_{lv}$.

In terms of these operatos, the ground state wave function is defined as

$$|\Phi_o\rangle = \hat{d}_{l_1} \cdots \hat{d}_{l_N} |\,\mathrm{vac}\,\rangle \;. \tag{3.9}$$

In the electron-hole pair separated by $\vec{R}_s$ becomes

$$|\Psi_{l+s,l}\rangle = \hat{a}^{+}_{l+s}\, \hat{d}^{+}_{l} |\Phi_o\rangle \;. \tag{3.10}$$

Stationary eigenstates with quasi-momentum $\vec{k}$ are formed as

$$|\Psi_{s,\vec{k}}\rangle = N_c^{-1/2} \sum_{l} e^{i\vec{k}\vec{R}_l} |\Psi_{l+s,l}\rangle \tag{3.11}$$

where N_c is the number of elementary cells in the crystal. One now obtains the wave function of an exciton with momentum $\vec{k}$ as a linear combination of symmetry adapted e-h pair functions with different separation.

$$|\Psi_{\vec{k}}\rangle = \sum_{s} \Omega_{s,\vec{k}} |\Psi_{s,k}\rangle . \tag{3.12}$$

Now taking advantage of the fact that the H.F. Solutions have been obtained, write the Hamiltonian as

$$\hat{H} = \langle \Phi_0 | H | \Phi_0 \rangle + \hat{H}_e + \hat{H}_h + \hat{H}_{eh} , \tag{3.13}$$

where

$$\hat{H}_e = \sum_{lm} \langle w_c(x-l) | \hat{F} | w_c(x-m) \rangle \, \hat{a}_l^+ \hat{a}_m , \tag{3.14}$$

$$\hat{H}_h = - \sum_{l,m} \langle w_v(x-l) | \hat{F} | w_v(x-m) \rangle \, \hat{d}_m^{\dagger} \hat{d}_l , \tag{3.15}$$

and

$$\hat{H}_{e-h} = - \sum_{l_1,l_2,l_3,l_4} \left\{ g\left(\begin{smallmatrix} l_1 l_2 \\ c\ v \end{smallmatrix} \middle|\middle| \begin{smallmatrix} l_3 l_4 \\ v\ c \end{smallmatrix}\right) - g\left(\begin{smallmatrix} l_2 l_1 \\ v\ c \end{smallmatrix} \middle|\middle| \begin{smallmatrix} l_3 l_4 \\ v\ c \end{smallmatrix}\right) \right. \hat{a}_{l_1}^+ \hat{a}_{l_4} \hat{d}_{l_3}^+ \hat{d}_{l_2} \tag{3.16}$$

Since the Wannier functions are not eigenfunctions of the Fock-operator, $\hat{F}$, switch to a Bloch representation to calculate the matrix elements of $\hat{H}_e$ and $\hat{H}_h$. This gives

$$\langle \Psi_{r,\vec{k}} | \hat{H}_e | \Psi_{s,\vec{k}} \rangle = N_c^{-1} \sum_{\vec{k}} e^{i\vec{k}(\vec{R}_r - \vec{R}_s)} \, \varepsilon_{c,\vec{k}} \tag{3.17}$$

and

$$\langle \Psi_{r,\vec{k}} | \hat{H}_h | \Psi_{s,\vec{k}} \rangle = N_c^{-1} \sum_{\vec{k}} e^{i\vec{k}(\vec{R}_r - \vec{R}_s)} (-\varepsilon_{v,\vec{k}-\vec{K}}) \tag{3.18}$$

Using

$$\langle \Psi_{r,\vec{k}} | E_k | \Psi_{s,\vec{k}} \rangle = N_c^{-1} E_{\vec{k}} \sum_{\vec{k}} e^{i\vec{k}(\vec{R}_r - \vec{R}_s)} , \tag{3.19}$$

and applying the resolvent method for the perturbation $\hat{H}_{e-h}$ (setting $E_{HF} = 0$), one obtains

$$\Psi_{\vec{k}} = \left[E_k - (\hat{H}_e + \hat{H}_h) \right]^{-1} \hat{H}_{eh} \Psi_{\vec{k}} . \tag{3.20}$$

Substituting Eq. (3.20) into Eq. (3.11), multiplying from the left with $\Psi_{n,\vec{k}}^{+}$ and integrating, one obtains

$$\Omega_{r,\vec{K}} = \sum_{s,t} \langle \Psi_{r,\vec{K}} | [E_{\vec{K}} - (\hat{H}_e + \hat{H}_h)]^{-1} | \Psi_{s,k} \rangle \langle \Psi_{s\vec{K}} | \hat{H}_{e-h} | \Psi_{t\vec{K}} \rangle \Omega_{t\vec{K}} . \tag{3.21}$$

The matrix elements of the e-h interaction become

$$\langle \Psi_{s\vec{K}} | \hat{H}_{e-h} | \Psi_{t\vec{K}} \rangle = -\sum_u e^{i\vec{K}\vec{R}_u} \left[g\left(\begin{smallmatrix} s+u & o \\ c & v \end{smallmatrix} \middle\| \begin{smallmatrix} u & t \\ v & c \end{smallmatrix}\right) - g\left(\begin{smallmatrix} o & s+u \\ v & c \end{smallmatrix} \middle\| \begin{smallmatrix} u & t \\ v & c \end{smallmatrix}\right) \right]$$

$$= V^{(v,c)}(\vec{R}_s, \vec{R}_t, \vec{K}) . \tag{3.22}$$

Substituting into Eq. (3.21), we arrive at a system of homogeneous linear equations for the determination of $\Omega_{t\vec{K}}$:

$$\Omega^{(v,c)}_{r\vec{K}} = \sum_{s,t} G^{(v,c)}(\vec{R}_r, \vec{R}_s, E_{\vec{K}}) V^{(v,c)}(\vec{R}_s, \vec{R}_t, \vec{k}) \Omega^{(vc)}_{t,\vec{K}} , \tag{3.23}$$

where

$$G^{(v,c)}(\vec{R}_r, \vec{R}_s, E_{\vec{K}}) = N_c^{-1} \sum_{\vec{K}} \frac{e^{i\vec{k}(\vec{R}_r - \vec{R}_s)}}{E_{\vec{K}} - (\varepsilon_{c,\vec{k}} - \varepsilon_{v,\vec{k}-\vec{K}})} . \tag{3.24}$$

The $E_{\vec{K}}$'s are determined by finding the zeroes of the determinant $D = |\underline{\underline{G}}\,\underline{\underline{V}} - \underline{\underline{1}}|$, where $\underline{\underline{1}}$ is the unit matrix and the wave function is obtained by solving the system of equations (3.23) for the normalized values of $\Omega_{r,\vec{K}}$.

Since the Hamiltonian applied does not contain spin dependent terms, take linear combinations of the e-h determinants to form eigenfunctions of a definite spin multiplicity. One obtains

$${}^{M}V^{(vc)}(\vec{R}_s, \vec{R}_t, \vec{K}) = \sum_u e^{-i\vec{K}\vec{R}_u} \left[g\left(\begin{smallmatrix} s+u & o \\ c & v \end{smallmatrix} \middle\| \begin{smallmatrix} u & t \\ v & c \end{smallmatrix}\right) - 2\delta_M g\left(\begin{smallmatrix} o & s+u \\ v & c \end{smallmatrix} \middle\| \begin{smallmatrix} u & t \\ v & c \end{smallmatrix}\right) \right],$$

$$\delta_M = \begin{cases} 1, & \text{singlets} \\ 0, & \text{triplets} . \end{cases} \tag{3.25}$$

Since ${}^{M}V$ is treated only in first-order perturbation, the screening effects were added by using

$$\epsilon^{-1}(r) = \epsilon_0^{-1} + \left[\frac{\epsilon_0^{-1}}{G_0} \right] e^{-Qr} , \tag{3.26}$$

where Q^{-1} plays the role of a characteristic "breakdown" length for dielectric effects. In the work of Ref. /4/ $Q = a^{-1}$ where a is the lattice constant. Since the precise value of ϵ_0 belonging to the backbones of PDA's is not known, calculations for $\epsilon_0 = 2,3$, and 5 were performed. The value of $\epsilon_0 \sim 3$ gave the binding energy of the e-h pair reduced by 0.35 - 0.4 eV. The difference between the PTS and TCDU were nearly independent of screening. These results suggest that that first singlet absorption maximum of PTS is around 2.1 - 2.2 eV, and TCDU around 1.7 - 1.8 eV. The band width is ~2.9 eV for PTS and ~3.5 eV for TCDU. The corresponding triplet solutions of 0.87 eV and 0.56 eV for PTS and TCDU, respectively.

IV. EXCITON SUPERCONDUCTOR MODEL

High-temperature has long been a cherished goal, which if achieved would have great technological importance. Currently the highest temperature at which superconductivity has been observed is approximately 25°K. These superconductors have the electrons interact via a phonon process. The critical temperature T_c at which superconductivity occurs is

$$k_B T_c = 1.14\hbar\,\omega_p e^{-1/\lambda} \quad . \tag{4.1}$$

Now if a semiconductor organic solid with a few electrons in the conduction band were to interact with the exciton field the ω_p would be replaced by ω_{ex}. The frequence $\omega_{ex} >> \omega_p$ and one has the hope of a room temperature superconductor. We shall outline a model /5/ which gives these results.

The Hamiltonian for our system of electrons interacting via a two-body potential V is

$$\hat{H} = \sum_{\vec{k}\sigma} \epsilon_{\vec{k}} \hat{n}_{k\sigma} + (1/2) \sum_{\substack{\vec{k},\vec{k}',\vec{q} \\ \sigma,\sigma'}} \langle \vec{k}+\vec{q}, \vec{k}'-\vec{q} | V | \vec{k}\vec{k}' \rangle \hat{C}^+_{\vec{k}+\vec{q},\sigma} \hat{C}^+_{\vec{k}'-\vec{q},\sigma'} \hat{C}_{\vec{k}'\sigma'} \hat{C}_{\vec{k}\sigma} \equiv \hat{H}_0 + \hat{H}_{int}, \tag{4.2}$$

where $\vec{q}$ is the change in momentum of the electrons after the interaction, $\epsilon_{\vec{k}}$ is the energy of the electrons described by a plane wave $\vec{k}$, the $\hat{C}_{\vec{k}}$'s are destruction operators of the electronic state with wave vector $\vec{k}$, $\hat{n}$ is the number operator, and σ represents the spin of the electron. In the Hartree-Fock approximation one linearizes the interaction term with respect to a given state $|0\rangle$, i.e., one approximates

$$\hat{C}_1^+\hat{C}_2^+\hat{C}_3\hat{C}_4 \approx \langle 0|\hat{C}_1^+\hat{C}_4|0\rangle \hat{C}_2^+\hat{C}_3 - \langle 0|\hat{C}_1^+\hat{C}_3|0\rangle \hat{C}_2^+\hat{C}_4 + \langle 0|C_2^+C_3|0\rangle \hat{C}_1^+\hat{C}_4 - \langle 0|\hat{C}_2^+\hat{C}_4|0\rangle \hat{C}_1^+\hat{C}_3 \quad . \tag{4.3}$$

The state $|0\rangle$ is then determined self-consistently in terms of the eigenstates of the linearized Hamiltonian.

The next step is to introduce a modified zero-order Hamiltonian

$$H'_0 = H_0 + H_\chi - \mu N \ , \tag{4.4}$$

where

$$H_\chi = \sum_{\vec{k},\sigma} \chi_{\vec{k}} \hat{n}_{\vec{k}\sigma} \tag{4.5}$$

is the Hartree-Fock potential, and the term

$$\mu N = \mu \sum_{\vec{k},\sigma} \hat{n}_{\vec{k}\sigma} \tag{4.6}$$

where μ is the chemical potential, just shifts the energy. H'_{int} then becomes

$$H'_{int} = H_{int} - H_\chi \tag{4.7}$$

This gives $H' = H'_0 + H'_{int} = H - \mu N$. Thus the Hartree-Fock approximation is equivalent to requiring that the elementary excitation spectrum of H'_0 is to first order unaffected by the residual interaction H'_{int}.

In order to build an effective Hamiltonian that can explain superconductivity one mixes the N and N $\pm$ 2 states together; so one breaks the conservation of the number of particle and adds operators of the form $\hat{C}^+_{\vec{k}\sigma}\hat{C}^+_{-\vec{k}\sigma'}$ to H'_0 and subtracts the same from H'_{int}. At this point it is convenient to introduce new field operators, which are extensions of the Nambu /6/ formalism, so that parallel and antiparallel spins are included.

Considering the following field operators,

$$\hat{\Psi}^+_{\vec{k}} = (\hat{C}^+_{\vec{k}\uparrow}, \hat{C}_{-\vec{k}\downarrow}, \hat{C}^+_{\vec{k}\downarrow}, \hat{C}_{-\vec{k}\uparrow}) \quad \text{and} \quad \hat{\Psi}_{\vec{k}} = \begin{pmatrix} \hat{C}_{\vec{k}\uparrow} \\ \hat{C}^+_{-\vec{k}\downarrow} \\ \hat{C}_{\vec{k}\downarrow} \\ \hat{C}^+_{-\vec{k}\uparrow} \end{pmatrix} , \tag{4.8}$$

one has

$$\hat{\Psi}_{\vec{k}}((1/2)\mathbb{1}\otimes\hat{\sigma}_3)\Psi_{\vec{k}} = (1/2)(\hat{C}^+_{\vec{k}\uparrow}\hat{C}_{\vec{k}\uparrow} - \hat{C}_{-\vec{k}\downarrow}\hat{C}^+_{-\vec{k}\downarrow} + \hat{C}^+_{-\vec{k}\uparrow}\hat{C}_{-\vec{k}\uparrow} - \hat{C}_{\vec{k}\downarrow}\hat{C}^+_{\vec{k}\downarrow})$$

$$= (1/2)(\hat{n}_{\vec{k}\uparrow} + \hat{n}_{-\vec{k}\downarrow} + \hat{n}_{-\vec{k}\uparrow} + \hat{n}_{\vec{k}\downarrow} - 2). \tag{4.9}$$

The convention for the outer product of the Pauli spin matrices and the 2 x 2 unit matrix is

$$\mathbb{1} \otimes \vec{\sigma}_3 = \begin{pmatrix} 1 & 0 & 0 & 0 \\ 0 & -1 & 0 & 0 \\ 0 & 0 & 1 & 0 \\ 0 & 0 & 0 & -1 \end{pmatrix} , \qquad \vec{\sigma}_3 \otimes \mathbb{1} = \begin{pmatrix} 1 & 0 & 0 & 0 \\ 0 & 1 & 0 & 0 \\ 0 & 0 & -1 & 0 \\ 0 & 0 & 0 & -1 \end{pmatrix} \tag{4.10}$$

The operator $(1/2)\mathbb{1} \otimes \vec{\sigma}_3$ is defined as $\underline{\underline{\tau}}_3$.

The 4 x 4 operators which give the singlet and the three triplet states are the following:

$$(1/2)\vec{\sigma}_3 \otimes \vec{\sigma}_1 = \hat{\underline{S}}_o \qquad \text{for } S_z = 0,\ S = 0 , \tag{4.11}$$

$$(1/2)i\vec{\sigma}_3 \otimes \vec{\sigma}_2 = \hat{\underline{S}}_1 \qquad \text{for } S_z = 0,\ S = 1, \tag{4.12}$$

$$(1/2\sqrt{2})i(\vec{\sigma}_1 \otimes \vec{\sigma}_2 + \vec{\sigma}_2 \otimes \vec{\sigma}_1) = \hat{\underline{S}}_2 \qquad \text{for } S_z = 1,\ S = 1 \tag{4.13}$$

$$(1/2\sqrt{2})i(\vec{\sigma}_1 \otimes \vec{\sigma}_3 - \vec{\sigma}_2 \otimes \vec{\sigma}_1) = \hat{\underline{S}}_3 \qquad \text{for } S_z = -1,\ S = 1 . \tag{4.14}$$

The above operators have been normalized (their inner products with themselves is unity). Another relation is

$$\underline{\underline{\tau}}_3 \hat{\underline{S}}_i \underline{\underline{\tau}}_3 = -1/4\ \hat{\underline{S}}_i \quad , \tag{4.15}$$

which will be used below.

With the above definitions one can write H_o' as

$$H_o' = \sum_{\vec{k}} \hat{\psi}^+_{\vec{k}} \left\{ \tilde{\epsilon}_{\vec{k}} \underline{\underline{\tau}}_3 + \sum_{i=o}^{3} \phi_{i\vec{k}} \hat{\underline{S}}_i \right\} \hat{\psi}_{\vec{k}} + \sum_{\vec{k}} \tilde{\epsilon}_{\vec{k}} \quad , \tag{4.16}$$

where

$$\tilde{\epsilon}_{\vec{k}} = \epsilon_{\vec{k}} + \chi_{\vec{k}} - \mu \tag{4.17}$$

and the $\phi_{i\vec{k}}$ are the "gaps" belonging to the above four cases. H_{int}' becomes

$$\begin{aligned} H_{int}' = (1/2) \sum_{\vec{k},\vec{k}',\vec{q}} \langle \vec{k}+\vec{q}, \vec{k}'-\vec{q} | V | \vec{k}\vec{k}' \rangle (\hat{\psi}_{\vec{k}+\vec{q}} \underline{\underline{\tau}}_3 \hat{\psi}_{\vec{k}}) (\hat{\psi}^+_{\vec{k}'-\vec{q}} \underline{\underline{\tau}}_3 \hat{\psi}_{\vec{k}'}) \\ - \sum_{\vec{k}} \psi^+_{\vec{k}} \left[\chi_{\vec{k}} \underline{\underline{\tau}}_3 + \sum_{i=o}^{3} \phi_{i\vec{k}} \hat{\underline{S}}_i \right] \hat{\psi}_{\vec{k}} . \end{aligned} \tag{4.18}$$

The Green's matrix elements can be written as

$$G_{o\alpha\beta}(\vec{p},t) = -i \langle o | \hat{T} [\hat{\psi}_{\vec{p}\alpha}(t) \hat{\psi}^+_{\vec{p}\beta}(o)] \,|\, 0 \rangle , \tag{4.19}$$

where $\hat{T}$ is the time-ordering operator and

$$\hat{\psi}_{\vec{p}}(t) = e^{iH_o' t}\,\hat{\psi}_{\vec{p}}(o)e^{-iH_o' t} \quad . \tag{4.20}$$

For example,

$$G_{o11}(\vec{p},t) = -i\langle 0|\hat{T}[\hat{C}_{\vec{p}1}(t)\hat{C}^{\dagger}_{\vec{p}1}(o)]|0\rangle \ .$$

In order to remove the infinite c-number term for H_o', it is convenient to define $\underset{\sim}{G}_o(\vec{p},t\to o)$ as

$$G_{o11}(\vec{p},t=o) = \lim_{t\to o^-} G_{o11}(\vec{p},t) \ , \tag{4.21}$$

$$G_{o22}(\vec{p},t=o) = \lim_{t\to o^+} G_{o22}(\vec{p},t) \quad ,$$

$$G_{o33}(\vec{p},t=o) = \lim_{t\to o^-} G_{o33}(\vec{p},t) \ ,$$

$$G_{o44}(\vec{p},t=o) = \lim_{t\to o^+} G_{o44}(\vec{p},t) \quad ,$$

Using the above definitions, one obtains

$$\underset{\sim}{G}_o(\vec{p},p_o) = \frac{\left[p_{o1}+\tilde{\epsilon}_{\vec{p}}\underset{\approx}{\tau}_3+\sum_{j=o}^{3}\Phi_{j\vec{p}}\hat{\underset{\sim}{S}}_j\right]e^{i\delta p_o\tau_3}}{p_o^2-\tilde{\epsilon}_{\vec{p}}^2-\sum_{i=o}^{3}\Phi_{ip}^2+i\delta} \quad , \tag{4.22}$$

which gives the BCS-tpye definitions for the energy $E_{\vec{p}}$,

$$E_{\vec{p}}^2 = \tilde{\epsilon}_{\vec{p}}^2 + \sum_{i=o}^{3}\Phi_{i\vec{p}}^2 \quad . \tag{4.23}$$

The restriction determining the chemical potential μ is given by

$$\begin{aligned}\langle 0|\sum_{\vec{p}\sigma}\hat{n}_{\vec{p}\sigma}|0\rangle &= \sum_{\vec{p},i}(-i)\left[G_{oii}(\vec{p},t=o)\right]\ (-1)^{i+1} \\ &= 2\sum_{\vec{p}}(-i)\ \mathrm{Tr}\left[\underset{\approx}{\tau}_3\ \underset{\sim}{G}_o(\vec{p},t=o)\right] \equiv N_o \ .\end{aligned} \tag{4.24}$$

Now

$$-i\underset{\sim}{G}_o(\vec{p},t=o) = -i\int \underset{\sim}{G}_o(\vec{p},p_o)\ dp_o/2\pi \ = \frac{(E_p-\tilde{\epsilon}_p)\underset{\approx}{\tau}_3-\sum_{i=o}^{3}\Phi_{i\vec{p}}\hat{\underset{\sim}{S}}_i}{2E_{\vec{p}}} \ . \tag{4.25}$$

Substituting Eq. (4.25) into Eq. (4.24) and changing the sum to an integral, gives

$$\int_{(\epsilon_{\vec{p}}+\chi_{\vec{p}})<\mu} \left[1 - \tilde{\epsilon}_{\vec{p}}/E_{\vec{p}}\right] d^3p/(2\pi)^3 = 2\int v_{\vec{p}}^2 \; d^3p/(2\pi)^3 = N_o \; . \qquad (4.26)$$

μ must be set so that Eq. (4.26) is valid.

Consider the lowest-order diagrams of the self-energy corrections, i.e.

one finds for the self-energy Σ ,

$$\Sigma(p) = -i\,\underset{=}{\tau}_3 \int \langle \vec{p},\vec{p}\,'|V|\vec{p},\vec{p}\,'\rangle \,\mathrm{Tr}[\underset{=}{\tau}_3 \underset{-}{G}_o(p')]\; d^4p/(2\pi)^4$$
$$+\, i\int \langle \vec{p}\vec{p}\,'|V|\vec{p}\vec{p}\,'\rangle \underset{=}{\tau}_3\, \underset{-}{G}_o(p')\, \underset{=}{\tau}_3\; d^4p/(2\pi)^4 - \left[\chi_{\vec{p}}\,\underset{=}{\tau}_3 + \sum_{i=o}^{3} \phi_{i\vec{p}}\, \underset{-}{\hat{S}}_i\right]. \qquad (4.27)$$

Since $\Sigma(p) = o$ in the generalized Hartree-Fock model by construction, a BCS-type gap equation is obtained by noting that $\underset{=}{\tau}_3$ and $\underset{-}{\hat{S}}_i$'s are linearly independent.

$$\phi_{i\vec{p}} = -(i/4)\int d^4p'/(2\pi)^4 \langle \vec{p}\vec{p}\,'|V|\vec{p}\vec{p}\,'\rangle\, \phi_{i\vec{p}\,'}/(p_o'^2 - E_{\vec{p}\,'}^2) \qquad (4.28)$$

$$= -(1/4)\int d^3p'/(2\pi)^3 \langle \vec{p}\vec{p}\,'|V|\vec{p}\vec{p}\,'\rangle\, \phi_{i\vec{p}\,'}/2E_{\vec{p}\,'} \; .$$

An alternative way to the generalized HF scheme is that of self-consistent perturbation theory. The general form of $\Sigma(p)$ is

$$\Sigma(p) = [1 - Z(p)]\,p_o\,\mathbb{1} + \chi_p\,\underset{=}{\tau}_3 + \sum_{i=o}^{3} \phi_{i\vec{p}}\,\underset{-}{\hat{S}}_i \; . \qquad (4.29)$$

In the approximation of the first-order self-consistent perturbation theory $Z(p) = 1$ and $\Sigma(p)$ is given by the equation above. The quantities χ and ϕ in Eq. (4.29) are functions of the four-momentum. This generality is required to treat retardation effects.

We replace the Coulomb and exchange contribution by the matrix elements of an effective screened Coulomb interaction $V_c(p-p') = V(q)/K(q)$, $K(q)$ being the wave-vector- and frequency dependent dielectric function. To simplify the algebra further the Bloch states are approximated by plane-wave states so that the matrix elements of the screened Coulomb interaction are given by $4\pi e^2/[|q|^2 K(q)]$. The screened Coulomb part of the self-energy $\Sigma(p)$ is then

$$\Sigma^{c}(p) = i\int d^4p'/(2\pi)^4 \underline{\underline{\tau}}_3\, \underline{G}_o(p')\underline{\underline{\tau}}_3\; 4\pi e^2/|\vec{q}|^2\kappa(q),\; (q=p-p'). \tag{4.30}$$

As a small number of electrons are added to the conduction band, polarization of the valence band will occur. Since only a comparatively small number of electrons is in the conduction band, a rigid-band approximation can be made. This implies that $\tilde{\epsilon}_{\bar{p}}$ remains the same. Assuming for the moment that the interaction is spin independent (which it is not), the lowest-order dressed exciton contribution to the self-energy becomes

$$\Sigma^{ex}(p) = i\int d^4p'/(2\pi)^4\, \underline{\underline{\tau}}_3\, \underline{G}_o(p')\underline{\underline{\tau}}_3 \sum_{\lambda}(\bar{g}_{pp'\lambda})^2\, D_\lambda(p-p')\;, \tag{4.31}$$

where the screened exciton Green's function $D_\lambda(q)$ is given by

$$D_\lambda(q) = \frac{2\Omega_{q\lambda}}{q_o^2 - \Omega_{q\lambda}^2/\kappa(q) + i\delta}\;. \tag{4.32}$$

Here λ is a polarization index and $\bar{g}_{pp'\lambda}$ is the screened coupling constant $g_{pp'\lambda}/\kappa(q)$. In the following, the polarization index will be dropped and $\Omega_{q\lambda}$ will be approximated by a single dispersionless exciton frequency ω_{ex}.

The Fröhlich interaction substituted for the exciton-electron interaction gives

$$g_q = -(i\hbar\,\omega_{ex}/|\vec{q}|)\left[\frac{\hbar}{2m\omega_{ex}}\right]^{1/4}(4\pi\alpha)^{1/2}\;, \tag{4.33}$$

where the dimensionless coupling constant α is given by

$$\alpha = \frac{e^2}{2\sqrt{\hbar/2m\omega_{ex}}}\left[\frac{1}{\hbar\omega_{ex}}\right](1 - 1/\epsilon_\infty). \tag{4.34}$$

ϵ_∞ is the optical dielectric constant. We obtain for $(\bar{g}_q)^2$

$$|(\bar{g}_q)|^2 = \frac{4\pi e^2}{|\vec{q}|^2\kappa(q)}\;\hbar\,\omega_{ex}(1/2)(1 - 1/\epsilon_\infty), \tag{4.35}$$

Substituting into Eq. (4.31), Eqs. (4.32) and (4.35), one obtains

$$\Sigma(p) = \Sigma^{c}(p) + \Sigma^{ex}(p) = i\int d^4p'/(2\pi)^4\, \underline{\underline{\tau}}_3\, \underline{G}_o(p')\, \underline{\underline{\tau}}_3\, V_{eff}(p-p') \tag{4.36}$$

where

$$V_{eff}(p-p') = \frac{4\pi e^2}{|q|^2 K(q)} \left[\frac{(\hbar\omega_{ex}^2/K(q))[1 - 1/\varepsilon_\infty]}{q_o^2 - \omega_{ex}^2/K(q) + i\delta} + 1 \right] . \qquad (4.37)$$

From Eq. (4.37) we see that for certain concentractions and $q_o^2 < \omega_{ex}^2/K(q)$ this effective interaction between electrons due to the screened Coulomb interaction and the exchange of virtual dressed excitons can become attractive corresponding to an overscreening of the Coulomb repulsion between the electrons. The attractive region is responsible for the superconductivity.

V. CONCLUSIONS

As can be seen by this paper the quasiparticle called an "exciton" is a very useful construct in describing the physics and chemistry of organic solids. It is not only a good definition to some excitations of the N-body system but can be used in describing the polarization field as was the case for the superconductivity model.

ACKNOWLEDGEMENT

The author would like to express his sincere thanks to Professor J.J. Ladik for many helpful discussions.

REFERENCES

/1/ Frenkel, J., Phys. Rev. 37 (1931) 17.
/2/ Wannier, G.H., Phys. Rev. 52 (1937) 191.
/3/ "Excitons, Their Properties and Uses", Reynolds, D.C. and Collins, T.C. (Academic Press, Inc.,New York) 1981.
/4/ Suhai, S., Phys. Rev. B (submitted).
/5/ Collins, T.C., Seel, M., Ladik, J.J., Chandrasekhar, M. and Chandrasekhar, H.R., Phys. Rev. B27 (1983) 140.
/6/ Nambu, Y., Phys. Rev. 117 (1960) 648.

THEORY OF THE ELECTRONIC STRUCTURE AND OPTICAL PROPERTIES OF ORGANIC SOLIDS: COLLECTIVE EFFECTS*

A. Barry Kunz

Department of Physics and Materials Research Laboratory
University of Illinois at Urbana-Champaign
Urbana, Illinois 61801, U.S.A.

Abstract: In this series of lectures we briefly consider two complimentary approaches to the study of organic solids: The method of simulation by finite clusters of molecules, and the methods of energy band theory. In both cases, the initial starting point is the Hartree-Fock method, which, as expected, turns out to be inadequate for any reasonable level of quantitative accuracy. Solids, being essentially infinite sized systems, restrict our choice of correlation methods to those which are size consistent. We are furthermore interested in properties such as optical excitation and need to be able to obtain the finite difference between extensive total energies. This further restricts our choices. Methods based upon ordinary Rayleigh-Schrodinger Perturbation-Theory are chosen and extensive results for solid CH_4 are used as an illustration.

*This research has been funded in part by the U. S. Navy Office of Naval Research, ONR-N-0014-81-K-0620, in cooperation with the Department of Physics, Michigan Technological University, and by the National Science Foundation, DMR-80-20250 in cooperation with the Materials Research Laboratory of the University of Illinois.

I. Introduction.

Theoretical studies on the electronic structure of three dimensional solids have largely excluded the organic or molecular solids. The vast majority of existing calculations have been performed for the solid rare gases.[1] More complicated molecular solids, such as those with two or more atoms per molecu-

J. Ladik et al. (eds.), Quantum Chemistry of Polymers - Solid State Aspects, 83–99.

lar unit, or more than one molecular unit per unit cell, have been largely ignored. The principal exception to this tendency has been for solid H_2. It should be further noted, that the interest in solid H_2 stems largely from interest in its possible transformation to a monatomic metal, exhibiting high temperature BCS type superconductivity.[2] In addition, a recent resurgance of interest in the solid rare gases has been generated by speculations that "low" pressure (~0.3 M bar) metal phases of solid Xe might exist.[3] In addition, some theoretical studies on solid N_2[4] and H_2O[5] exist. In addition, quite a few studies on properties of polymer systems exist.[6] There are probably many reasons for the neglect of this interesting and technologically interesting class of solids. Several of the reasons are likely related to the complicated and at times ill-defined crystal structure of such systems and the associated difficulties in constructing adequate theoretical models. A second and perhaps more serious problem relates to the question of which approach one might use to determine the electrical structure. As an example, the spectrum of solid CH_4 has been determined over an energy range of 8 to about 35 eV. The fundamental spectral region of from threshold (>8.5 eV) to about 14 eV shows marked similarity in both solid and gas phase. It is generally conceded that the gas phase spectra in this energy region is dominated by transitions from the bonding to antibonding bound state orbitals or to Rydberg series like transitions.[7] It seems reasonable to expect that the crystalline spectrum is likely to be similarly dominated by transitions to bound rather than free final states. That is we do not expect the contributions from energy band theory to play a major role in the low lying excitations of solid CH_4. On the other hand the spectral region above 14 or so eV may well be dominated by band to band transitions and this may account for the apparent differences between the high lying spectrum of gas phase CH_4 and solid CH_4. Similar considerations apply to many other molecular solids.

The previous theoretical study on solid methane lends credibility to this argument, as the calculation of Piela, Pietronero, and Resta finds a band gap in excess of 27.2 eV for solid methane.[8] It is not likely that this result is quantitatively accurate as these authors used a very abbreviated basis set in their calculation and found the conduction results to be highly sensitive to the virtual basis set. A further source of error in this early study is the use of the Hartree-Fock approximation uncorrected for any correlation corrections. Similar studies by Mickish and Kunz on the somewhat similar solid rare gases have found that the Hartree-Fock method consistently overestimates the band gap of these systems by about 4 or 5 eV.[1] In addition all band methods are inaccurate, in that, all neglect the formation of local excited states called excitons.

We believe new approaches are needed if one is to truly interpret the electronic structure of such systems as solid Methane. Recent theoretical results of Kunz and Flynn have demonstrated that it is possible to include the effect of electron-hole interaction and exciton formation without violating Bloch's theorem in calculations of the optical properties of such divergent solids as LiF and Mg or Ca. This is accomplished by means of a degenerate perturbative calculation using the $\vec{K}$-conserved one body valence to conduction band excitations as a basis.[9] This model retains the periodic symmetry of the lattice avoiding complications introduced by the use of finite cluster models to describe the local excitations. These finite cluster models, nonetheless, are useful and accurate approximates as we shall see. The formulationm of the problem in this way by Kunz and Flynn causes one to wish to begin with Hartree-Fock descriptions of the solid since a well defined Many-Body wavefunction is needed. The Hartree-Fock model neglects all correlations and the limited basis set used to describe excitonic effects does not describe properly the relaxation or polarization properties of the system. In these lectures, the correlation effects are incorporated by means of a simple Many Body Perturbation Theory Model (MBPT). The necessary theoretical methods are described in Section II. The numerical calculations are described in Section III, and conclusions are given in the final section.

II. Theoretical Development

The initial step in this development is the choice of the Hartree-Fock method. This choice is largely determined by the need to perform extensive correlation calculations in addition to the initial Hartree-Fock study. To facilitate development, we employ variants on the familiar Linear-Combination-of-Atomic-Orbitals method (LCAO). In the case of cluster calculations, these AO's are first rotated into molecular orbitals (MO's) spanning the entire cluster, and in the case of the band calculations, the AO's are rotated into MO's spanning the crystallographic unit cell. This rotation into MO's is advantageous because for unit cells of ever increasing size or complexity, an adequate description in terms of AO's yields rather large secular determinants. The LCMO scheme reduces substantially the size of the secular determinant. This method was first introduced by Piela *et al.*[8] for studies on solid methane. In such a simple case the basis set for the occupied orbitals is reduced from 9 to 5 orbitals. Furthermore, the MO's may contain polarization functions in them and therefore yield far greater accuracy than a much larger set of AO's.

The essential features of this approach is this. Let each unit cell be divided into molecules (real ones or not), and we devise a basis set to represent the MO's of these molecules. The primitive basis set used are spherical-harmonic Gaussian Type Orbitals (GTO's) centered about different origins, and have the form

$$\chi_i(\vec{r}-\vec{R}_i) = \exp\,[-Z_i(\vec{r}-\vec{R}_i)^2]\; Y_\ell^{\,m}\,(\theta,\phi). \tag{1}$$

The $\vec{R}_j$ are the origins about which these functions are centered and need not be an actual nuclear site, the $Y_\ell^{\,m}$ are the usual spherical harmonics. The orbital exponent Z_i is chosen by energy minimization. The MO's in turn are just linear combinations of these GTO's,

$$\phi_j\,(\vec{r}-\vec{R}_\alpha) = \sum_i \; a_i^j\, \chi_i(\vec{r}-\vec{R}_i), \tag{2}$$

The $\vec{R}_\alpha$'s are the locations of the molecules in the system. From the MO's, one forms Bloch orbitals which span the entire system:

$$\psi_j(\vec{k},\vec{r}) = N^{-1/2} \sum_\alpha \; e^{i\vec{k}\cdot\vec{R}_\alpha}\, \phi_j(\vec{r}-\vec{R}_\alpha). \tag{3}$$

The MO's (Eq. (2)) or the Bloch functions (Eq. (3)) form the basis by which we solve the HF problem or its extensions.

The first point is that the Hartree-Fock equation need be solved self-consistently. For a finite molecular cluster, this is achieved by conventional iterative means. However, the infinite periodic system imposes special difficulties. These are simply that the occupied canonical Bloch orbitals are infinite in number and therefore enumerating the contribution of each orbital to the Fock operator imposes a strain on ones computer budget. Two options are available. The first is to use a finite mesh in $\vec{k}$- space and use some form of quadrature to construct the Fock operator. The second is to rotate into a basis set of local orbitals.[10,11,12] In the early stages, both methods were tried with negligible differences in numerical result between them. However, at the current stage of our code development, the local orbitals method enjoys a substantial speed advantage.

The intent of the present study is to obtain spectroscopic information and hence we need examine the meaning of the energy bands. The occupied bands are the negative of the ionization energy for that band for the state of wave vector $\vec{k}$. The virtual bands are similar representations for the electron affinities. In this event one is assuming the use of the Hartree-Fock eigenvalue and also of Koopmans' theorem as is usually done. The essential physics here refers to ionization properties, not to excitation properties of the n-electron system.

In order to improve upon the Hartree-Fock results one must include correlation corrections. In doing this, initially the author will maintain the same physical definition for the energy bands as in the Koopmans' case. That is, the bands now become quasi-particle bands in which the energy of an occupied level is the negative of the energy needed to create it, and the energy of the virtual states are the negative of the energy recovered in creating it. This is in keeping with the earlier usage of the electronic polaron model and its extensions as discussed by Pantelides _et al._[13,14,15]

It is now necessary to discuss correlation corrections. The first problem is that of size consistency (The total energy is an extensive quantity).[16,17] In fact the total energy of an infinite solid is infinite and only the energy/molecule is finite. Unfortunately, the energy change upon ionization is also finite and the energy change/molecule vanishes. That is the energy difference is still finite. Similar considerations apply to excitations of the n-electron system. A simple classical way to view this is to realize that the size of the wave created by hurling a brick into a pond is largely independent of the size of the pond. Therefore we must establish a size consistent framework for the system total energies in such a way that formally we can obtain differences in extensive quantities, cancelling the infinities before we compute finite differences. Alternately, we may reduce the size of the systems so that total energy determinations are possible.

Let us work in a local representation here. This is appropriate since many molecular solids are filled shell systems. For notational simplicitiy, designate the Wannier function $W_{iN}(\vec{r})$ as the ith Wannier function about site $\vec{R}_N$. Form a complete set of Wannier orbitals describing the ground state of the neutral, N-electron solid in the Hartree-Fock limit. We will use them to generate the ion states as well. For a system of N-electrons the Hamiltonian is

$$H = \sum_{i=1}^{n} - \frac{\hbar^2}{2m} \nabla_i^2 - \sum_{i=1}^{N} \sum_{I=1}^{M} \frac{Z_I e^2}{|\vec{r}-\vec{R}_I|} + \frac{1}{2} \sum_{i=1}^{N} \sum_{j=1}^{N} {}^{1} \frac{c^2}{|\vec{r}_{ij}|} \qquad (4)$$

The electronic has mass m, and is charge e, Z_I is the atomic number of the nucleus at site I. The ith electron has coordinate $\vec{r}_i$ and the Ith nucleus has coordinate $\vec{R}_I$. In terms of Wannier Functions, in the single determinant limit, the energy of the system is

$$E_N = \sum_{i_j}^{(N)} < w_{iI}| - \frac{\hbar^2}{2m} \nabla^2 - \sum_{J=1}^{M} \frac{Z_J e^2}{|\vec{r}-\vec{R}_j|} |w_{iI}> \tag{5}$$

$$+ \frac{1}{2} \sum_{\substack{iI \\ jJ}}^{(N)} [<w_{iI}w_{jJ}|\frac{e^2}{r_{12}}|w_{iI}w_{jJ}> - < w_{iI}w_{jJ}|\frac{e^2}{r_{12}}|w_{jJ}w_{iI}>]$$

The symbol (N) on the summation implies sums over all states in the occupied N electron space. To keep the physics of the energy bands discussed earlier we need to look at the N-1 and N+1 electron system next.

Let the ground state of the N-electron Hartree-Fock system be designated as $|N>$ and let α^+_{pI}, α_{pI} create or destroy a Wannier function at site I with other quantum numbers p. We adopt the conventive that quantum numbers i, j, k etc., refer to occupied orbitals, a, b, c to virtual orbitals and o, p, q to either/both. A Slater determined of the N-1 body system is

$$|N-1;\, jB > \equiv \alpha_{jB}|N > \tag{6}$$

This will by symmetry adapted later. The energy expectation value of this state is simply

$$E^{jB}_{N-1} = E_N - < w_{jB}|F(N)|w_{jB}> \tag{7}$$

Here F(N) is simply the N-electron ground state Hartree-Fock operator. Similarly one may obtain the off diagonal matrix elements between two states $|N-1,\, iA>$ and $|N-1,\, jB>$. These are:

$$D^{iAjB}_{N-1} = <w_{jB}|F(N)|w_{iA}>. \tag{8}$$

One may project on the state $|N-1,\, jb>$ to form a proper translational invariant Bloch function, $\psi_j^{N-1}(\vec{k})$:

$$\psi_j^{(N-1)}(\vec{k}) = \sum_B \frac{1}{\sqrt{M}} e^{i\vec{k}\cdot\vec{R}_B} |N-1;\, jB> \tag{9}$$

In terms of eqs (5)-(9) one may construct a band structure in terms of Wannier-functions and Slater determinante for the occupied orbitals. These are yet uncorrelated. One may treat the N+1 body states similarly. Furthermore, recognizing that E_N in eq (11) is infinite and also irrelevant, since energy changes are needed, we proceed to define E_n as 0, and thus simplify our computation.

A framework is needed in order to simply correlate this problem since the simple single Slater determinants $|n-1,\, jB>$ are highly degenerate, and within a band, the $\psi_j(\vec{k})$ are nearly degenerate. Consider the problem in a general framework initially. H is a Hamiltonian,

$$H = H_o + V. \tag{10}$$

We assume that the eigenstates of H_o are known as

$$H_o\phi_i = w_i\phi_i. \tag{11}$$

The projector onto a given eigenstate of H_o say ϕ_i is P_i and is given as

$$P_i = |\phi_i\rangle\langle\phi_i|.$$

Furthermore a projector onto the first n states say is $\overline{P}$ and

$$\overline{P} = \sum_{i=1}^{n} |\phi_i\rangle\langle\phi_i| = \sum_{i=1}^{n} P_i. \tag{12}$$

Assume we order our eigenfunction of H_o so that the states of interest lie in the range 0 to n. Furthermore no other states are degenerate with these states. Now let us solve the desired equation:

$$H\psi = E\psi = (H_o+V)\psi. \tag{13}$$

Let us chose a w_j for $1 \leqslant j \leqslant n$, then:

$$(1-\overline{P})(H_o-w_j)\psi = (1-\overline{P})(E-w_j-V)\psi \tag{14}$$

One may commute $(1-\overline{P})$ with H_o-w_j and proceed to see

$$\psi = \overline{P}\psi + (H_o-w_j)^{-1} (1-\overline{P}) (E-w_j-V)\psi. \tag{15}$$

Furthermore;

$$P_i\psi = \pi_i\phi_i, \ \pi_i = \langle\phi_i|\psi\rangle,$$

so that

$$\overline{P}\psi = \sum_{k=1}^{n} \pi_k\phi_k = \phi. \tag{16}$$

Therefore

$$\left[1 - (H_o-w_j)^{-1} (1-\overline{P})(E-w_j-V)\right]\psi = \phi. \tag{17}$$

If one defines

$$T = \{1 - (H_o-w_j)^{-1} (1-\overline{P})(E-w_j-V)\}^{-1}, \tag{18}$$

then

$$\psi = t\phi. \tag{19}$$

Furthermore one can show that

$$(E-w_i)\,\pi_i = \sum_{k=1}^{n} \pi_k \overline{V}_{ik}, \tag{20}$$

where

$$\overline{V}_{ik} = \langle\phi_i|VT|\phi_k\rangle. \tag{21}$$

Eqs. (20) and (21) define a perfectly good algebraic eigenvalue equation for the system energies. To proceed further, one expands the inverse appearing in T. That is,

$$\begin{aligned} VT &= N + V\frac{1}{H_o-w_j}(1-\overline{P})(E-w_j-V) + \ldots \\ &= \overline{V}, \end{aligned} \tag{22}$$

or

$$\overline{V}_{ik} = V_{ik} + \sum_{a=N+1}^{\infty} \frac{a}{w_j-w_a} V_{ia} V_{ak}. \tag{23}$$

The structure of the eigenvalue problem defined by Eqs (20), (21) and (23) is now clear. The matrix elements to lowest approximation are similar to those of second order R.S.P.T. and this is clearly a size consistent approach. If all the eigenvectors in the first n are degenerate, one recovers normal degenerate perturbation theory. Consider our problem, where we use Wannier functions, this framework makes our approach clear. First correlate the single Slater determinant of Wannier functions, then proceed with Bloch symmetry projection to remove the degeneracy. The N-body wavefunction has proper Bloch symmetry for closed band systems. By using a proper choice of A in the Adams-Gilbert local orbital formulation called A^W one may obtain Wannier orbitals.[11] The actual choice of A^W is not important, only that such exist. Then

$$[F + PA^WP]\, w_{iI} = \varepsilon_{iI} w_{iI} \tag{24}$$

The first order Fock-Dirac density matrix is ρ. From this one constructs a zero order Hamiltonian. For a system of M-electrons, H_o is defined as

$$H_o = \sum_{i=1}^{M} [F(\vec{r}_i) + \rho_i A_i^W \rho_i] \tag{25}$$

and then the perturbation, V becomes

$$V \equiv H - H_o \tag{26}$$

$$E(N) = E_N + \sum_{iI>jJ}^{(N)} \sum_{aA>bB} \frac{|V_{iIjJ}^{aAbB}|^2}{\varepsilon_{iI}+\varepsilon_{jJ}-\varepsilon_{aA}-\varepsilon_{aB}}. \tag{27}$$

Here the summation (N) indicates all Wannier orbitals in the N-electron state. The matrix element is simply

$$V_{iIjJ}^{aAbB} = \langle w_{iI} w_{jJ} | \frac{e^2}{r_{12}} | w_{aA} w_{bB} \rangle - \langle w_{iI} w_{jJ} | \frac{e^2}{r_{12}} | w_{bB} w_{aA} \rangle \tag{28}$$

The N-body orbitals will be used to describe both the N-1 and N+1 body states. Brillouin's theorem is not valid for such states. Consider first the N-1 body problem. Let Wannier orbital w_{iB} be deleted from the N-body ground state. Then to second order one finds that

$$E(N-1,iB) = E_n - D_{N-1}^{iBiB} + \sum_{jI}^{(N+1)} \sum_{aA} \frac{|F(N-1;iB)_{jI}^{aA}|^2}{\varepsilon_{jI}-\varepsilon_{aA}}$$
$$+ \sum_{kK>jJ}^{(N-1)} \sum_{aA>bC} \frac{|V_{kK,jJ}^{aAbB}|^2}{\varepsilon_{kK}+\varepsilon_{jJ}-\varepsilon_{aA}-\varepsilon_{bC}}. \tag{29}$$

In Eq (19) the V is still as defined in Eq (18) and F(N-1;iB) is obtained by deleting terms referring to orbital w_{iB} from F(N). Therefore

$$F(N-1;iB)_{jI}^{aA} = \langle w_{jI} | F(N-1,iB) | w_{aA} \rangle. \tag{30}$$

One proceeds in like fashion for the N+1 electron case, adding W_{CB} to the N electron state.

$$E(N+1;cB) = E_N + D_{N+1}^{cBcB} + \sum_{jJ}^{(N+1)} \sum_{aA} \frac{|F(N+1,cB)_{jJ}^{aA}|^2}{\varepsilon_{jJ}-\varepsilon_{aA}}$$
$$+ \sum_{jJ>kK}^{(N+1)} \sum_{aA>dD} \frac{|V_{jJkK}^{aAdD}|^2}{\varepsilon_{jJ}+\varepsilon_{kK}-\varepsilon_{aA}-\varepsilon_{\alpha D}} \tag{31}$$

In Eq (21) V remains as in Eq (18) and F(N+1;cB) is obtained by adding terms referring to orbital w_{cB} to the N-electron Fock operator.

One may obtain the physically interesting energy differences from these expressions. The ionization potentials are defined by E(N) - E(N-1;iB). This difference called here Δ_{iB} is given as

$$\Delta_{iB} = D^{iBiB}_{N-1} + -\sum_{jJ}^{(N-1)} \sum_{aA} \frac{|F(N-1,iB)^{aA}_{jJ}|^2}{\varepsilon_{jJ} - \varepsilon_{aA}}$$

$$+ \sum_{jJ\neq iB}^{(N)} \sum_{aA>cC} \frac{|V^{aAcC}_{iBjJ}|^2}{\varepsilon_{iB}+\varepsilon_{jJ}-\varepsilon_{aA}-\varepsilon_{cC}} - \sum_{\substack{jJ>kK \\ \neq iB}} \sum_{aA} \frac{|V^{iBaA}_{jJkK}|^2}{\varepsilon_{jJ}+\varepsilon_{kK}-\varepsilon_{iB}-\varepsilon_{aA}} \tag{32}$$

Likewise the electron affinity terms are obtained by letting $\Delta_{CB} = E(N+1;CB)-E(N)$. Then

$$\Delta_{cB} = D^{cBcB}_{N+1} + \sum_{jJ}^{(N+1)} \sum_{aA} \frac{|F(N+1,\ CB)^{aA}_{jJ}|^2}{\varepsilon_{jJ} - \varepsilon_{aA}} + \sum_{iI}^{(N+1)} \sum_{aA>dD\neq cB}$$

$$\frac{|V^{aAdD}_{iIcB}|^2}{\varepsilon_{iI} + \varepsilon_{cB} - \varepsilon_{aA} - \varepsilon_{dD}} - \sum_{iI>jJ}^{(N+1)} \sum_{aA\neq cB} \frac{|V^{cBaA}_{iIjJ}|^2}{\varepsilon_{iI} + \varepsilon_{jJ} - \varepsilon_{cB} - \varepsilon_{aA}} \tag{33}$$

It is these formulas we will use in this study.

One final piece is needed to complete this theory. This is to include the actual effect of electron-hole interaction upon excitation. An accurate method of doing this for both tightly bound or loosely bound excitations has been recently given by Kunz and Flynn.[9]

The essential point is to use the Hartree-Fock bands as a basis set after incorporation of correlation corrections into the band energies. The Fock ground state $|N\rangle$ is then used to describe schematically the process. Let $\alpha_v(\vec{k})$ annihilate a valence electron of wavevector $\vec{k}$ and let $\alpha^+_c(\vec{k})$ create a conduction electron of wavevector $\vec{k}$. Consider the state then:

$$|N,\vec{k}\rangle = \alpha^+_c(\vec{k})\ \alpha_v(\vec{k})\ |N\rangle \tag{34}$$

It is only states like this which can be reached from the ground state via optical processes. Furthermore all such ground states $|N,\vec{k}\rangle$ correspond to the same total crystal wavevector; that of the ground state. The most general excited state that one may access is then $|N,\ E\rangle$, where

$$|N,e\rangle = \sum_{\vec{k}} a_{\vec{k}}\ |N,\vec{k}\rangle \tag{35}$$

In this sum, the ground state $|N\rangle$ is excluded because it differs in parity from the excited state. By finding the $a_{\vec{k}}$ and $\langle N,\ E|H|N,\ E\rangle$, one may determine the spectrum of the solid including electron-hole interaction. This is achieved by means of a CI calculation among the states $|N,\ \vec{k}\rangle$. The formation of such exciton states is not an extensive property and size consistency is not a problem as demonstrated by Kunz and Flynn.[9] Exact implementation of such an infinite CI is of course impossible and we use a finite number of states, some 270 configurations. A second approximation is made as well. This is to truncate the coulomb interaction at the boundary of a unit cell. This is not unreasonable for tightly bound excited systems as in the case of CH_4 particularly since the large lattice constant (11.14 au) encloses a substantial volume in a unit cell. The dominant consequence of this is to allow the formation of only a single bound exciton, not an entire Rydberg series below the bands. However when the coefficients $a^{\dagger}_{k}$ are used to evaluate the optical response one finds substantial adjustment over the Hartree-Fock results. These changes are due to the redistribution of oscillator strength to the bottom of the conduction band due to the inclusion of electron-hole interaction.

The alternate approach employed is to use a finite molecular cluster simulation. This is also done using the method of local orbitals. In this case one partitions the system into the cluster and the environment. The environment imposes itself on the cluster by means of a bounding potential. The methods of doing this are well represented in the literature, and a general approach is given by Kunz and Klein[18] which need not be repeated here. Correlation is imposed using the technique of this section and in particular eqs. (20), (21) and (23) as needed. For non-degenerate states of course, these reduce to ordinary second order RSPT. Most cases considered here are not degenerate in the cluster limit, but for those cases for which degeneracy is a problem, we have found the full approach to be very powerful.[19]

III. Results for Solid CH_4

A Gaussian basis set was first developed for the CH_4 molecule in free space and then reoptimized for the crystal to allow accurate description of the energy bands, occupied and virtual. It was found easy to obtain accurate valence bands, but that the conduction bands were quite sensitive to the choice of outer orbital. The variational principal applies to the solution of the one particle states in a LCMO formalism, and the selection of the basis is quite easy. In practice, the conduction bands are found to be stable against small changes in basis set. The valence structure here agrees well with that obtained after corrections to formalism by Piela *et al.*[8,20,21] The conduction bands are in very poor agreement however. This is due to the far too restrictive basis set employed in the Piela *et al.*[8] calculation of the virtual bands. In performing this calculation, some idealizations are needed. A lattice constant of 11.14 au, in agreement with Piela *et al.* is used and the C sub lattice is fixed as a fcc one as per experiment. The four H's form in tetrahedra about the C in a unit cell. In the real world, the tetrahedra do not allign from one cell to another but have orientational disorder. We, as did Piela *et al.* fix the H's in an ordered fcc lattice as well. The current calculation uses the same geometry as does Piela *et al.* The C-H distance is obtained computationally from Beck[22] and for a lattice constant of 11.14 au, the equilibrium constant, is essentially the same C-H distance as in the free molecule.

Although the Hartree-Fock band results overestimate any reasonable band gap, they do reduce the Piela gap by about 13.6 eV however,[8] and one need add correlation along the lines suggested in Section II. In performing the correlation correction computations, the author deviates from the ideals expressed in the preceding section to the extent that instead of solving for a set of rather complicated, orthogonal Wannier functions as implied by the derivations, one approximates these by a set of local orbitals. In obtaining these the unit on which localization occurs is the CH_4 molecule is used, and also the appropriate multicenter localization.[23] These orbitals are quite local, the valence-valence overlaps being 0.03 or less here. First order overlap corrections are made in the inter molecular terms for further precision. Due to the procedure adopted, *all* orders of overlap in the *large* intra molecular overlaps are included exactly. The inclusion of these corrections is essential if one wishes to achieve quantitative accuracy. In evaluating the perturbation sums, d orbitals on the C atom and p orbitals on the H atoms were added to the band structure basis set. The effect of the several contributions to Eqs (32) and (33) are given in Table 1. There we call the second term on the right hand sides of Eqs (32) and (33) the relaxation and the sum of the second and third terms, which come from two electron virtual excitations, clearly represent correlation terms.

The energy bands for CH_4 including correlation are shown in Figure 1. The density of electron states is also seen in these figures. As is clear from these figures, the band gap is indirect and from Γ_{15v} to $X_{5'C}$. The direct gap is at the X point and is $X_{5'v}$ to $X_{5'C}$. The correlated indirect gap is found to be 13.0 eV, and the correlated direct gap is found to be 13.3 eV.

Finally, one computes the position of the exciton levels in CH_4. This is accomplished using the method given in Section II which has been more fully described in Ref. 9. In this calculation the coulomb interactive is treated as a one molecule interaction. The effective electron-hole interaction is here computed to be 5.4 eV. This is the value of the V_0 discussed in

Table 1

Contributions to the ionization potential and electron affinity of solid CH_4 are shown as a function of lattice parameter. Results are given for the correlation correction and the relaxation correction. Results are in eV.

a	11.14 au	10.50 au	10.00 au
valence correlation	0.1 eV	0.2 eV	0.4 eV
valence relaxation	1.2 eV	1.2 eV	1.2 eV
conduction correlation	-.7 eV	-.8 eV	-.9 eV
conduction relaxation	~0.0 eV	~0.0 eV	~0.0 eV
net gap change	-2.0 eV	-2.2 eV	-2.5 eV

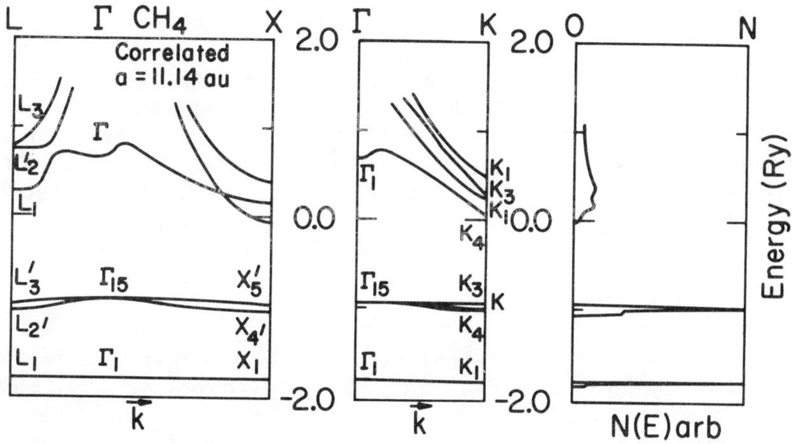

Figure 1 The correlated band structure of solid CH_4 and density of states in shown for lattice parameter = 11.14 au.

Ref. 9. Using this value, the exciton is found to be at 10.9 eV. The optical spectrum of Koch and Skibowski[24] does find a spectral peak at 11.0 eV and this may well be our exciton. A more quantitative analysis of the optical response is not possible at this time because the highest valence and lower conduction bands are of like symmetry and the techniques developed in Ref. 9 and currently available do not permit a calculation of the optical response for the case in which the band to band transitions are dipole forbidden only the positions. Therefore, the author reluctantly contents himself with using only the $\vec{k}$ conserved joint density of states in comparison with the measured reflectance spectrum shown in Figure 2.[24] As is clear from this figure, even if one were to include the exciton at 11.0 eV, a fair degree of discrepancy remains. This is largely at low energy. This discrepancy is expected. A similar result is seen in the free CH_4 molecule and is due to the mobility of the H nuclei and their large zero point motion. The excited CH_4 molecule can lower its energy by about 1.6 eV by relaxing from ideal T_D geometry to D_{2h} geometry. Due to large zero point motion it may be possible to excite from the ground state T_D geometry directly into the relaxed, distorted D_{2h} geometry directly. This certainly appears to be the case in the free molecule and a discussion of this is being prepared by Beck and Kunz.[25] If one assumes the same type of Jahn-Teller distortion is present in the solid, a distorted exciton line would then appear at about 9.3 eV. This is shown as a dotted line in Figure 1. Since the first experimental peak in solid CH_4 lies at 9.6 eV, this inclusion greatly enhances the comparison of theory and experiment. In addition the low energy continuous spectrum between about 12 and 14 eV would be enhanced in strength by the redistribution of oscillator strength due to exciton formation as was seen in LiF.[9]

Large scale cluster calculation for bulk CH_4 (13 molecules or 65 atoms) and for the CH_4 surface (9 molecules or 45 atoms), including all electrons and correlation via the perturbative route, have been recently performed by Beck.[26] These calculations are for the excitons alone and tend to confirm the energy band results qualitatively and quantitatively. The specific details of the perturbation treatment for large systems is well described in the literature.[27]

IV. Conclusions

The essential conclusions are few and simple. These are one can construct a satisfactory, self-consistent Hartree-Fock band structure for molecular solids, including the conduction bands, if one carefully optimizes the basis set. If one wishes to obtain quantitative comparisons with experiment, the inclusion of correlation corrections is essential. Furthermore, in describing the ion states in terms of the neutral system orbitals corrections termed relaxation corrections are needed. It is seen here, using a Wannier basis, how such arise and may be included. It is also seen that inclusion of electron-hole interaction is needed if one is to quantitatively study the optical spectrum. In addition, due to the light mas of H one need also be prepared to include Jahn-Teller distortion if one is to be fully quantitative.

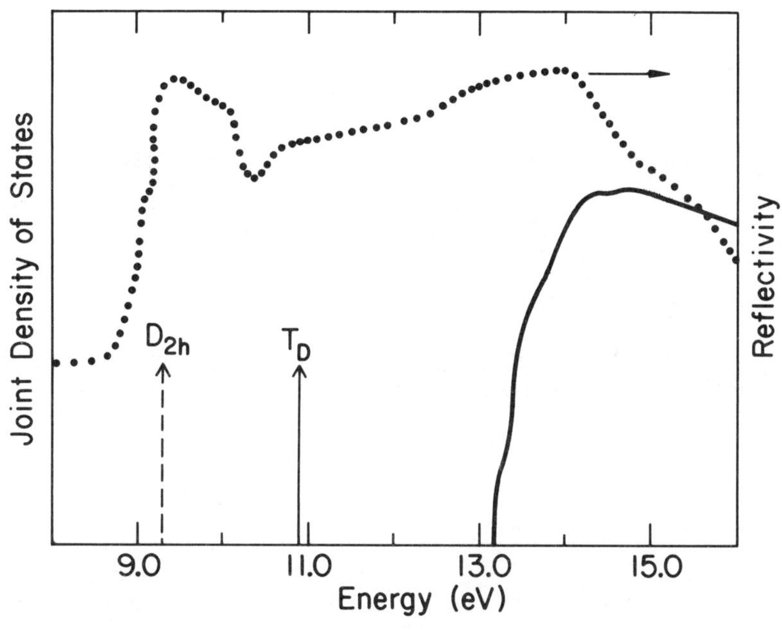

Figure 2 The optical joint density of state is shown for solid CH_4 along with the T_D geometry exciton position and probable D_{2N} geometry exciton. The optical reflectivity of Ref. 24 is also shown

References

1. A. B. Kunz and D. J. Mickish, Phys. Rev. B 8, 779 (1973); and references contained therein.

2. A. Zunger, J. Phys. Chem. Solids 36, 229 (1975); G. Pastori-Parraviccini, F. Piccini and L. Resca, Phys. Stat. Sol. 60, 801 (1972); E. L. Pollack, T. A. Bruce, G. V. Chester and J. A. Krumhansl, Phys. Rev. B 5, 4180 (1972); J. A. Krumhansl and S. Y. Wu, Phys. Rev. B 5, 4155 (1972); D. M. Gray and A. B. Gray, Phys. Rev. B 19, 5329 (1979); L. Kumar, H. J. Monkhorst and J. Oddershede, Int. J. Quant. Chem. 12, 145 (1978); C. Friedli and N. W. Ashcroft, Phys. Rev. B 16, 662 (1977).

3. D. A. Nelson and A. L. Ruoff, Phys. Rev. Lett. 42, 383 (1979); M. Ross and A. K. McMahon, Phys. Rev. B 21, 1658 (1980); A. K. Ray, S. B. Trickey, R. S. Weidman and A. B. Kunz, Phys. Rev. Lett. 45, 933 (1980); A. K. Ray, S. B. Trickey and A. B. Kunz, Sol. St. Comm. 41, 351 (1982).

4. E. Huler and A. Zunger, Phys. Rev. B 12, 5878 (1975); A. Zunger, Mol. Phys. 28, 713 (1976).

5. E. Huler and A. Zunger, Chem. Phys. 13, 433 (1975).

6. See for example, Quantum Theory of Polymers, J-M Andre, J. Delhalle and J. Ladik, eds. D. Reidel Co. Dordrecht (1978).

7. R. W. Ditchburn, Proc. Roy. Soc. 229A, 44 (1955); P. H. Metzger and G. R. Cook, J. Chem. Phys. 41, 642 (1964).

8. L. Piela, L. Pietronero and R. Resta, Phys. Rev. B 7, 5321 (1973).

9. A. B. Kunz and C. P. Flynn, J. Phys. C. 16, 1659 (1983); Phys. Rev. Lett. 50, 1524 (1983).

10. W. H. Adams, J. Chem. Phys. 34, 89 (1961); 37, 2009 (1962).

11. T. L. Gilbert in Molecular Orbitals in Chemistry, Physics and Biology, P. O. Lowdin and B. Pullman, eds. (Academic Press, 1964).

12. A. B. Kunz, Phys. Stat. Sol. (b) 36, 301 (1969); Phys. Rev. B 7, 5364 (1973); B 8, 1690 (1973).

13. Y. Toyazawa, Progr. Theoret. Phys. (Kyoto) 12, 421 (1954).

14. A. B. Kunz, Phys. Rev. B 6, 606 (1972).

15. S. T. Pantelides, D. J. Mickish and A. B. Kunz, Phys. Rev. B 10, 2602 (1974).

16. D. J. Thouless, The Quantum Mechanics of Many-Body Systems, (Academic Press, 1961).

17. E. R. Davidson and D. W. Silver, Chem. Phys. Lett. 52, 403 (1977).

18. A. B. Kunz and D. L. Klein, Phys. Rev. B 17, 4614 (1978).

19. P. Goalwin, P. Keegstra and A. B. Kunz, unpublished research results.

20. A. B. Kunz, Phys. Rev. B 9, 5330 (1974).

21. L. Piela, L. Pietronero and R. Resta, Phys. Rev. B 9, 5332 (1974).

22. D. R. Beck, unpublished research results.

23. A. B. Kunz, Phys. Rev. B 26, 2056 (1982); B. 26 2070 (1982).

24. B. H. Lombos, P. Sauvageaw and C. Sandorfy, Chem. Phys. Lett. 1, 382 (1967); E. E. Koch and M. Skibowski, Chem. Phys. Lett. 9, 29 (1972).

25. D. R. Beck and A. B. Kunz, ONR report, unpublished.

26. D. R. Beck, ONR report, unpublished.

27. R. J. Bartlett and I. Shavitt, Chem. Phys. Lett. 50, 190 (1977); R. J. Bartlett, J. Shavitt and G. D. Purvis, III, J. Chem. Phys. 71, 281 (1979); R. J. Bartlett and G. D. Purvis, III, Ann. of the N.Y. Academy of Sciences 367, 62 (1981).

CALCULATION OF THE MECHANICAL AND OPTICAL PROPERTIES OF POLYETHYLENE INCLUDING ELECTRON CORRELATION EFFECTS

Sándor Suhai

German Cancer Research Center
Im Neuenheimer Feld 280, D-6900 Heidelberg, FRG
and Chair for Theoretical Chemistry, Friedrich-Alexander-University Erlangen-Nürnberg,
Egerlandstr. 3, D-8520 Erlangen, FRG

ABSTRACT

The microscopic longitudinal elastic modulus, deviations from Hooke's law, the maximum of the retractive force and tensil strength are calculated for single infinite polyethylene chains using a first principle crystal orbital method and including electron correlation effects by perturbation theory. The ultraviolet spectrum of polyethylene is interpreted in terms of singlet charge transfer excitons.

1. CALCULATION OF THE GROUND STATE ENERGY

The investigation of the response of macromolecules to external mechanical forces or to electromagnetic fields may basically contribute to our understanding of the structural and functional properties of these systems. The starting point of all studies of this kind is the proper description of the equilibrium (ground) state of the molecule without external fields. In our a priori calculations, the ground state energy is obtained in two steps: as a zeroth order approximation the Hartree-Fock(HF) contribution is calculated by the ab initio crystal orbital method (1,2) and electronic correlation effects are included by perturbation theory afterwards.

J. Ladik et al. (eds.), Quantum Chemistry of Polymers - Solid State Aspects, 101–110.

1.1 Zeroth-order HF many-electron wavefunction

To obtain HF wavefunctions of appropriate quality for infinite systems, the following problems have to be solved:
i) Choice of symmetry adapted one-particle states: one has the possibility to use either infinitely extended Bloch-states or spatially more or less localized Wannier functions (WF's) as basis set to expand the many-electron wave function. The superiority of this latter set is unambiguos only in perturbation theory since it leads to relatively large matrix eigenvalue problems at the one-particle level and also the localization of the resulting WF's is not always optimal. We prefer, therefore, the use of Bloch-type basis functions, which are not only easier to calculate but provide at the same time also the eigenvalues of the Fock operator needed for the construction of the Green function. The price of this choice is, on the other hand, that a rather sophisticated procedure has to be applied to obtain properly localized WF's for the purpose of perturbation theory.

ii) Both the Bloch and Wannier functions are expanded in a Gaussian-type atomic basis set. Three such basis sets of increasing size have been applied in these calculations the minimal STO-3G (3), the extended 6-31G(4) and a third one containing d-type polarization functions on carbon and p-functions on hydrogen (6-31G**) (5). The polarization functions turn out to be especially important in calculating the correlation contribution to the ground state energy.

iii) The infinite lattice sums have to be truncated in such a way that both the balance between the repulsive and attractive terms and the translational symmetry are preserved (6). Since the elementary cell of polyethylene (PE) has no dipole moment, the purely electrostatic interactions converge very fastly (the leading term is the quadrupole-quadrupole interaction) and the long range tail of the lattice sums is determined by the exchange potential. An early cut-off of this contribution leads, however, even in an otherwise correct truncation scheme to numerical instabilities (7). In fact, depending on the basis set applied, no physically reasonable HF solutions can be obtained for PE until the 4th to 5th interacting neighbors are included and for a proper convergence eight neighbors were needed in these calculations.

iv) For substantial alongations, the symmetry adapted (restricted) HF solution does not represent a physically proper reference state for perturbation theory. Fortunately, spin-polarized (unrestricted HF, UHF) states could be found in the region of 30-40 percent elongation (ε) and with the help of their density matrices, the UHF solution could be traced back until $\varepsilon \gtrsim 0.10$. The HF proved to be, however, triplet stable for $\varepsilon \lesssim 0.10$.

1.2 Correlation energy calculations

The Møller-Plesset (MP) partitioning scheme (8) of the Rayleigh-Schrödinger perturbation theory provides a size-consistent scheme for the calculation of electron correlation effects in polymers (9). Earlier studies with this method (10-13) have shown that physically interpretable results can be expected even in second order if the atomic basis set is large enough. The following three methodological problems have to be considered for these calculations:

i) The matrix elements of the perturbation operator can be calculated only with the help of WF's in an economic way. To obtain spatially well localized WF's, we have applied a variational optimization procedure to the free phase of the Bloch functions before Fourier-transforming them.

ii) The correlation energy is especially sensitive to short-range effects, therefore, all virtual excitations involving states of large quasi-momentum (or equivalently, high-lying energy bands in the reduced zone scheme) have to be included into the calculation.

iii) Though the RHF-based perturbation theory is more economic, the UHF-type reference functions are indispensable in the region where the improper dissociation properties of the RHF states become observable.

2. MECHANICAL PROPERTIES

Different experimental and theoretical procedures for the investigation of the deformation properties of PE have been discussed in a recent paper (13) in more detail. In this contribution, we present shortly only those new theoretical results which throw some more light for the role of various electronic effects in determining the longitudinal elastic modulus and the deviations from Hooke's law in PE.

As a first step, we have optimized the geometrical structure of the polymer both at the HF and HF+MP levels using the previously mentioned atomic basis sets. Since the changes within the methylene group affect to only a small extent the energetic processes taking place within the carbon chain, in the case of extended basis sets as well as for correlation calculations the coordinates of the methylene group were fixed at the HF/STO-3G values. In stretching the polymer, we constrained it to preserve planarity and reoptimized its structural parameters for different values of ε.

2.1 Microscopic Young's modulus

From the total energy of the polymer per elementary cell as a function of the stretch we can extract the longitudinal elastic modulus given by $Y=(1/A)\partial F_z/\partial l$, where l is the length and A the effective cross-sectional area of a monomer unit and F_z is the force per monomer acting along the polymer axis z: $F_z=\partial E_{tot}/\partial l$. The values obtained for Y are shown in Table 1.

TABLE 1. Theoretical values of the longitudinal elastic modulus of PE obtained with the HF and HF+MP methods using different atomic basis sets (in GPa units).

Atomic basis	HF	HF+MP
STO-3G	417	362
6-31G	339	303
6-31G**	334	276

The corresponding experimental values determined by different types of measurements scatter between 240 and 360 GPa (13). The longitudinal acoustic mode, corresponding to accordionlike vibrations of the planar PE chain can be best compared with the theoretically investigated deformations. The Raman spectroscopical measurements result for this mode Y=290 GPa (14).

2.2 Deviations from Hooke's law

We expect from theoretical considerations that a simple Hookean model cannot properly describe the mechanical behaviour of macromolecules in the region of large strains arising from covalent bond deformation. Furthermore, it can also be expected that correlation effects become increasingly important during such deformations. Fig 1. shows the retractive force acting on a PE chain at different elongations. We can observe that the deviation from

Hooke's law starts at $\varepsilon \sim 0.05$ and it becomes really substantial in the region of $\varepsilon \sim 0.1-0.2$. The use of an UHF reference function modifies the results to a large extent in the region of $\varepsilon \sim 0.2-0.4$. The maximum of the retractive force is reached at $\varepsilon = 0.38$ for HF-MP (f=0.814 mdyn) and at $\varepsilon = 0.31$ for UHF-MP (f=0.631 mdyn), respectively. The corresponding maxima of the microscopic tensil strength are at 44.6 and 34.2 GPa, respectively.

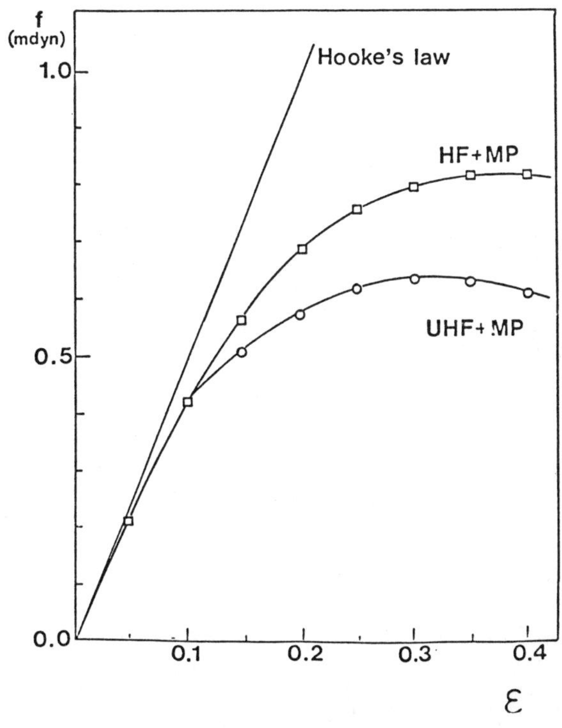

FIGURE 1. The retractive force acting on a single PE chain at different elongations.

3. CALCULATION OF THE UV SPECTRUM

For the theoretical description of the optical properties in covalently bound macromolecules neither of the classical (Frenkel- or Wannier-type) exciton models is appropriate. A physically more sound basis for the solution of this problem is provided by Takeuti's charge transfer exciton theory (15) within the framework of the one particle Green function approximation. An a priori implementation of this procedure has been described in more detail in Ref.16. It can be summarized in the following steps:

i) As starting point, the wavefunction of an excited electron-hole (e-h) pair is obtained in the form

$$\Psi_{l+s,l} = a^{+}_{l+s}\, d^{+}_{l}\, \Phi_V\,, \tag{1}$$

where Φ_V is the wavefunction (Slater determinant) of the completely filled valence band, d^{+}_{l} creates a hole in the cell $\vec{R}_l$ and a^{+}_{l+s} creates an electron at $\vec{R}_l+\vec{R}_s$. ($|\vec{R}_s|$ is the separation of the e-h pair.) All one-particle states are described by WF's localized at the corresponding cell.

ii) A symmetry adapted projection is formed with exciton quasi-momentum $\vec{K}$:

$$\Psi_{s,\vec{K}} = N_c^{-1/2} \sum_{l} e^{i\vec{K}\vec{R}_l}\, \Psi_{l+s,l}\,. \tag{2}$$

(N_c stands for the number of elementary cells.)

iii) The complete excitonic wavefunction is formed as a linear combination of excitons with proper quasi-momentum but with different spatial extensions:

$$\Psi_{\vec{K}} = \sum_{s=1}^{N} \Omega_{s,\vec{K}}\, \Psi_{s,\vec{K}}\,. \tag{3}$$

The corresponding Schrödinger equation

$$\hat{H}\,\Psi_{\vec{K}} = E_{\vec{K}}\,\Psi_{\vec{K}} \tag{4}$$

provides the exciton dispersion $E_{\vec{K}}$. It can be solved by applying the resolvent formalism if we separate the Hamiltonian in the form

$$\hat{H} = \hat{H}_e + \hat{H}_h + \hat{H}_{e,h} \equiv \hat{H}_0 + \hat{H}_{e,h}\,. \tag{5}$$

The Schrödinger equation takes in this case the form

$$\Psi_{\vec{K}} = \{E_{\vec{K}} - (\hat{H}_e + \hat{H}_h)\}^{-1} \cdot \hat{H}_{e,h} \cdot \Psi_{\vec{K}} \tag{6}$$

and the substitution of $\Psi_{\vec{K}}$ in terms of symmetry adapted excitonic 'basis functions' $\Psi_{s,\vec{K}}$ leads to the matrix equation

$$\underline{\Omega}_{\vec{K}} = \underline{\underline{G}}(E_{\vec{K}}) \cdot \underline{\underline{V}}(\vec{K}) \cdot \underline{\Omega}_{\vec{K}} \qquad (7)$$

$\underline{\Omega}_{\vec{K}}$ is here a vector with elements $\Omega_{s,\vec{K}}$, $\underline{\underline{G}}(E_{\vec{K}})$ is the one-particle Green matrix of the periodic crystal and $\underline{\underline{V}}(\vec{K})$ contains the matrix elements of the e-h interaction(16). The single-particle energies entering the calculation of $\underline{\underline{G}}(E_{\vec{K}})$ play a crucial role in determining the exciton levels. On the other hand, previous experience has shown that the reorganization of pair correlation bonds in the course of creating an electron and a hole substantially influences these quantities(10). Using the electron polaron model (9,10), we calculated them for PE besides the HF also at the correlated quasiparticle level (Table 2).

TABLE 2. Comparison of different theoretical and experimental quantities characterising the ground state electronic structure of PE(in eV).

	Hartree-Fock[a]	Quasi-part.[b]	Exp.[c]
Ionization potential	10.2	8.5	7.6-8.8
Valence band width	6.2	5.4	6.0
Electron affinity	3.2	1.8	0.6-1.2
Conduction band width	4.3	3.8	-
Fundamental gap	13.4	10.3	8.8

a) RHF-method using 6-31G** atomic basis set.
b) Electron polarons from RHF+MP theory and 6-31G** basis.
c) Ref.17.

With the help of the resulting Green matrix we have solved the system of homogeneous linear equations (7) for singlet excitons assuming different maximal e-h separations for $\Psi_{\vec{K}}$ in Equ.(3). The resulting $E_{\vec{K}}$ curve (Fig. 2.) shows that the Frenkel exciton (N=0) contributes only a small part (0.32 eV) to the whole exciton binding energy (E_b=1.63 eV). The importance of the charge transfer type excitations can be understood if we look at Fig. 3.

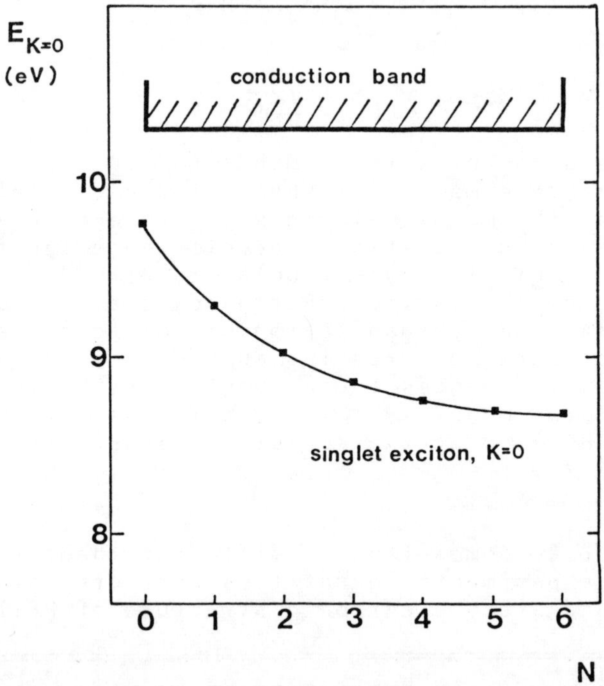

FIGURE 2. The energy of the K=0 singlet exciton in PE measured from the top of the valence band. N is the maximum of allowed e-h separations measured in C_2H_4 units.

showing the weight of e-h pairs with increasing separation in the total excitonic wavefunction.

The comparison of the theoretical absorption edge at 8.67 eV in Fig. 2 with the strong absorption observed in PE in the region of 7.6-8.8 eV (17) suggests a description of the optical properties in terms of charge transfer excitons in this polymer.

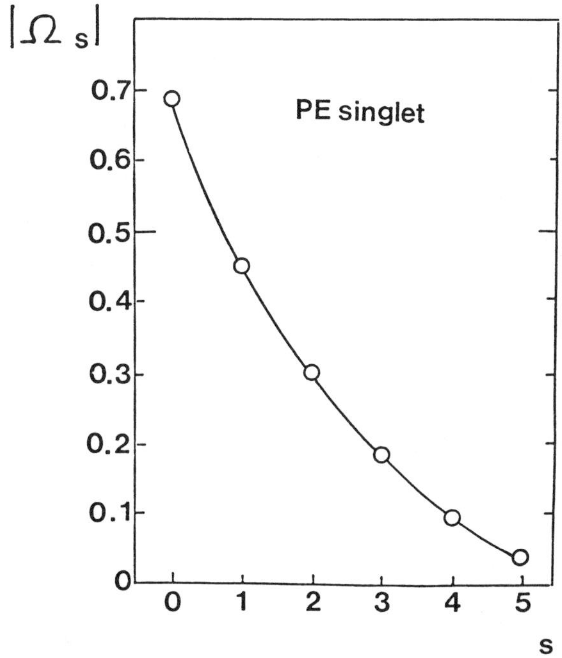

FIGURE 3. The weight of e-h pairs with radius $|\vec{R}_s|$ in the wavefunction of the K=0 singlet exciton of PE (s is measured in C_2H_4 units).

REFERENCES

1. André,J.-M., Gouverneur,L., and Leroy,G. 1967 Int.J.Quant.Chem. 1,427.
2. Del Re,G.,Ladik, J., and Biczo, G. 1967,Phys.Rev. 155,967.
3. Hehre, W.J., Stewart, R.F., and Pople, J.A. 1969, J.Chem.Phys. 51,2657.
4. Ditchfield,R., Hehre,J.W., and Pople, J.A. 1971, J.Chem.Phys. 54,726.
5. Gaussian 76, Binkley,R.A., Hariharan, P.C., Seeger, R., Whiteside,R.A. and Pople,J.A., QCPE No.368, Bloomington, Indiana.

6. Suhai, S. 1980, J.Chem.Phys. 73,3843;to appear.
7. Suhai, S.,Bagus,P.S., and Ladik,J. 1982, Chem. Phys.Lett. 68,467.
8. Møller,C. and Plesset, M.S. 1934, Phys.Rev. 46,618.
9. For a review see the contribution of J.Ladik in this volume, p. 337
10. Suhai,S. 1983, Phys.Rev.B27, 3506.
11. Suhai,S. 1983, Chem.Phys.Lett. 96,619.
12. Suhai,S. 1983, Int.J.Quant.Chem.23,1239.
13. Suhai,S. 1983, J.Polym.Sci.:Polym.Phys.Ed. 21,1341.
14. Strobl,G.R. and Eckel,E. 1976, J.Polym.Sci.:Polym. Phys.Ed. 14,913.
15. Takeuti,Y.1957, Progr.Theor.Phys.(Kyoto) 18,421.
16. Suhai,S.1983, Phys.Rev.B (in press); for a short review see also the contribution of T.C.Collins in this volume, p. 57.
17. Less,K.J. and Wilson,E.G. 1973, J.Phys.C:Solid State Phys. 6,3110.

SYNTHESIS AND PROPERTIES OF CONDUCTING BRIDGED MACROCYCLIC METAL COMPLEXES

Michael Hanack

Institut für Organische Chemie der Universität
Tübingen, Auf der Morgenstelle 18,
D-7400 Tübingen / West-Germany

1. INTRODUCTION

Organic conductors which are based on organometallic compounds have gained more importance recently. New types of low-dimensional polymeric compounds, most containing metallomacrocycles, have been synthesized and extensively investigated [1,2].

Metal complexes containing macrocycles e.g. the phthalocyanine (PcH_2) [1], can be assembled in three different ways: As ladder-polymers [3-5] and as plane-polymers [5,6]. Especially interesting is the third mode of assembling, the stacked arrangement. Elements for this arrangement are planar or nearly planar metallomacrocyclic systems (see Fig. 1a), containing π-electrons, the central metal-atom being surrounded by four coordinating nitrogen atoms. This arrangement leads to low-dimensional compounds with the possibility of electron delocalization by π-π-overlap of the macrocycles, or, in few cases by direct metal-metal contact, resulting after partial oxidation (doping) in highly conducting materials.

J. Ladik et al. (eds.), Quantum Chemistry of Polymers - Solid State Aspects, 111–136.

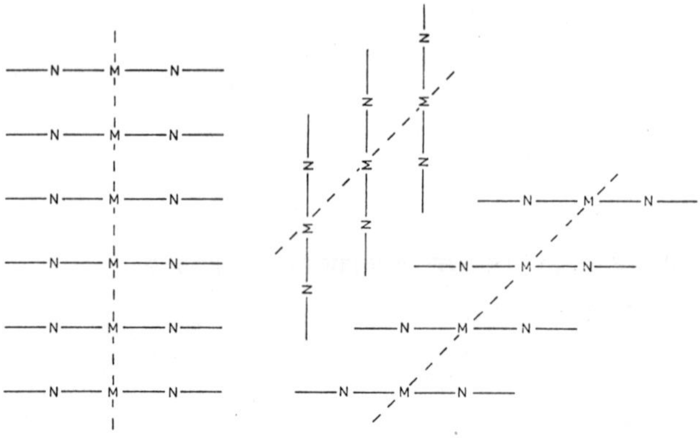

Fig.1 a b

In addition to phthalocyanine (PcH_2) [1], (tetrabenzotetra-azaporphyrin), octaethylporphyrin ($OEPH_2$) [2], tetramethylporphin ($TMPH_2$) [3], tetrabenzoporphyrin ($TBPH_2$) [4], dihydrodibenzotetraaza[14]annulene ($taaH_2$) [5] and hemiporphyrazine (HpH_2) [6] (see Fig. 2) are examples of macrocycles which have been investigated in efforts to construct low-dimensional conductors [6-16].

Fig.2

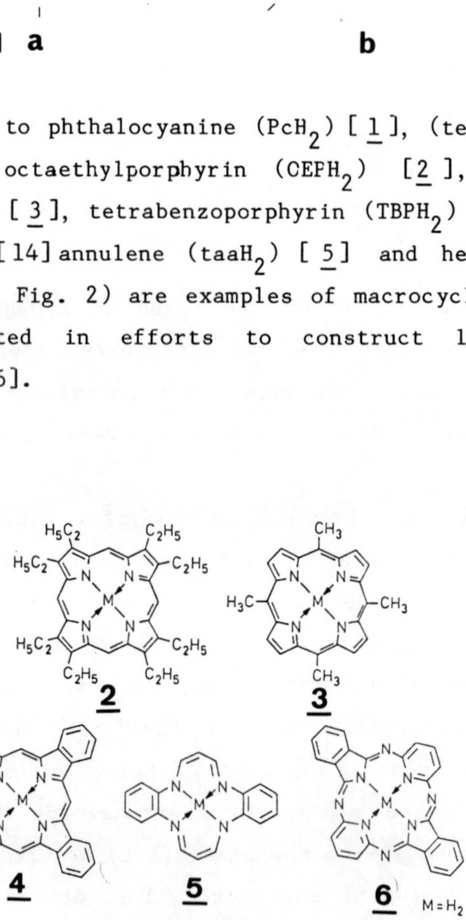

The metallomacrocycles, e.g. phthalocyaninatometals, on the other hand, only very seldom crystallize in the stacked arrangement shown in Fig. 1a. Generally the packing is different, as is illustrated in Fig. 1b corresponding schematically to the so-called α- or ß-modification and is not favorable for the formation of a conduction band by π-π-overlap of the macrocycles. Stacked metal complexes in general have been reviewed recently [2,8]. PcNi, for example crystallize as shown in Fig. 1b. It is an insulator with a conductivity $< 10^{-11}$ S/cm (polycrystalline samples). Doping with iodine leads to $PcNiI_x$ (x = 0 - 1.74) with I_3^- (and less often I_5^-) as the counter-ion. Oxidative doping with iodine results in highly increased conductivities of the polycrystalline material. The iodinated metal complex of the stoichiometry $PcNiI_{1.0}$ forms stacked structures analogous to Fig. 1a as shown by crystal structure determination [7,12,13]. The PcNi-macrocycles are staggered by 39.5° with an interplanar distance of 324.4 pm [7], and with I_3^- as counter-ions leading to a charge distribution of $(PcNi)^{+0.33}$ $(I_3^-)_{0.33}$. The stacks of PcNi are surrounded by four parallel channels which contain linear chains of disordered I_3^- ions. In the stacking direction, the PcNiI-crystals show a conductivity of σ_{RT} = 260 - 750 S/cm and behave as a metal down to 100 mK.

The reason for the high conductivity of $PcNiI_{1.0}$ is still not fully understood. It appears that one of the essential prerequisites for a facile charge mobility is indeed a molecular stacking as shown in Fig. 1a with planar rings containing π-electrons, and with distances allowing the intermolecular interactions necessary for establishing a conduction pathway. Another criterion for high conductivity, though less well interpreted, is the formal partial oxidation (incomplete charge transfer, mixed valence, non-integral oxidation state) by the dopant (halogen). In general only the macrocycles are affected by the dopant through the formation of π-radical cations. The central metal atom plays an inferior role, as for example in (PcNi)I where the central Ni^{2+} ion does not change in formal oxidation state.

2. BRIDGED MACROCYCLIC METAL COMPLEXES

As pointed out earlier, metallomacrocycles and many of the halogen doped systems do not crystallize in stacks, thereby not meeting one of the prerequisites for high conductivity. In addition segregated stacks are often not formed; the stacking repeat distances and the donor-acceptor distances can not be controlled in any way. By using oxidizing dopants other than halogens, e.g. quinones, integrated stacks can be formed leading to insulators [1] for which a possible arrangement is shown in Fig. 3.

Fig.3

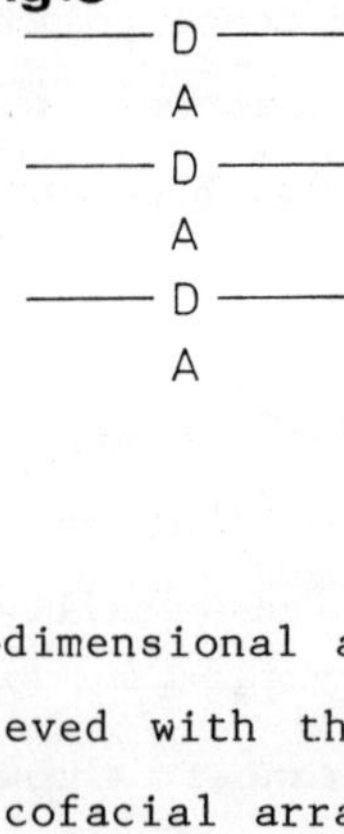

A stacked quasi-one-dimensional arrangement of the metallomacrocycles can be achieved with the Kenney-Marks approach in which a face to face or cofacial arrangement can be achieved by linking the central metal atoms via bridging ligands (e.g. O^{2-}) through strong covalent bonds according to Fig. 4 ("chip and stick"-arrangement).

Fig.4

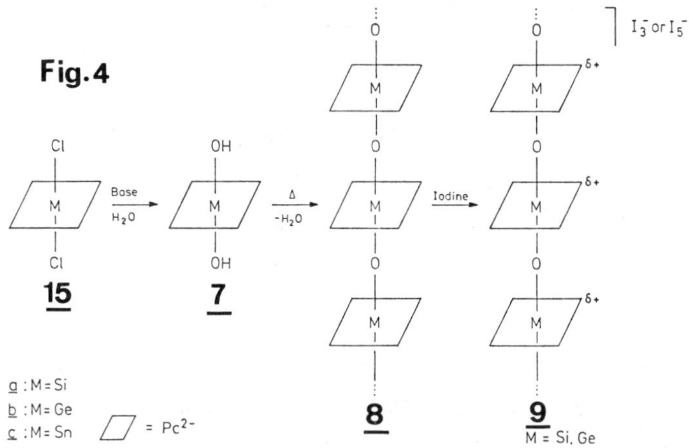

The polymetalloxanes [8a-c] (M = Si, Ge, Sn) are formed by condensation of phthalocyaninatometaldihydroxides [7a-c] (M = Si, Ge, Sn), as demonstrated for the first time by Kenney et al [17,18]. The $[PcMO]_n$-polymers [8a-c] and the corresponding halogen-doped polymetalloxanes $[(PcMO)I_y]_n$ [9], which led to a new class of electrically conductive compounds were investigated in detail by T. J. Marks and his group [19-21]. Both [8a] and [8b] have high thermal and chemical stabilities. In fact, [8a] can be dissolved in conc. H_2SO_4 and in HSO_3CH_3 and can be recovered without change.

Estimation of the average molecular weights for [8a-c] by IR end-group analysis, tritium labelling- and laser light scattering experiments (in H_2SO_4 for [8a]) yields a degree of polymerization of 70 - 140 subunits for [8a] and somewhat lower values for [8b] and [8c]. Longer polymerization times and high tempertatures seem to increase the molecular weights [20].

Partial oxidation of the $[PcMO]_n$-polymers [8a-c] was carried out using iodine and bromine as dopants. $PcSi(OH)_2$ [7a] does not absorb detectable quantities of I_2 under the same conditions. That the doping of $[PcMO]_n$ [8a,b] with halogens is an oxidation process was demonstrated by UV/VIS-, EPR-, Resonance Raman- and transmission IR-spectroscopy. The formation of $(PcMO)^{y/3+}(1/3\ I_3^-)_{y\ n}$ (M = Si, Ge, y ≤ 1.1) and the corresponding bromine compounds for M = Si is indicated [21]; phthalocyanine π-radical cations and chains of I_3^- counter-ions are generated. The doping is inhomogeneous, and a single limiting phase of y ∼ 1.1 is observed.

The room-temperature conductivities of polycrystalline samples of $[(PcSiO)I_y]_n$ and $[(PcGeO)I_y]_n$ for various stoichiometries are given in Table 1. The conductivities are dependent upon the iodine uptake and are several powers of ten higher than the conductivities of nondoped [8a] and [8b]. From variable temperature conductivity data it was concluded, that the room-temperature conductivities are inversely proportional to the interplanar ring-ring spacings, i.e. $\sigma_{Si} > \sigma_{Ge}$.

Table 1 Pressed-Powder Electrical Conductivity Data for $[(PcMO)I_y]_n$ Doped Polymers at Room Temperature[a]

Compound	Room-temperature conductivity (S/cm)
$[PcSiO]_n$	$5,5 \cdot 10^{-6}$
$[(PcSio)I_{0,12}]_n$	$1,4 \cdot 10^{-3}$
$[(PcSiO)I_{0,71}]_n$	$3,2 \cdot 10^{-1}$
$[(PcSiO)I_{1,55}]_n$	1,4
$[PcGeO]_n$	$2,2 \cdot 10^{-10}$
$[(PcGeO)I_{0,14}]_n$	$2,6 \cdot 10^{-6}$
$[(PcGeO)I_{0.56}]_n$	$1,9 \cdot 10^{-2}$
$[(PcGeO)I_{1,08}]_n$	$1,1 \cdot 10^{-1}$

[a] From reference [21].

Recently the syntheses of the peripherally alkylated μ-oxo-polymers $[R_4PcMO]_n$ [10a-d], R = t-Bu, Tms; M = Si, Ge, Sn (see Fig. 5) were carried out in our laboratory [22]. The presence of the bulky substituents in these compounds gives rise to high solubilities in common organic solvents like chloroform, methylene chloride, toluene, etc. The polymeric μ-oxo(tetraalkylphthalocyaninato)metal IVB derivatives [10a-d] are accessible via thermal condensation of the corresponding monomeric dihydroxy compounds [12a-d] in high vacuum or in boiling 1-chloronaphthalene.

Fig.5

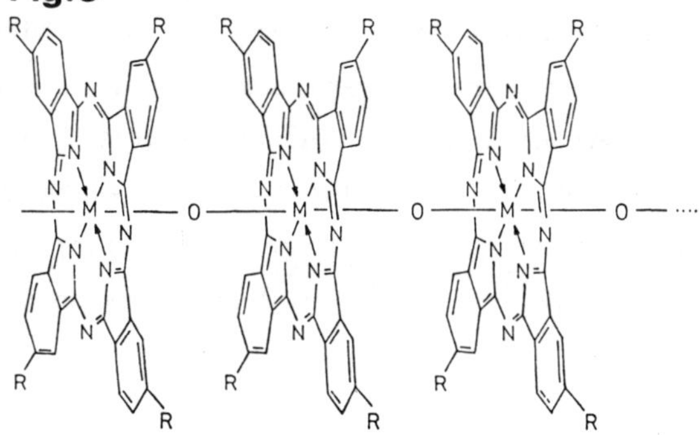

10

R_4PcMX_2 ⟶ $R_4PcM(OH)_2$ ⟶ $[R_4PcMO]_n$

11a-d   12a-d   10a-d

a R = t-Bu, M = Si, X = Cl

b R = t-Bu, M = Ge, X = Cl

c R = Tms, M = Ge, X = Cl

d R = t-Bu, M = Sn, X = I

The polymers [10a-d] were characterized by their IR-, far-IR-, UV-VIS-spectra, microanalytical data and TG/DTA measurements. The chain length of [10a] was estimated by infrared end-group analysis to have a minimum average value of $\bar{n}$ = 10. More accurate data were obtained via tritium labeling techniques, which indicated a degree of polymerization of $\bar{n}$ = 25 [23].

The undoped materials [10a-d] show electrical conductivities which are similar to those of the peripherally unsubstituted polymers [8a-c].

The polymers [10a-d] were doped following known heterogeneous procedures with iodine vapor (solid-gas phase), iodine solutions (solid-liquid) or by grinding of both components (solid-solid). The properties of these materials were compared with the solids obtained by homogeneous doping in solution (liquid-liquid). The samples are thermally stable up to 140°C. The dc-dark conductivities of the doped polymers [10a-c] reach limiting values of 10^{-3} S/cm (M = Si) and 10^{-4} S/cm (M = Ge) and show no significant variation with the doping procedure.

A change in the bridging ligand was made in an attempt to get some information on the mechanism of the electrical conductivity of chalcogen bridged group IV B metals. Sulfur instead of oxygen was used in order to see if the better d-orbital overlap has an influence on the conductivity and on the doping. Polymeric phthalocyaninato-μ-thiogermanium(IV) $[PcGeS]_n$ [13] was obtained in two independent ways [24]; by reacting $PcGe(OH)_2$ [7b] with H_2S under a pressure of 80 bar ($8 \cdot 10^6$ Pa) at 130°C and through the reaction of [7b] with an excess of triphenylsilanethiol (Ph_3SiSH) in refluxing chlorobenzene. The sulfur-bridged polymer was characterized by elemental analysis, IR/FIR and mass spectroscopy. Further characterization was accomplished by chemical means: [13] can be quantitatively hydrolyzed by NH_3/pyridine (1:1) to $PcGe(OH)_2$ [7b] and by HCl/THF (1:6) to $PcGeCl_2$. Evolving H_2S was precipitated with $CdCl_2$ as CdS.

In comparison with $[PcGeO]_n$ [8b], polycrystalline samples of [13] do not show a pronounced increase in electrical conductivity ($\sigma_{RT} = 2 \cdot 10^{-9}$ S/cm). The doping of [13] with iodine under various conditions leads, in contrast to the analogous $[PcGeO]_n$ [8b], to decomposition by cleaving the germanium-sulfur bond. Doping with bromine leads to $PcGeBr_2$.

Polymeric stacks of bridged metallofluorophthalocyanines, $[PcMF]_n$ [14], with M = Al, Ga, Cr (shown schematically in Fig. 6) were first prepared by Kenney et al [25]. $[PcGaF]_n$ crystallizes in stacks of nearly eclipsed PcGa-macrocycles connected by linear Ga-F-Ga bridges with interplanar distances of 387 pm as shown by X-ray analysis [26].

Fig.6

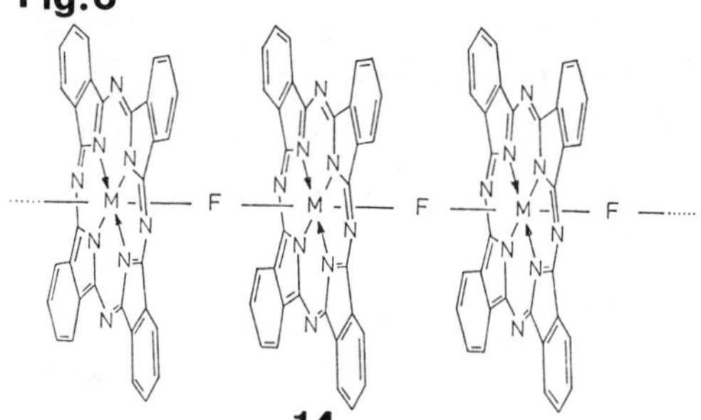

M = Al, Ga, Cr

14

All known $[PcMF]_n$-polymers can be doped with iodine to yield $[(PcMF)I_y]_n$, where y, depending upon M, varies from 0.012 to 3.3. The doped polymers were characterized by Raman- and IR-spectroscopy, the former indicating the presence of I_3^- and I_5^- as counter-ions [27]. The conductivity of $[(PcMF)I_y]_n$ rises with increasing iodine content, the highest conductivity being observed for $[(PcAlF)I_{3.3}]_n$ (σ_{RT} = 5 S/cm, activation energy 0.17 eV), which was prepared from sublimed $[PcAlF]_n$.

Bridging ligands other than O^{2-} and S^{2-} have also been used in our laboratory to prepare polymeric stacked metallomacrocycles. Reaction of $PcSiCl_2$ [5a] with a molar amount of bis-bromomagnesiumacetylene [16] in boiling THF gave the polymer [17a] in nearly quantitative yields.

$$\underset{[15a]}{PcSiCl_2} + \underset{[16]}{BrMgC{\equiv}C\text{-}MgBr} \xrightarrow{THF} \underset{[17a]}{[PcSiC{\equiv}C\text{-}]_n}$$

It was characterized through Raman- ($\nu_{C\equiv C}$ = 2150 cm^{-1}) and MS-spectroscopy as well as by chemical methods [28]. The reaction of $PcSiCl_2$ with the appropriate di-Grignard-compound gave $[PcSiC{\equiv}C\text{-}C{\equiv}C\text{-}]_n$ [18] and $[PcSiC{\equiv}C\text{-}C_6H_4\text{-}C{\equiv}C]_n$ [19]. [17], [18] and [19] are insulators with powder conductivities 10^{-12} S/cm. Doping with iodine leads to decomposition of the polymers.

The basic problem with the reaction of phthalocyaninato IVB dichlorides [15] with mono- and bis-Grignard reagents of acetylene and its derivatives is the poor solubility of [15] which results in a heterogeneous reaction mixture. These disadvantages led us to the synthesis of peripherally substituted phthalocyaninato derivatives which are known to be soluble in organic solvents:

$$\underset{[\underline{11}]}{R_4PcMX_2} + \underset{[\underline{16}]}{BrMgC{\equiv}CMgBr} \longrightarrow \underset{[\underline{20}]}{[R_4PcMC{\equiv}C]_n}$$

M = Ge, Sn

R = t-Bu, Tms

The powder conductivities of about 10^{-12} S/cm are similar to those of the peripherally unsubstituted derivatives [17]. Doping leads to gross decomposition of the polymers [29].

3. BRIDGED MACROCYCLIC TRANSITION METAL COMPLEXES

In the bridged macrocyclic systems mentioned so far, the conduction pathway, with a few exceptions has been described as originating from π-π-overlap of the stacked metallomacrocycles which are separated by small distances. Only a minor role has been attributed to the central metal ion.

The generation of a conduction pathway by π-π-overlap of the stacked metallomacrocycles, however, is not the only possibility allowing charge migration in bridged metal complexes. Another pathway can be brought about by bridging groups which allow electron migration from one metal atom to the other. For this purpose organic bridging ligands, containing (conjugated) π -electrons, allowing delocalization along the chain are necessary. To retain a stacked, one-dimensional arrangement of a bridged polymer such as [21] shown in Fig. 7, the bridging ligand L, axial to the metal M, must be linear or at least have a sufficiently small M-L-M angle. Additionally, the central metal atom must be a transition metal with a preferred octahedral coordination e.g. Fe, Ru, Co, Mn or Cr. The π-electron containing bridging group L may be cyanide or a polyatomic organic molecule, such as pyrazine, bipyridine, etc. The structural flexibility of such bridged macrocyclic transition metal complexes provides a unique possibility to study systematically the physical properties of such compounds and to develop new conducting materials.

In the construction of the polymers [21] the three possible units: the macrocycle, the central metal atom M and the bridging ligand L have each been varied.

The bridging ligand L in Fig. 7 can be axially linked to the central transition metal of the macrocycle in three different ways:

A. By two σ-bonds as shown in Fig. 7A. A possible bridging molecule for this type would be the acetyliddianion $C{\equiv}C^{2-}$.

B. With two coordinative bonds as shown in Fig. 7B. The bridging ligand here can be e.g. pyrazine (pyz), bipyridine (bpy) or 1,4-diisocyanobenzene (dib).

C. By one σ- and one coordinative bond, a possible bridging ligand would be the cyano group CN^-.

The type of binding of the bridging ligand L in combination with the specific electronic-interactions of the equatorial coordinated macrocycle is expected to have a major impact on a possible electron migration along the central metal-ligand-metal spine.

Fig.7

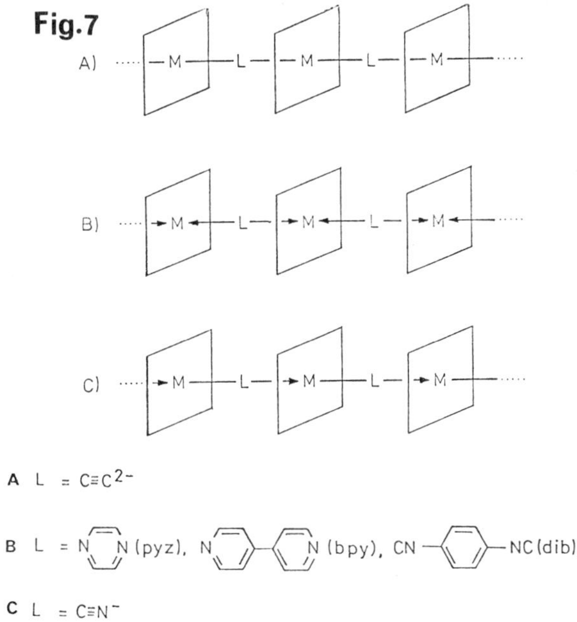

Type A. Two σ-Bonds to the Central Metal Atom.

In principle, the acetylenic group (-C≡C-) can be used as a bridging ligand to link units of the metallomacrocycle through σ-bonds. Such a polymer could exhibit interesting properties concerning conductivity. Extended Hückel Molecular Orbital (EHMO)-calculations have been carried out for a model substance [22] with the dianion of tetraazaporphin (TAP^{2-}) as the macrocycle, Fe^{2+} as the central metal, and the dianion of acetylene $(C\equiv C)^{2-}$ as the bridging ligand (Fig. 8) [30,31]. The EHMO-calculations and, particularly, group-theory considerations indicate that the resulting $13a_1$ crystal orbital in this type of polymer should be extremely concentrated along the central axis via the bridging group. This crystal orbital would be only partially filled therefore crossing the approximate Fermi level. In this way virtually metal-like electrical conductivity would be expected in the one-dimensional chain.

Fig.8

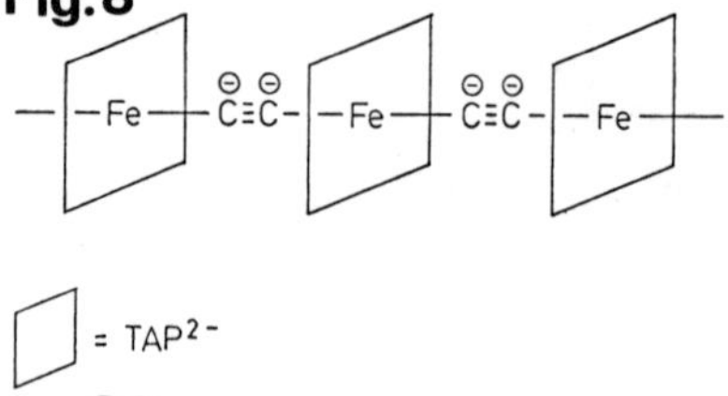

22

Since $Li_2[PcFe(C\equiv C-C_6H_5)_2]\cdot 7THF$ has been reported in the literature [32] it was considered to be a monomeric model for the polymeric structure [22]. The same approaches used in preparation of the monomeric model have, so far, failed to produce the polymer.

Type B. Two coordinative bonds to the central metal atom

Our investigations in the field of polymeric bridged metallomacrocycles were initiated with phthalocyaninatoiron(II), PcFe [23] (Fig. 7B, M = Fe), as a useful macrocyclic system [33]. In PcFe the iron ion is unsaturated electronically and also has additional sites for coordination. PcFe reacts to form a series of new monomeric $PcFeL_2$ adducts [24a-e] (Fig. 9) with L = pyrazine (pyz), 2-methyl- (Mepyz), 2,6-dimethyl- (Me_2pyz), 2-ethyl- (Etpyz), or 2-chloropyrazine (Clpyz).

Fig.9

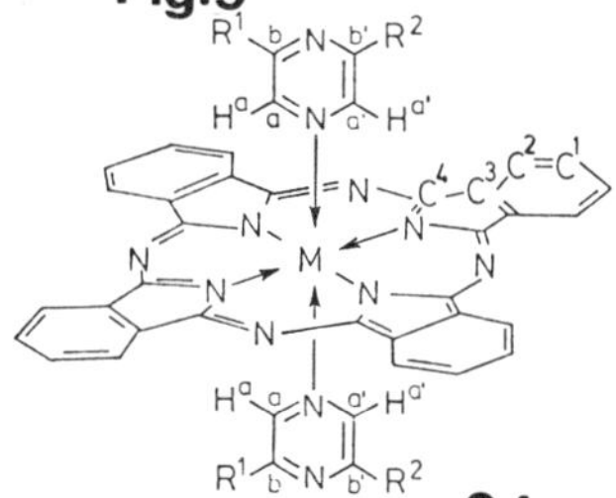

24

The pyrazine-bridged [$PcFe(pyz)]_n$ [<u>25</u>] (Fig. 10) can be easily synthesized. Solid ß-PcFe suspended in a boiling solution of pyrazine in chlorobenzene reacts to yield insoluble $[PcFe(pyz)]_n$. A second synthetic route to $[PcFe(pyz)]_n$ is available through simple extraction of monomeric $PcFe(pyz)_2$.

Fig.10

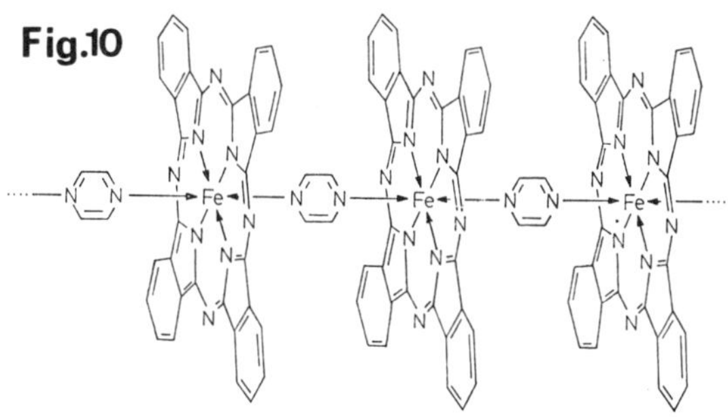

25

Using the same or comparable methods the polymers shown in Fig. 7B with Pc^{2-} and M = Fe^{2+}, Co^{2+}, L = bpy and dib have been prepared. The polymers are stable up to 240°C (for M = Fe and L = pyz) and up to 120°C for M = Co and L = pyz. The polymeric structure for $[PcFe(pyz)]_n$ [<u>25</u>] for example has been proven by TG-DTA-measurements, Mößbauer-spectroscopy and magnetic measurements (the compound is diamagnetic). From IR-studies the degree of polymerisation was estimated to be >20. For [$PcCo(pyz)]_n$ the polymeric structure was deduced from its IR-, NIR- and ESR-spectra [34].

$[PcFe(pyz)]_n$ shows a conductivity of $2 \cdot 10^{-5}$ $S \cdot cm^{-1}$ at room temperature with an activation energy of 0.40 eV (pressed pellets). As shown in table 2 the conductivity of [25], which is 10^7 times higher than the conductivity of $PcFe(pyz)_2$, does not change considerably if the phthalocyanine ring is substituted in the peripherical positions as in [26] and [27]. The possibility for interleaving of the parallel Pc-rings in [25] from different chains appears unlikely, when comparing the conductivities of [25] with that of the ring-substituted polymers [26] and [27]. [25], [26] and [27] (Table 2) show conductivities of the same order of magnitude.

TABLE 2 Conductivities at RT of differently substituted $[PcFe(pyz)]_n$ polymers.

	$\sigma_{RT}(S \cdot cm^{-1})$
$[PcFe(pyz)]_n$ [25]	$2 \cdot 10^{-5}$
$[(CH_3)_8PcFe(pyz)]_n$ [26]	$9 \cdot 10^{-6}$
$[(CH_3O)_8PcFe(pyz)]_n$ [27]	$5 \cdot 10^{-6}$

The conductivity does not change either if instead of pyz (leading to a distance of 700 pm in between the Pc-rings) dib is used as the bridging ligand (distance Pc-Pc = 1150 pm). $[PcFedib]_n$ [28] shows a conductivity in the same order $\sigma_{RT} = 2 \cdot 10^{-5}$ $S \cdot cm^{-1}$.

PcFe reacts with 1,4-diazabicyclo[2.2.2]octane (dabco) in a melt to form the monomer $PcFe(dabco)_2$ [29] or in a solvent ($CHCl_3$) with formation of the polymer $[PcFe(dabco)]_n$ [30]. In contrast to the polymers [25], [26], [27] and [28] the bridging ligand in [30] has no π-electrons. Its conductivity was found to be four powers of ten lower ($\sigma_{RT} = 1\cdot10^{-9}$ S$\cdot$cm^{-1}) than that of [25], [26], [27] and [28] but practically the same as the conductivity of the monomer PcFe(dabco)$_2$ [29]. ($\sigma_{RT} = 1\cdot10^{-10}$ S$\cdot$cm^{-1}).

In addition to Fe and Co, Ru was also used for the first time as the central metal atom. [PcRu(pyz)]$_n$ [31] shows a room temperature conductivity (powder, pressed pellets, 1 kbar) of $1\cdot10^{-5}$ S$\cdot$cm^{-1}, therewith not exceeding the conductivity of [PcFe(pyz)]$_n$ [25].

Polymers of the type shown in Figure 7 with pyrazine as the bridging ligand L can also be prepared using 5.10.15.20-tetraphenylporphine (TPP^{2-}) or dihydrodibenzo[b,i]-1,4,8,11-tetraaza[14]annulene (taa^{2-}) [5] as the macrocycle forming [TPPFe(pyz)]$_n$ [32a] and [Fetaa(pyz)]$_n$ [32b]. Both are formed as insoluble, diamagnetic, airstable solids by reacting TPPFe or Fetaa respectively with pyrazine under various conditions. For the first time tetrabenzporphine, TBP, was also used as the macrocycle.

As in case of the afore mentioned phthalocyaninato-µ-oxo- and fluorine polymers [PcMO]$_n$ [8] and [PcMF]$_n$ [14] the phthalocyaninatoironpyrazine polymer [PcFe(pyz)]$_n$ [25] is also dopable with iodine and yields compounds with the stoichiometry [PcFe(pyz)I$_y$]$_n$ [33] (y = 0 to 2.6) [35]. The compositions of the doped pyrazine polymers were established both by microanalysis and TG/DTA. The thermal stability of [PcFe(pyz)I$_y$]$_n$ [33] is therefore in the same range as for iodine doped [(PcSiO)I$_y$]$_n$ and [(PcAlF)I$_y$]$_n$ [26,27]. ^{57}Fe Mößbauer spectroscopy on [PcFe(pyz)I$_y$]$_n$ [33] unambigiously shows, that doping does not destroy the polymeric structure of the compound. Isomer shifts and quadrupole splitting of PcFe(pyz)$_2$, [PcFe(pyz)]$_n$, and [PcFe(pyz)I$_y$]$_n$ are nearly identical, supporting hexacoordinated iron ions in all three compounds. The presence of I_5^-, as indicated by the resonance Raman results, clearly demonstrates that oxidation takes place in the linear chain compound $[PcFe(pyz)]_n$.

The highest conductivities, measured for compressed polycrystalline samples, was found for $[PcFe(pyz)I_{2.54}]_n$ which shows a room temperature conductivity of $\sigma_{RT} = 0.2\ S\cdot cm^{-1}$. The temperature dependence of the doped polymer conductivity can be fit to a model of thermal activation. The derived activation energies E_a decrease with increasing conductivity.

Several other type B-polymers (see Fig. 7) have been doped with iodine, leading also to stable iodine containing compounds.

From these experiments it is evident, that type B transition metal polymers are dopable with iodine, leading two stable compounds of conductivities which are in the same range as $[(PcMO)I_y]_n$ and $[(PcMF)I_y]_n$. In spite of the rather large interplanar spacings in a single chain of the type B-polymers doping leads to a large increase in electrical conductivity.

Type C. One σ- and one coordinative bond to the central metal atom.

For the first time phthalocyaninatometal polymers with a bridging ligand linked by one σ- and one coordinative bond to the central metal have been synthesized. These are the (μ-cyano)-phthalocyaninatometal compounds shown in Fig. 11. The most interesting property of these type of (μ-cyano)phthalocyaninatometal compounds is, that with Co and Fe as the central metal atom, they exhibit DC room temperature conductivities around

10^{-2} S•cm^{-1} without doping, therewith showing conductivities which are in the same range as the iodine doped $[PcMO]_n$ (M = Si, Ge) and $[PcML]_n$ (M = Fe, Ru, L = pyz, dib).

Fig.11

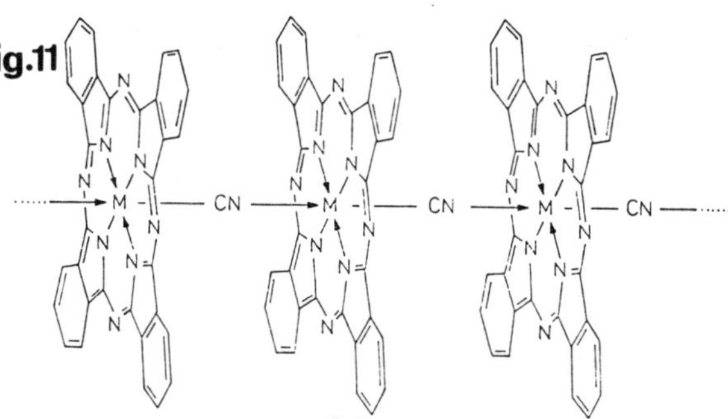

35

a: M = Co c: M = Mn
b: M = Fe d: M = Cr

Compounds of this type have been synthesized in addition to Co and Fe with Mn and Cr as the central metal atoms (Fig. 11) using the methods shown in Fig. 12 [36].

Fig.12 Syntheses of Monomeric and Polymeric Phthalocyaninatometal(III) Cyano-Derivatives

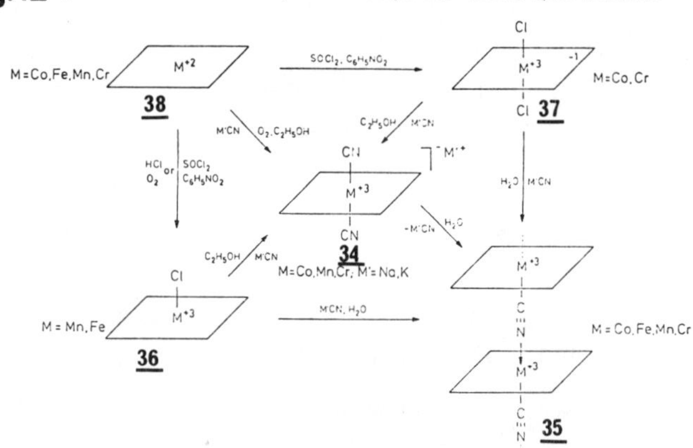

The polymers [35a-d] (Fig. 11) were characterized by IR and far-IR spectroscopy, magnetic measurements, thermogravimetrical and microanalytical analyses and chemical decomposition. The CN valence frequencies of [35a,c,d] compared with the monomeric complexes [34], (Table 3) are shifted about 30 cm^{-1} (M = Co) and 20 cm^{-1} (M = Cr, Mn) to higher energy. This increase in CN-valence frequency is evidence for the presence of a cyano bridge [37] in [35a-d]. The polymers exhibit thermal stabilities up to 250°C and are insoluble in noncoordinating solvents.

The polymeric structure of [35] (Fig. 11) is destroyed when the material is treated with a competing ligand. This confirms the proposed chain structure [38] thus being exemplified for $[PcCoCN]_n$: When $[PcCoCN]_n$ is treated with bases (L) such as pyridine (py), 2-methylpyrazine (mepyz), piperidine (pip), and n-butylamine (ba), monomeric complexes are obtained with the composition PcCo(CN)L [39] (Fig. 13).

Fig.13

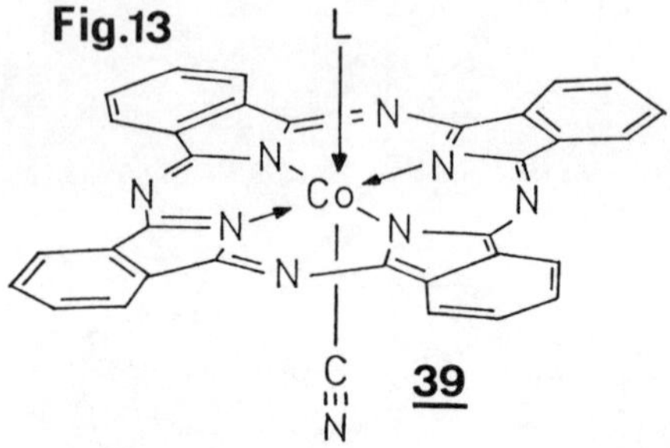

39

$[PcCoCN]_n$ [35a] exhibits electrical conductivities around 10^{-2} $S \cdot cm^{-1}$ (Table 4) without additional doping, which are in the same range as those obtained for doped samples of the μ-oxo and μ-fluoro polymers $[R_4PcMX]_n$. When the polymeric structure of $[PcCoCN]_n$ was destroyed by treatment with a competing ligand

TABLE 3 Infrared and Thermoanalytical Data for Monomeric and Polymeric Cyano Complexes.

	ν_{CN} [cm^{-1}] a)	Dissociation range of the cyano group [°C] b)
$NaPcCo(CN)_2$ $5 \cdot H_2O$	2130	280 - 310
$Na(H_3C)_8PcCo(CN)_2 \cdot 3\ H_2O$	2132	170 - 300
$KPcMn(CN)_2 \cdot 2$ EtOH	2114	250 - 380
$NaPcCr(CN)_2$ EtOH,$2H_2O$	2133	310 - 455
PcCo(py)CN	2147	180 decomp.
$[PcCoCN]_n$	2158	220 - 300
$[(H_3C)_8PcCoCN]_n$	2142	150 - 270
$[PcFeCN]_n$	2133	170 - 230
$[PcMnCN]_n$	2133	250 c)
$[PcCrCN]_n$	2150	170 - 300

a) Nujol mull

b) Heating rate 2°/min. N_2

c) Endothermic weightloss.

e.g. py, pip, ba, mepyz to form PcCo(L)CN, [39], the conductivity was diminished by 6 - 10 orders of magnitude (Table 4).

TABLE 4 Dc-Dark Conductivities for Polymeric Phthalocyaninatometal(III) Cyano-Derivatives

Compound	$S \cdot cm^{-1}$	Activation Energy eV
$[PcCoCN]_n$	$2 \cdot 10^{-2}$ b)	0.1
$[(H_3C)_8PcCoCN]_n$	$5 \cdot 10^{-5}$ b)	
$[PcFeCN]_n$	$6 \cdot 10^{-3}$ b)	0.1
$[PcMnCN]_n$	$1 \cdot 10^{-5}$ a)	
$[PcCrCN]_n$	$3 \cdot 10^{-6}$ a)	

a) Two-probe technique, 1 kbar.

b) Four-probe technique, 1 kbar (cf. ref. [41]).

The iron cyano polymer $[PcFeCN]_n$ [35b] shows a conductivity comparable to that of the cobalt polymer. Lower values for the (μ-cyano)phthalocyaninatomanganese and chromium compounds [35c,d] (Table 4) in the range of 10^{-5} - 10^{-6} $S \cdot cm^{-1}$ were obtained. $[PcCoCN]_n$ and $[PcFeCN]_n$ show a semiconducting behavior, the activation energy was found to be 0.1 eV [50].

No final explanation for the comparable high conductivity of $[PcCoCN]_n$ and $[PcFeCN]_n$ can be given at this time. EPR spectra obtained from polycrystalline samples of $[PcCoCN]_n$ exhibit two signals,one at the free electron value and the other at g = 2.3. While the signal at g = 2.0028 suggests an oxidation centered on the macrocyclic ligand, the weak signal at g = 2.3 is ascribed to the $g_\perp$ branch of a pentacoordinated PcCo(II) unit [40], which corresponds to a partial reduction of the back bone.

Acknowledgment

Financial support of the Bundesministerium für Forschung und Technologie and the Stiftung Volkswagenwerk is greatfully acknowledged. I thank Dr. Martin Mezger for his help preparing the manuscript.

References

[1] C.W. Dirk, K.F. Schoch, Jr. and T.J. Marks in Conductive Polymers (R.B. Seymour, ed.), Plenum Press, New York, 1981, p. 209.

[2] T.J. Marks and D.W. Kalina in Extended Linear Chain Compounds, (J.S. Miller, ed.), Vol. 1, Plenum Press, New York, 1982, p. 197.

[3] A.A. Berlin, L.G. Cherkashina and E.I. Balabanov, Vysokomol. Soedin, 4: 376 (1962).

[4] Y. Nose, N. Sera, M. Hatano and S. Kambara, J. Chem. Soc. Japan Ind. chem. Sec, 67: 1600 (1964).

[5] A.A. Berlin, Russ. Chem. Rev., 48: 1125 (1979).

[6] C.J. Schramm, D.R. Stojakovic, B.M. Hoffman and T.J. Marks, Science, 200: 47 (1978).

[7] C.J. Schramm, R.P. Scaringe, D.R. Stojakovic, B.M. Hoffman, J.A. Ibers and T.J. Marks, J. Am. Chem. Soc., 102: 6702 (1980).

[8] J.A. Ibers, L.C. Pace, J. Martinsen and B.M. Hoffman, Structure and Bonding, 50: 1 (1982).

[9] W.B. Euler, J. Martinsen, L.J. Pace, B.M. Hoffman and J.A. Ibers, Mol. Cryst. Liq. Cryst., 81: 231 (1982).

[10] B.M. Hoffmann and J.A. Ibers, Acc. Chem. Res., 16: 15 (1983).

[11] S.K. Wright, C.J. Schramm, T.E. Phillips, D.M. Scholler and B.M. Hoffman, Synth. Metals, 1: 43 (1979/80).

[12] T.E. Philips, R.P. Scaringe, B.M. Hoffman and J.A. Ibers, J. Am. Chem. Soc., 102: 3435 (1980).

[13] J. Martinsen, L.J. Pace, T.E. Philips, B.M. Hoffman and J.A. Ibers, J. Am. Chem. Soc., 104: 83 (1982).

[14] Y.-M. Wuu, S.-M. Peng and H. Chang, J. Inorg. Nucl. Chem., 42: 839 (1980).

[15] W.E. Hatfield in Conductive Polymers, (R.B. Seymour, ed.), Plenum Press, New York, 1981.

[16] R. Fay and M. Hanack, unpublished results.

[17] R.D. Joyner and M.E. Kenney, J. Am. Chem. Soc., 82: 5790 (1960).

[18] R.D. Joyner and M.E. Kenney, Inorg. Chem., 1: 717 (1962).

[19] T.J. Marks, K.F. Schoch and B.R. Kundalkar, Synth. Met., 1: 337 (1979/80).

[20] C.A. Dirk, T. Inabe, K.F. Schoch, jr. and T.J. Marks, J. Am. Chem. Soc., 105: 1539 (1983).

[21] B.N. Diel, T. Inabe, J.W. Lyding, K.F. Schoch, Jr., C.R. Kannewurf and T.J. Marks, J. Am. Chem. Soc., 105: 1551 (1983).

[22] J. Metz, G. Pawlowski and M. Hanack, Z. Naturforsch., 38b: 378 (1983).

[23] T. Inabe, J. Metz, M. Hanack and T.J. Marks, manuscript in preparation.

[24] K. Fischer and M. Hanack, Chem. Ber., 116: 1860 (1983).

[25] J.P. Linsky, T.R. Paul, R.S. Nohr and M.E. Kenney, Inorg. Chem., 19: 3131 (1980).

[26] R.S. Nohr and K.J. Wynne, J. Chem. Soc. Chem. Commun., 1981: 1210.

[27] K.J. Wynne, R.S. Nohr, Mol. Cryst. Liq. Cryst., 81: 243 (1981).

[28] K. Mitulla and M. Hanack, Z. Naturforsch., 35b: 111 (1980).

[29] M. Hanack, J. Metz and G. Pawlowski, Chem. Ber., 115: 2836 (1982).

[30] M. Hanack, F.F. Seelig and J Strähle, Z. Naturforsch., 34a: 983 (1979).

[31] F.F. Seelig, Z. Naturforsch., 34a: 986 (1979).

[32] R. Taube and H. Drevs, Z. Anorg. Allg. Chem., 429: 5 (1977).

[33] O. Schneider and M. Hanack, Angew. Chem., 92: 391 (1980); Angew. Chem. Int. Ed. Engl., 19: 392 (1980).

[34] a) O. Schneider and M. Hanack, Angew. Chem., 94: 68 (1982).
b) O. Schneider, J. Metz and M. Hanack, Mol. Cryst. Liq. Cryst., 81: 273 (1982).
c) J. Metz and M. Hanack, Nouv. J. Chim., 5: 541 (1981).
d) O. Schneider and M. Hanack, Chem. Ber., 116: 2088 (1983).

[35] B.N. Diel, T. Inabe, N.K. Jaggi, J.W. Lyding, O. Schneider, M. Hanack, C.R. Kannewurf, T.J. Marks and L.H. Schwartz, J. Am. Chem. Soc., in press.

[36] A. Datz, J. Metz, O. Schneider and M. Hanack, Synth. Metals, in press.

[37] K. Nakamoto, Infrared and Raman Spectra of Inorganic and Coordination Compounds, Wiley, New York, 1978.

[38] J. Metz and M. Hanack, J. Am. Chem. Soc., 105: 828 (1983).

[39] O. Schneider and T.J. Marks, unpublished.

[40] J. Metz, Dissertation, University of Tübingen 1983.

[41] L.J. van der Pauw, Philips Technische Rundschau, 20: 230 (1958).

CARRIER GENERATION, RECOMBINATION, AND TRANSPORT IN ORGANIC CRYSTALS

Martin Pope

New York University, Chemistry Department, New York

Charles E. Swenberg

Armed Forces Radiobiology Research Institute, Bethesda

Various mechanisms of carrier generation in organic crystals are reviewed. These include band-to-band transitions, auto-ionization, and direct charge-pair generation. Geminate recombination is discussed, including the effect of temperature and electric field on the initial charge distribution. Carrier transport is reviewed, using crucial experimental results. An analysis of various transport mechanisms is presented, including one that may be relevant.

Three processes that are of major importance in the phenomenon of photoconductivity are carrier generation, carrier recombination, and carrier transport. These will be discussed in that order, as they relate to homomolecular polycyclic aromatic hydrocarbon (PAH) crystals. It is surprising that the mechanisms of carrier generation and carrier transport in well-characterized crystalline compounds are still subjects for heated dispute.

I. PHOTOGENERATION OF CARRIERS IN THE BULK

Among the homomolecular PAH compounds, anthracene has long been the "hydrogen-atom" for theoretical calculations. It now appears that there are distinct differences not only in the quantitative response of various members of the PAH family to light, but there may be qualitative differences as well. As one proceeds from naphthalene to pentacene for example, the optical absorption spectrum shows a significant increase in the contribution of states that have ionic character (1). This increase in

J. Ladik et al. (eds.), Quantum Chemistry of Polymers - Solid State Aspects, 137–164.

ionic character is paralleled by an increase in the quantum efficiency of photogeneration. However, there is not yet unanimity on the association of this increased ionic character with a specific mode of ionization.

Historically, the first conflicted picture to be resolved was whether the photogeneration of carriers proceeded as a result of a direct transition from the valence band to the conduction band (band-to-band, BB) (2, p. 470). In anthracene, it was known that the absorption spectrum was essentially explicable in terms of transitions to bound, neutral, Frenkel exciton states; the BB transition would thus have to be weak since the final states would be buried in the Frenkel exciton spectrum. An alternative hypothesis was that carrier generation requires the excitation of a Frenkel exciton that could dissociate if its energy was degenerate with that of a pair of uncorrelated carriers. This process is referred to as auto-ionization (AI). It does not follow that all AI transitions lead to completely uncorrelated carrier pairs. Most of the emitted electrons will thermalize within the Coulomb capture radius of the geminate positive ion, forming a transient charge-transfer state that can either decay to the ground state or dissociate by the absorption of ambient energy. Following Jortner (3), the wavefunction of such an AI state can be written as

$$\psi_E = a_E \phi_A + \Sigma_i b_i \phi_i + \int C_{E'} \phi_{E'} dE' \quad (1)$$

where $\phi_{E'}$ are the crystal continuum states, and ϕ_i are the sets of excited vibrational levels of the ground state and of other bound electronic states that are degenerate, or nearly so with the vibrational state localized at ϕ_A. Anticipating the recent work of Bounds and Siebrand (4), an example of another type of bound electronic state that would be degenerate with the ϕ_A would be a charge-transfer (CT) state, in which the hole and electron are on different molecules, but still correlated with one another. These CT states could ionize completely by absorbing thermal energy, or energy from an external electric field or auxiliary light source. The decision as to whether BB transitions or an AI transition was the proper description for the photoconduction process was made on the basis of studies using polarized light, but it is now evident that these experiments would not distinguish between the AI mechanism or that in which a dissociable CT exciton was excited. This may be seen as follows: for the purposes of this argument, assume that all three processes namely AI, BB transitions, and direct CT formation occur. If one classifies bound states as being either Frenkel states or CT states, then the maximum efficiency of carrier production, ϕ^- can be expressed as

$$\Phi^- \propto \frac{k_F \{[x(1-y)\ \eta_{AI} + (1-x)]\ \eta_{CT} + \eta_{AI}\ xy\} + \alpha}{k} \tag{2}$$

where k is the overall optical absorption coefficient, being equal to $k_F + \alpha$, where k_F is the absorption coefficient to a CT state or to a Frenkel state that can autoionize with an efficiency η_{AI}. Here x represents the fraction of states that lead to AI and y represents the fraction of AI states that lead to carriers without passing through an intermediate CT state. The efficiency of ionization of the CT state is given as η_{CT}. In either case, AI or direct CT formation, an intermediate CT state can be formed; this conclusion follows from the observations (5,6) that there is a temperature and field-dependent ion-pair dissociation process that precedes complete ionization. This field and temperature dependent process of dissociation is referred to as the Onsager mechanism (7) which will be discussed shortly. The term α in equation (2) is the absorption coefficient to a plane wave state (BB transitions) in which the ionization efficiency is unity. If BB transitions dominate, then α dominates the numerator and it follows that $\phi^- \propto \alpha/k$. Since ϕ^- is found to be about 10^{-4} in anthracene (it is much higher in tetracene and pentacene, (8)) this implies that $k = k_F + \alpha >> \alpha$ or $k_F >> \alpha$. Thus Φ^- would vary inversely with the absorption coefficient k_F. It is possible to check this prediction by using polarized light to excite the crystal. In this way, k may be varied by a factor of 7 at a particular wavelength (constant energy), thus providing a technique for changing k without introducing at the same time the additional complication of changing the energy of excitation. This experiment was carried out by Geacintov and Pope (9), and Braun and Hornig (10), and it was found that there was essentially no change in Φ^- as k was varied. This observation ruled out BB transition in anthracene. This polarization experiment cannot, however, distinguish between the processes of AI and direct CT formation. It is probably the case that both mechanisms are operative in the sense that for some materials and in some energy ranges, one or the other process (or perhaps both) will dominate. Thus, in the subthreshold energy region for photoconduction (i.e. $h\nu < E_g$ where E_g is the band gap and $h\nu$ is the photon energy), the contribution of AI is small, while there is incontrovertible evidence that direct CT exciton formation does take place. This will be discussed shortly. On the other hand for excitation energies exceeding the vacuum level, the observation that electrons can be photoemitted with the maximum kinetic energy (11,12) permitted by the Einstein photoelectric equation is proof that no intermediate CT state need be excited to produce free carriers. Furthermore, it has been found by Chance and Braun (5), that the activation energy E_a for carrier photogeneration in the energy region $5.4 < h\nu < 6.2$ eV is constant at about 0.06 eV and the quantum yield at 6.2 eV is about 4×10^{-4}. This quantum yield should be compared with that of photoemission at 6.2 eV (13), which is about 3×10^{-5}. From geometric considerations, it may be appreciated that roughly one-fourth of the total number of free electrons will emerge from the surface, so the actual AI yield should be at least 10^{-4} at this energy, which is approximately what is observed experimentally in photoconductivity. In addition, the constancy of E_a over the energy region $5.4 < h\nu < 6.2$ eV is consistent with the maintenance of the same mechanism or mechanisms of photogeneration over this energy range and with the position (5) that AI also plays a major role at energies less than the ionization energy of the crystal. It thus appears reasonable to expect that there are excitation energies for which the photoconductivity mechanism in a given PAH crystal changes from direct CT exciton generation to AI.

The CT state in anthracene was first treated theoretically by Hernandez and Choi (14), who concluded that the direct optical transition to this state would have an oscillator strength of 10^{-4}, and that it would be buried in a region of high oscillator strength Frenkel exciton transitions, making its spectroscopic detection difficult. The first indirect evidence for the generation of CT excitons was provided by Pope and Burgos (15), and this was based on photoemission studies, particularly in tetracene. The first optical absorption evidence for the existence of a CT state in a homomolecular PAH crystal was presented by Tanaka (16), who worked with 9,10 dichloroanthracene. These results were substantiated by Abbi and Hanson (17). The proposition that direct CT generation was the primary mechanism for photoconductivity in all PAH crystals was put forth in a series of theoretical papers by Bounds and Siebrand (4,18), and more recently by Bounds, Petelenz and Siebrand (19,20). Experimental evidence for direct excitation of CT exciton states in tetracene and pentacene has been provided by Sebastian, Weiser and Bässler (21). More recently, the CT exciton transition was detected in anthracene by Sebastian, Weiser, Peter, and Bässler (22). All of the experimental studies were based on the electric field modulation of optical absorption.

We will first discuss the theoretical work of Bounds, Petelenz and Siebrand (BPS). The fundamental premise of this study (based on anthracene) is that the direct excitation of the CT exciton is made possible by the coupling of this low oscillator strength state, with that of the high oscillator strength ($f \simeq 1$) S_3 (^1B_u) Frenkel state located at 4.63 eV. The S_3 state thus lies about 0.5 eV above the anthracene band gap $E_g \sim 4.1$ eV, and all optical excitations in the energy region up to 4.63 eV would produce CT states (albeit, vibrationally excited); in other words, there would be an energy of activation for photoconductivity even when the optical energy $h\nu > E_g$. This is in fact observed (5). The energy E_{CT} of CT states of varying separation distance between the charges was calculated by BPS, using the expression

$$E_{CT}(\ell_1 k_1, \ell_2 k_2) = E_g + W_c(\ell_1 k_1, \ell_2 k_2) + \Delta W_{eh}(\ell_1 k_1, \ell_2 k_2) - \Delta W_e - \Delta W_h, \quad (3)$$

where $k_2\ell_2$ is the molecule and unit cell respectively from which the electron is removed and $k_1\ell_1$ is the final state of the excess electron; E_g is the band gap, W_c is the Coulomb energy of the two carriers, and ΔW are apparent polarization energies, the subscript eh referring to the electron-hole pair and the separate e and h subscripts to the electron and hole respectively. The apparent polarization energy is defined as

$$\Delta W = P + W_M + W_R \quad , \quad (4)$$

where P is the polarization energy of a charge, which includes charge-induced dipole and induced dipole-induced dipole interactions, W_M is the charge-multipole energy and W_R is the lattice relaxation energy; the P term is the largest in magnitude. The results of these calculations for anthracene are shown in Fig. 1.

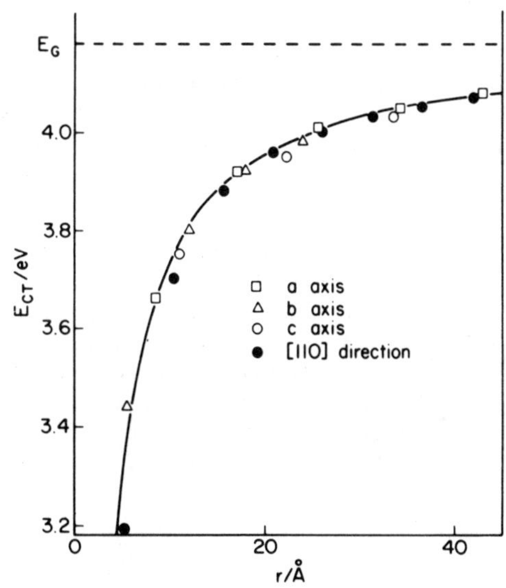

Fig. 1. Energy of an electron-hole pair in anthracene as a function of separation in four crystal directions. The solid line represents the energy the pair would have in an isotropic dielectric with ε = 3.23 for a Coulombic potential. Bounds et al (19).

These data may be described qualitatively but not quantitatively by a Coulomb expression of the form

$$E_n = E_g - \mu e^4/8(h\varepsilon\varepsilon_o n)^2 \quad , \qquad (5)$$

where μ is the effective mass of the electron-hole pair, $\varepsilon\varepsilon_o$ the dielectric constant parameters, and n is a principal quantum number, related to the distance r between electron and hole. Equation (5) is overly simple, and when properly corrected, yields the results shown in Table I. The CT state values shown in

Table I. Energy Levels E_n of CT States (in eV) (19)

n	Quasi-Localized Model[a]	Wannier Model[b]
1	3.20	3.46
2	3.70	3.72
3	3.88	3.85
4	3.96	3.93
5	4.00	3.98

[a] Along the [110] direction, calculated using eq. (3).
[b] Calculated using an empirical correction to the Coulomb potential.

Table I can be related to the activation energies E_a found for carrier generation in anthracene by the relationship $E_a = E_g - E_n$; quantum efficiencies of carrier generation $\phi_o(h\nu)$ were also calculated, and the results of these are shown in Fig. 2. The agree-

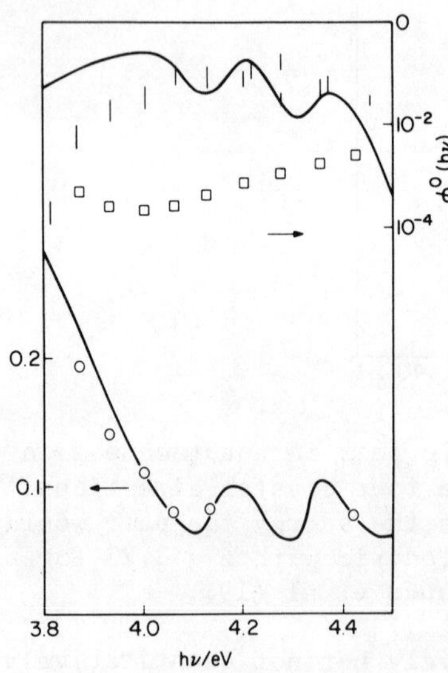

Fig. 2. Calculation (lines) and obsereved activation energies of carrier generation (bottom) and quantum yields of electron-hole pair formation (top). From Bounds et al. (19). Circles and squares taken from (6) and bars from L. E. Lyons and K. A. Milne, J. Chem. Phys., 65 1474 (1976).

ment for the E_a values is good, but not so good for the $\phi_o(h\nu)$ calculations. The explicit assumption is made here that while it is possible to excite vibrational states of CT excitons for each value of E_n, only the 0-0 vibrational state is available for thermal dissociation, the excess energy being dissipated in a time shorter than dissociation or recombination. It was possible for BPS to estimate the contribution of CT states to the overall absorption spectrum; in anthracene, this contribution was small but significant (23), while in tetracene and pentacene, the CT contribution was dominant.

Considerable experimental support for the CT exciton mechanism for carrier generation has been provided by work on electric field modulated absorption spectroscopy. As described Sebastian et al (21), the electric field induced change in absorption coefficient k for excitation of a Frenkel exciton may be given to first order by the expression

$$\Delta k \propto (\Delta\bar{p})\ F^2\ \partial k/\partial E + \ldots + \qquad (6)$$

where $\Delta\bar{p}$ is the average change in molecular polarizability, F is the applied electric field, and $\partial k/\partial E$ is the differential change of absorption coefficient with the optical excitation energy. For the excitation of a pure CT state the change in the absorption coefficient is given by

$$\Delta k \propto (1/3\ qr)^2\ F^2\ \partial^2 k/\partial E^2 + \ldots \qquad (7)$$

where qr is the magnitude of the dipole moment of the CT state. In Fig. 3 is shown the absorption and electroabsorption spectra of pentacene; for excitation energies above 1.83 eV it is evident that Δk varies directly with $\partial^2 k/\partial E^2$. Similar studies were made on anthracene by Sebastian et al (22), although it proved to be more difficult to detect the CT states. These authors correlated the energies of these peaks in the CT exciton spectra with the size of the CT exciton using a Coulombic equation similar to that given by equation (8); this is shown in Fig. 4. The extrapolated value of E_g proved to be 4.4 eV, in contrast with the accepted value of about 4.1 eV; they justified this discrepancy by identifying their CT values with the vertical (Franck-Condon) transitions from a Frenkel ground state to vibrationally excited CT states. Furthermore, in opposition to a basic premise of the BPS treatment, they postulated that the excess vibrational energy for any particular CT state could be used to further separate the members of the ion-pair state. This postulate, in a sense, combines the processes of CT exciton absorption and ballistic carrier separation. The identification of the structures found by Sebastian et al (22), in the electroabsorption spectrum with the sequence of CT states of different separations r_{CT} has been

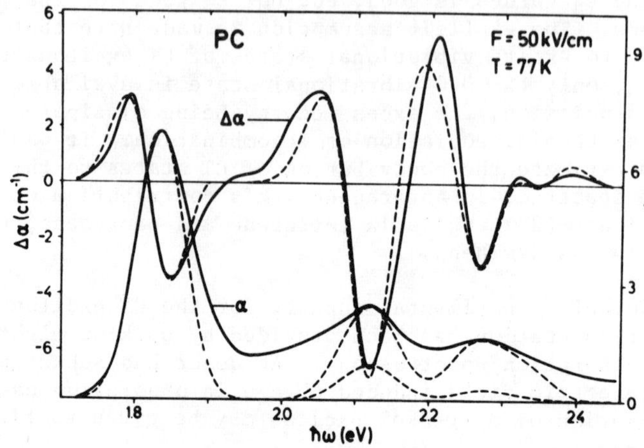

Fig. 3. Absorption and electroabsorption spectra of pentacene films. Solid curves refer to experiment, the dashed curves represent fits on the basis of Stark effect and charge transfer excitons. The lower absorption curves are calculated with the assumption of complete charge transfer. α is denoted by k in text. From Sebastian et al (21).

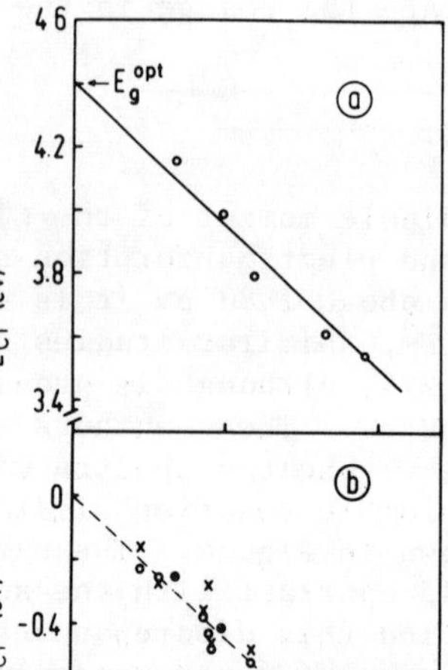

Fig. 4. (a) Charge-transfer energies versus reciprocal CT distance r_{CT}. (b) CT binding energies. Full dots are experimental, crosses and open circles are computed data from ref. 19 (O: point molecule treatment, x: submolecule treatment). From Sebastian et al (22).

questioned by Siebrand and Zgierski (24). These authors calculated the spectrum of CT states in anthracene and attributed the structure observed by SWPB (22), to the vibrational overtones of the nearest-neighbor CT state.

The AI mechanism of ionization coupled with the ballistic model of electron-hole separation has been discussed in detail by Silinsh et al (8), who studied tetracene and pentacene. The essential principles of this model are exposed in Fig. 5. Here,

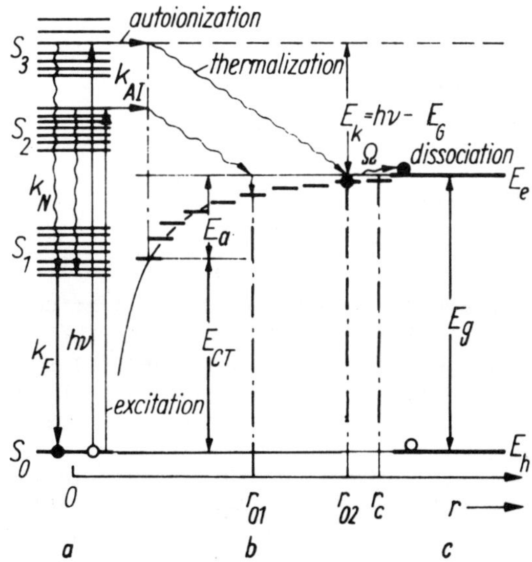

Fig. 5. Photogeneration stages in anthracene-type molecular crystals according to the ballistic model. a) neutral electronic states; b) bounded charge pair (CT) states; c) ionized states of the crystal. From Silinsh et al (8).

AI steps are succeeded by the escape of the hot electrons into the lattice, where they thermalize by acoustic phonon scattering, producing the series of bound CT states. Following the creation of the CT states, the mechanism of dissociation is the same as that of the direct CT exciton generation mechanism. Using a rather simple scattering theory, Silinsh et al (8), calculated the optical energy dependence of the thermalization distances, $r_o(h\nu)$, which in turn, correspond to specific separations of CT states. The agreement between their calculated and experimental values is shown in Fig. 6. The values of the energies of activation E_a were calculated using the expression

$$E_a = E_g - e^2/4\pi\varepsilon\varepsilon_o r_o(h\nu) \quad . \tag{8}$$

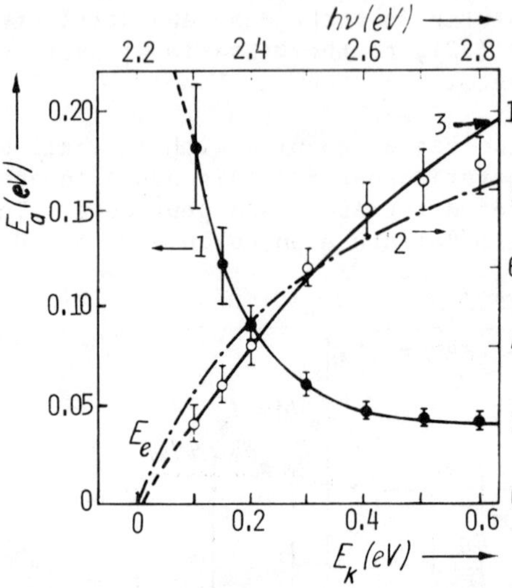

Fig. 6. Experimental zero-field thermalization length $r_o(h\nu)$ (open circles) and corresponding $E_a(h\nu)$ (filled dots) dependences on hν in pentacene; (1) $E_a(h\nu)$ curve; (2) $r_o(h\nu)$; calculated using $r_o = (D\tau_{th})^{\frac{1}{2}}$, see text; (3) empirical formula for r_o. From Silinsh et al (8).

As may be seen in Fig. 6 the agreement is satisfactory. Calculations were also made of the photoconductivity quantum efficiency and agreement was found between calculated (using ballistic theory and a modified Onsager theory) and measured values in the energy region $E(h\nu) > E_g$. In the region $E(h\nu) \sim E_g$, they found a discrepancy that they attributed to a contribution from direct CT state absorption. In the higher regions of photon energy, they found another discrepancy that they attributed to excitation to a different AI state.

In the cases considered above, the thermalization distance, or CT exciton size is determined by the energy of the absorbed photon. There is another interesting experimental result in which it appears that the efficiency of photoconduction is independent of photon energy (25). This work was carried out with X-metal-free phthalocyanine crystals embedded in a polymeric matrix. It was found that all the ionization proceeded from the first excited singlet state, S_1. In the direct CT exciton excitation picture, one might explain this result by the proposition that excitation to any state above S_1 results in a rapid internal conversion to S_1, which must have considerable CT character because the application of an external electric field leads to a significant fluorescence quenching from S_1. The quenching would be due to the field-induced dissociation of the CT states, which depletes the concentration of states that could recombine to form singlet-states that could fluoresce. If S_1 lies below the conducting level, as appears to be the case with β-phthalocyanine (2, p. 671) then the AI mechanism with ballistic carrier separation is not plausible, unless the initial state is vibrationally excited.

Summarizing the results presented herein regarding the relative roles of AI and direct CT exciton formation, it appears that there are materials and energy ranges in which one or the other, or both mechanisms prevail. This point has been made by Silinsh et al (8), and we favor this position. The actual determination of the relative roles of AI and direct CT exciton formation might be resolvable if photocurrent rise time measurements could be made. This suggestion was made by Bounds et al (20,23), who presented a theoretical analysis of what the photocurrent rise time should look like in anthracene, as a function of temperature and excitation wavelength. They chose anthracene as a model compound because the calculated values of the absorption coefficient to the CT state seem to be much smaller than the observed absorption coefficient to the Frenkel exciton states. Thus, if the AI plus ballistic model is operative, it should have a much better chance of being seen in anthracene than in pentacene, where the CT transitions appear to dominate Frenkel transitions in the energy region greater than that of S_1. If the ballistic mechanism is operative, Bounds et al (23) assume that only the nearest neighbor CT state will be excited, regardless of photon energy. The rise time of photoconductivity would thus be a constant, independent of photon energy. However, in the case of direct absorption to a CT state there should be an essentially instantaneous population of CT states of all radii. The risetime of conductivity should drop with increasing photon energy, due to the increased population of quickly ionizing large-radius CT exitons. These differences in the rate of formation of the CT state can be incorporated into the overall rate of formation of free carriers. It emerges that by using the parameters required to explain the action spectrum of the energy of activation, it is predicted that the rise time of the photocurrent should be constant at about 55 psec if the ballistic model is correct and should drop to 15 psec with increasing photon energy if the CT mechanism is correct. The assumptions made in deducing this expected behavior are arguable, and it seems that the experiment could be used to rule out direct CT absorption, but not AI.

II. GEMINATE RECOMBINATION

Following the initial step of carrier generation, there evolves a final configuration of a hole and an electron separated from each other by a distance that in some way is a function of the energy of the exciting light (except as indicated above).

These carriers have originated from the same excited molecule and are referred to as geminate pairs. These pair-states can dissociate as a result of a thermally activated diffusive separation of the two carriers and the application of an external field. The quantitative description of this photoionization process was given by Onsager (7). The application of the Onsager theory to organic crystals was carried out by Batt, Braun and Hornig (26,27), and by Chance and Braun (5,28). It was found that the theory was quite successful in explaining the observed results without introducing any parameter other than the dielectric constant of the medium. The Onsager theory gives the probability that a thermalized geminate charge pair will escape recombination, and will dissociate under the influence of an external electric field of strength F. Onsager assumed an isotropic system containing a low concentration of charge pairs (no interaction between charge pairs) in thermal equilibrium with a medium of dielectric constant ε. The relationship for the probability that the charge pair will dissociate is (29)

$$f(r,\theta) = \exp(-A)\exp(-B)\sum_{m=o}^{\infty}\sum_{n=o}^{\infty}\frac{A^m}{m!}\frac{B^{m+n}}{(m+n)!} \qquad (9)$$

which,when only terms linear in B are retained, gives

$$f(r,\theta) = [\exp(-A)](1 + AB) \qquad (10)$$

where $A = 2q/r$, $B = \beta r(1 + \cos\theta)$, $q \equiv e^2/8\pi\varepsilon\varepsilon_o kT$, $\beta \equiv eF/2kT$, r is an initial separation distance between the oppositely charged carriers and θ is the angle between the radius vector r and the applied field vector F.

The parameter 2q has the dimensions of distance and in the absence of any external field, $f(r,\theta)$ is equal to $\exp(-2q/r)$. The value of 2q is referred to as the Coulombic capture radius r_c, and corresponds to the distance at which the kinetic energy of the diffusing particle is equal to the Coulombic attractive potential energy. For an electron moving with thermal energy kT,

$$e^2/4\pi\varepsilon_o\varepsilon r_c = kT$$

or

$$r_c = e^2/4\pi\,\varepsilon_o\varepsilon kT = 2q \qquad (11)$$

The observed carrier quantum yield Φ can be expressed as the integral of $f(r,\theta)$ over space (28)

$$\Phi = \Phi_o \int g(r,\theta)\, f(r,\theta)\, d\tau \qquad (12)$$

where Φ_o is the primary quantum yield in carrier pairs, and $g(r,\theta)$ is the probability per unit volume of finding the ejected electron in a volume element $d\tau$ at r,θ. Assuming that g is spherically symmetric and independent of F, and that the Onsager function can be truncated at the linear term in F, then by setting $g(r,\theta)=g(r)$ one obtains from equations (10)-(12)

$$\Phi(F) = 4\pi\Phi_o\ (1+2\beta q) \int_0^\infty \exp\{-2q/r\}\ g(r)\ r^2\ dr \tag{13}$$

Onsager assumed that the electric field had no appreciable influence on $g(r,\theta)$. This assumption was modified by Silinsh et al (8), as will be discussed herein. It may be seen from equation (13) that in the low field limit, the integral is independent of F, implying that Φ is a linear function of F. Thus, at any given excitation énergy, a plot of the relative carrier yield against F should give a straight line with the ratio of slope (S) to intercept (I) of

$$S/I = 2\beta q/F = e^3/8\pi\varepsilon\varepsilon_o k^2T^2 = 3.38 \times 10^{-5}(298/T)\text{cm V}^{-1}, \tag{14}$$

using a mean value of $\varepsilon = 3.23 \pm 0.09$. Using anthracene, Chance and Braun (5) measured the quantum yield of photocarriers as a function of applied field. Their results are shown in Fig. 7 where the solid lines are theoretical and the points experimental. The fit is excellent.

The evaluation of the thermalization distance necessitates an assumption as to the form of $g(r,\theta)$. The assumption of spherical symmetry implies an isotropic medium. Although anthracene is not strictly isotropic, this assumption is not severe. The form of g(r) is however generally unknown, and may be deduced by the goodness of fit of various trial functions. In their original work, Batt et al (27), used a delta function for g(r), i.e.

$$g(r) = \delta(r-r_o)/4\pi r_o^2 \quad . \tag{15}$$

Inserting this into equation (13) and integrating, one gets

$$\Phi(F)/\Phi_o(1+2\beta q) = \exp(-e^2/4\pi\varepsilon\varepsilon_o kTr_o) \simeq \Phi(0)/\Phi_o \tag{16}$$

where $\Phi(0)$ is the carrier yield at zero applied field. The approximate equality on the right-hand side of equation (16) holds if the quantum yield measurements are made at low field strength ($\sim 10^4$V cm^{-1}). In this way, $\Phi(0)$ may be determined; βq is obtained from equation (14). Assuming Φ_o to be independent of temperature. Chance and Braun (28), obtained from equation (16) the energy of activation for carrier separation at zero applied field

$$E_a = e^2/4\pi\varepsilon_o\varepsilon r_o(h\nu) \tag{17}$$

By measuring E_a, at different excitation energies it is thus possible to calculate $r_o(h\nu)$, given g(r). Their results for

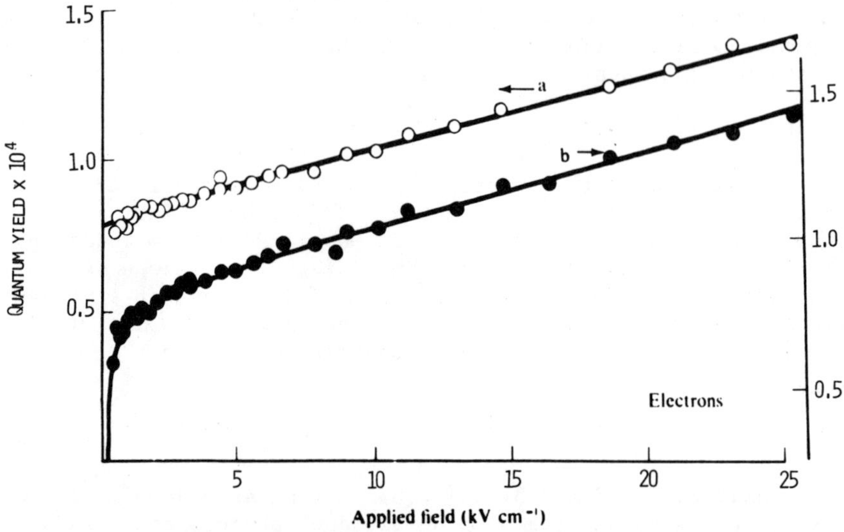

Fig. 7. Electron carrier quantum yields at 255 nm as a function of applied field. Curve a is for a previously unirradiated or "virgin" crystal sample. Curve b illustrates the reduced low-field yield that occurs when trapped holes, left behind in the experiments shown in curve a; are present in the excited volume of the crystal. Note that curves a and b actually superimpose at high fields but that curve b has been shifted arbitrarily for clarity of presentation. From Chance and Braun (5).

anthracene are shown in Fig. 8 where for the delta function approximation the average value of r is the effective thermalization distance, i.e.: $\langle r \rangle = r_o$. The constancy of $\langle r \rangle$ with photon energy in the two energy regions shown can be interpreted either by the AI mechanism or the direct CT exciton generation. Using the AI mechanism, one would assume that different, but distinct AI states were ionizing in each energy region. Using the direct CT exciton excitation mechanism, Sebastian et al (22), conjectured (within the context of the their assumption that the excess vibrational energy of the CT exciton is used to separate the carriers) that the maximum jump distance of both electron and hole are limited by the mean scattering length λ as determined by the electron-phonon interaction. This is taken to be between 17 and 20A; this is two units along the a axis or 3 units along the b axis. Thus, the maximum distance between hole and electron after the optical excitation would be λ, and the maximum distance that each carrier would jump after utilizing the excess vibrational energy would be λ making a total distance of 3 λ or 58Å. A different value of λ is assumed for the higher energy region in Fig. 8.

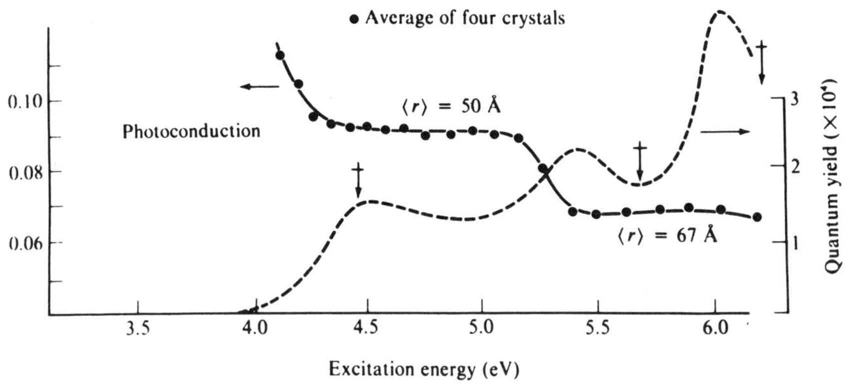

Fig. 8. Activation energy versus excitation wavelength. Each point represents the average of the results for four non-virgin crystal samples. The dashed line is the quantum yield extrapolated to zero applied fields for unpolarized light incident on the ab crystal face. The three vertical arrows indicate the location of peaks in the density of states in the valence band of anthracene. From Chance and Braun (28).

As has been mentioned, the Onsager equations (9) and (10) employ a thermalization distance r_0 that is independent of electric field and temperature. This assumption was modified by Silinsh et al (8), who attempted to produce a relationship in which $r_0 = r_0\ (h\nu, T, F)$. This modification takes into account electron drift during thermalization, which would introduce an anisotropic $g(r,\theta)$. The temperature dependence of r_0 in pentacene was determined by Silinsh et al (8) who observed a decrease in the photoconductive quantum yield above 200K, in the threshold energy region 2.3 to 2.5 eV. They attributed this to temperature activated scattering. The field dependence of r_0 appeared to be significant when $r_0 \simeq 40$Å. The magnitude of the drift was estimated by the expression

$$r_F = \bar{\mu} F \tau_{th} \tag{18}$$

where $\bar{\mu}$ is the mean electron mobility during thermalization, and τ_{th} is the thermalization time, calculated from the expression $r_0 = (D\tau_{th})^{\frac{1}{2}}$. During the thermalization time, the carriers are assumed to drift a distance r_F. The new carrier distribution is assumed to be shifted, but it maintains its δ-function character.

The Onsager expression is revised as follows:

$$\Phi[r_o(h\nu), r_F(F), F, T] = \Phi_o/2 \exp[-\beta r_F] \times \int_{-1}^{1} \exp[-2q/s - \beta s - \beta r_o\xi] \times \sum_{m=0}^{\infty}\sum_{m=1}^{\infty} \beta^{m+n} \frac{(s+r_F+r_o\xi)^{m+n}}{s^m} \frac{(2q)^m}{m!(m+n!)} d\xi \quad (19)$$

where $s = \sqrt{r_o^2 + r_p^2 + 2r_o r_F\xi}$, $\xi = \cos\theta$ and where the other parameters have been defined previously. The agreement between this equation and experiment shown in Fig. 9 is fair, and it provides support for the notion that AI plays a role in carrier generation even in materials like pentacene, where CT contributions play a major role in determining the absorption spectrum.

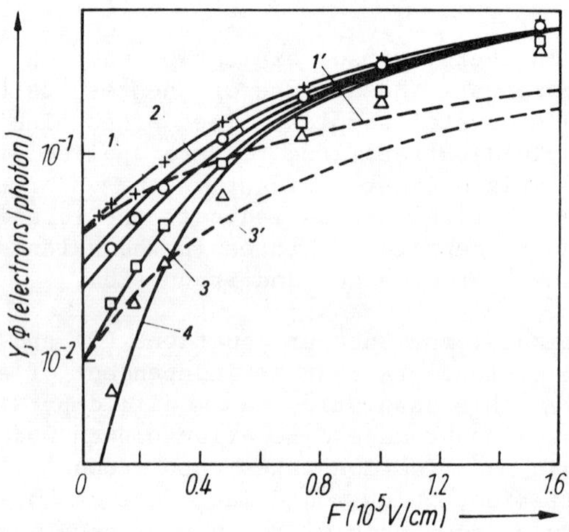

Fig. 9. Dependence of pulse photoconductivity quantum efficiency Y on F at $h\nu$ = 2.6 eV in pentacene at +T = 330; o 280, □ 250, Δ 200K and $\Phi(F)$ curves at corresponding T values: calculated according to equation (19) (curves 1 to 4) and according to equation (9) (curves 1' and 3'), taking $\Phi_o = 0.5$; r_o = 53Å and an average carrier mobility of 0.4 $cm^2 V s^{-1}$. From Silinsh et al (8).

III. CARRIER TRANSPORT

Charge transport in solids is generally considered in two extreme limits. In one, the carrier moves in a broad band with a relatively large mean free path, ℓ, i.e.: ℓ >> nearest neighbor distance. In this limit the mobility $\mu >> 1$ cm^2 V^{-1} s^{-1} and has a temperature dependence determined by phonon scattering, $\mu \propto T^{-n}$ $n > 1$ (30). At the other extreme, carriers are viewed as being highly localized with their motion consisting of a series of random hops between different molecular sites. For a strongly localized carrier, $\mu <<1$ cm^2 V^{-1} s^{-1} and $\mu \propto g(T)$ exp $(-E/kT)$ where E is the activation energy for hopping. The explicit dependence of g(T) on temperature and the intermolecular interaction energy depends on the particular polaron model, although it generally varies as T^{-m}, $m \geq \frac{1}{2}$. In contrast to carrier transport in wide bands as in Si and Ge, and hopping carrier transport as observed for example for electrons in orthorhombic sulfur (31), molecular crystals like anthracene and naphthalene have carrier mobilities that do not conform to either of these extreme transport limits. Carriers in these crystals have mobilities of the order of unity and have a temperature dependence of mobility that is band-like in some crystallographic directions; there are temperature regimes over which μ is almost insensitive to temperature variations and in the particular case of naphthalene the electron mobility along the c$'$-direction is fairly constant for T > 100K and shows a sharp increase for T < 100K. This is illustrated in Fig. 10.

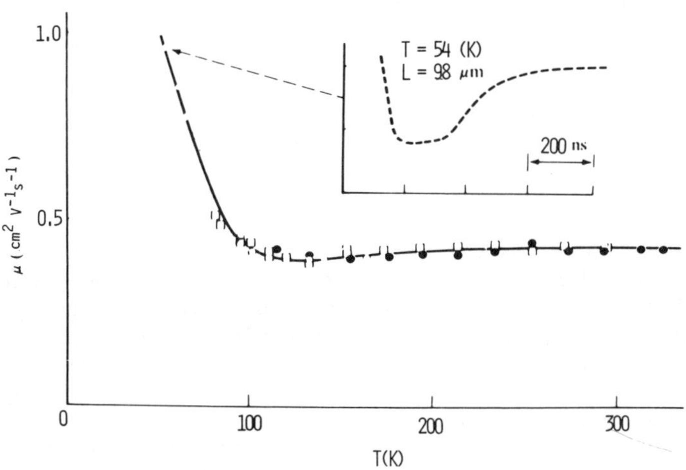

Fig. 10. The mobility, as measured by transient photoconductivity techniques, of electrons in the c$'$ direction of naphthalene from 54K to 324K. Insert illustrates a typical transit pulse. The different symbols represent measurement of different samples that have been normalized to 0.44 cm^2 V^{-1} s^{-1} at room temperature. From Schein et al (32).

The highly anisotropic behavior of carrier transport in organic crystals can be easily rationalized by the weak character of the intermolecular forces and the anisotropic nature of the crystal structure. However, it is this weak character of the intermolecular energies, which are comparable in magnitude with electron-phonon interaction energies (dynamic off-diagonal disorder) that makes a quantitative understanding of mobility in anthracene and similar solids difficult. Although the past five years has witnessed considerable experimental and theoretical activity there is still lacking a comprehensive theory of electrical transport in organic solids that is consistent with the experimental data. Thus our discussion will focus on what appears to be a promising theoretical model and on the presentation of the experimental data.

The different types of carrier transport can be categorized with respect to the order of magnitudes of the different interaction terms in the one-electron Hamiltonian:

$$\mathcal{H} = \mathcal{H}_o + \mathcal{H}_1 + \mathcal{H}_2 + \mathcal{H}_3 \tag{20}$$

$$\mathcal{H}_o = \Sigma_n \varepsilon a_n^+ a_n + \Sigma_\lambda \hbar \omega_\lambda (b_\lambda^+ b_\lambda + \tfrac{1}{2}) \tag{21}$$

$$\mathcal{H}_1 = \Sigma_{nm} J_{nm} a_n^+ a_m \tag{22}$$

$$\mathcal{H}_2 = \Sigma_\lambda \Sigma_n \Sigma_m Z_{nm\lambda} a_n^+ a_m (b_\lambda + b_{-\lambda}^+) \tag{23}$$

$$\mathcal{H}_3 = \Sigma_n \delta\varepsilon_n a_n^+ a_n + \Sigma_n \Sigma_m \delta J_{nm} (a_n^+ a_m) \tag{24}$$

Here $\mathcal{H}_o$ denotes the total energy of the system in the absence of interactions between molecules (labeled n and m) and the phonons (λ), ε is the molecular site energy, $\hbar\omega_\lambda$ is the phonon energy and J_{nm} is the interaction energy between molecules n and m. The interaction $\mathcal{H}_2$ describes the effects of lattice vibration on the carrier motion. Terms $Z_{nn\lambda}$ (n=m) designate the effects of dynamical diagonal disorder whereas when $n \neq m$ terms $Z_{nm\lambda}$ in $\mathcal{H}_2$ denote dynamical off-diagonal disorder interactions. The effects of static disorder on carrier motion are contained in the perturbing Hamiltonian, $\mathcal{H}_3$, where $\delta\varepsilon_n$ and δJ_{nm} are respectively the diagonal and off-diagonal components. The creation and annihilation operators have their usual meaning. Several possible types of limiting carrier transport behavior can be easily distinguished on the basis of the relative magnitudes of the perturbations ($\mathcal{H}_1$, $\mathcal{H}_2$ and $\mathcal{H}_3$) in the transport Hamiltonian.

<u>Case a</u>. If the perturbation $\mathcal{H}_1$ is very large compared to $\mathcal{H}_2$ and $\mathcal{H}_3$, then a good zero-order Hamiltonian for calculating transport properties is $\mathcal{H}_o + \mathcal{H}_1$. In this limit the eigenstates are the usual Bloch band states and the mobility tensor to first order in the relaxation time $\tau(k)$ is given as

$$\underset{\approx}{\mu} = (e/kT) \langle \tau(\underset{\sim}{k})\, \underset{\sim}{v}(\underset{\sim}{k})\, \underset{\sim}{v}(\underset{\sim}{k}) \rangle_o \tag{25}$$

where v(k) is the carrier velocity for momentum state k and the brackets denote an equilibrium ensemble average. Since $v(k) \propto T^{-p}$ where $p > 0$ (2, p. 347), it follows that band type motion implies $\mu \propto T^{-n}$ where $n > 1$. Hole mobilities for naphthalene (and anthracene) are in accordance with this type of temperature dependence as is evident in Fig. 11. The turnover in μ_{hole} at low temperature is caused by the presence of shallow hole traps since their effects on hole transport increase as the temperature is decreased. This band-like transport of holes in these materials is consistent with the fact that the calculated valence-bandwidth is relatively large ($\sim$ 0.1 eV) in all crystal directions.

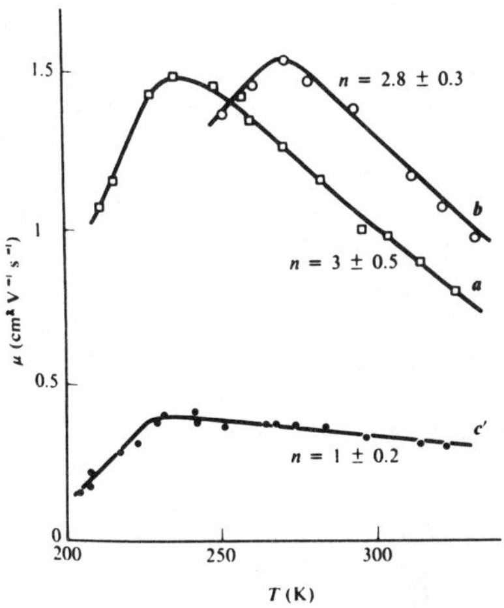

Fig. 11. The temperature dependence of hole mobilities for electric fields along the a, b and c$'$ axes for naphthalene mounted and studied in a pure N_2 atmosphere. The power dependence of $T(T^{-n})$ is indicated in each graph. From Berrehar and Schott (33).

Case b. Here, dynamical disorder effects are large compared to the transfer Hamiltonian $\mathcal{H}_1$ and the effects of static disorder are zero or very small. In this limit, the effects of lattice vibrations must be included in forming the initial basis states; this results in a renormalization of the carrier effective mass and alters the bare intermolecular interactions (J_{nm}). The quasiparticle after inclusion of lattice effects is called a polaron. Its motion through the lattice is generally viewed as a series of local hops between lattice sites. For hopping polaron transport (2, p. 356)

$$\mu = g(T) \exp\{-E/kT\}, \tag{26}$$

where the activation energy E depends on the polaron binding energy E_b and the nearest neighbor interaction energy J; g(T) is a weak function of temperature and its form depends on whether the transport between molecular sites is adiabatic or nonadiabatic. Regardless of the details of the polaron motion, $g(T) \propto T^{-n}$ where $n > \frac{1}{2}$ (2, p. 361).

Schein (34), has discussed the applicability of the polaron model for carrier transport in anthracene-like materials and has provided convincing arguments that regardless of the choice of polaron model the functional dependence of μ on T given by equation (26) is unable to account for the approximately constant (temperature independent) electron mobility along the c'-direction in naphthalene for T > 100K. The temperature dependence of $\mu_{cc'}$ for naphthalene is shown in Fig. 10. The slight decrease in the mobility at T ≈ 100K before the sharp rise at low temperatures (T < 100K) is now known to be due to crystal impurities (35). Much stronger evidence against the applicability of small polaron theories is provided by high electric field measurements. This can be seen as follows: in a localized model, the transfer of an electron between two molecular sites is permissible when the local configuration of atoms about a given site is such that the carrier has the same energy on either molecular site. Thus, if A denotes the local electron-phonon coupling constant (energy/distance) and Q_q is the generalized configurational coordinate for an electron at site q, then a Franck-Condon transition from q to q' occurs only when

$$AQ_q = AQ_{q'} + eF\, d_{qq'} \quad , \tag{27}$$

where $d_{qq'}$ is the physical distance between molecular sites q and q' and F is the electric field. Local configurations satisfying equation (27) are achieved as a result of thermal fluctuations and statistically, their probability of occurrence is proportional to exp $(-W_a/kT)$, where W_a is the minimum energy required to distort both molecules to a configuration such that equation (27) is valid. If BQ^2 denotes the elastic energy due to molecular distortion and $Q_o = A/2B$ is the change in the equilibrium position of the localized electron due to lattice vibrations, then the total distortional energy of the two neighboring molecules is (2, p. 354)

$$W(Q) = B\,\{\, Q_o + eFa/A - Q\}^2 + B\, Q^2 \tag{28}$$

where a is the nearest neighbor molecular separation. W(Q) has a minimum at $Q = Q_o/2 + eFa/2A$; thus

$$W_a(F) = \tfrac{1}{2}\, E_b + eFa/2 + (eFa)^2/8E_b \tag{29}$$

where we have used the relationship $E_b = A^2/4B = BQ_o^2$. The field dependence of μ follows directly from equation (29) by computing the net forward transition probability, ΔP, thus (2, p. 356)

$$\mu(F)=(ea^2/kT)\Delta P=(ea^2kT)\{\exp\{-W_a(-F)/kT\}-\exp\{-W_a(F)/kT\}\}$$

$$=(ea^2/kT)\exp\{-E_b/kT\}(2kT/eFa)\sinh(eFa/2kT)\exp\{-(eFa)^2/8E_bkT\}. \quad (30)$$

When the effects of the transfer energy J are included, the field independent factor $T^{-1}\exp(-E_b/kT)$ is appropriately modified and the functional form replacing it depends on whether the carrier motion is adiabatic or nonadiabatic (36). The important point to note about the derivation is its generality; the same field strength dependence in the mobility is expected for all forms of small polaron transport, exclusive of hot electron effects. Depending on the value of the polaron binding energy, E_b, the mobility $\mu(F)$ can either increase, decrease or remain unchanged for $eFa \leqslant 3kT$. The observation by Schein and McGhie (37), that $\mu_{c'c'}$ for anthracene and naphthalene at T = 100K is electric field-independent up to a field strength $F \approx 17V/\mu m$ can be made compatible with the predictions of small polaron theory only if $E_b \approx$ 21-35meV for anthracene and $E_b \approx$ 22-26meV for naphthalene. However these values for E_b are quite large and imply an activated temperature dependence for μ for T > 100K in contradiction to experiment. Furthermore the lack of temperature dependence in $\mu_{c'c'}$ for $100 < T \leqslant 325K$ cannot be rationalized within the context of a band model. In particular Schein (34), has shown that a temperature dependent carrier relaxation time for all types of scattering mechanisms is unable to account for the lack of a temperature dependence in $\mu_{c'c'}$ (electrons). Similar to naphthalene, the electron mobility along the c' direction for anthracene, deuterated anthracene and the layered compound As_2S_3 for T > 100K has been reported to be temperature independent(38). As a further illustration of the large differences in carrier transport along different crystallographic directions exhibited by anthracene-like crystals we note that electron mobility in the a-direction for naphthalene displays band-like motion with $\mu_{aa}(T) \propto T^{-1.53}$ (39). Support for this assignment is strengthened by noting that Schein et al (36) have rationalized the effects of deuteration on mobility in the context of the band model by assuming that the temperature dependence of the mobility arises solely in response to variations in phonon populations.

Case c. Effects of static disorder (the $\mathcal{H}_3$ term in equation (20)) are large. In this case, if the on-site static fluctuations are large compared to the average delocalizing effects of the average intermolecular transfer energy then charge localization ensues. Although this type of behavior is particularly important in the case of polymers and molecular glasses, it is unimportant for anthracene-like crystals since static disorder effects are very small in the excellent quality organic crystals usually employed.

Case d. Here, the perturbing terms $\mathcal{H}_1$ and $\mathcal{H}_2$ are of comparable magnitude, i.e.: the strength of the carrier interaction with lattice vibrations are of the same order of magnitude as are the intermolecular transfer energies, J_{nm}. It is this case that is thought to hold for most van der Waals molecular crystals like anthracene. Unfortunately, there is no theory that properly includes the effects of phonon interactions on carrier transport when phonon-electron interaction energies are comparable to intermolecular transfer energies and that also predicts mobilities consistent with experimental data. However, Sumi (40-43), gives a treatment of electrical conduction that appears to contain most of the essential elements for a comprehensive theory. In this theory, electron motion within the ab-plane is viewed as band-like with optical and acoustical phonons determining the carrier mean free path. Motion out-of plane is taken to be diffusive, due to the strong coupling to molecular rotational phonons (librations); electron motion in the ab-plane is assumed to be unaffected by librational modes. Hence for motion perpendicular to the ab-plane only the terms $Z_{nm\lambda}$ in the Hamiltonian $\mathcal{H}_2$ (see equation (23)) where n and m label molecules in nearest neighbor ab-plane that are along the same c axis direction are included. The effects of diagonal interactions (terms where n=m) and other dynamical off-diagonal interactions are neglected. Thus, the total nearest neighbor transfer energy, $J'_{n\alpha}$, for motion in the c'-direction, is taken as

$$J'_{n\alpha} = J + (\tfrac{1}{2}\gamma\, \hbar\omega_2)^{\frac{1}{2}} (b_{n\alpha} + b^{+}_{n\alpha} - b_{n+1,\alpha} - b^{+}_{n+1,\alpha}) \tag{31}$$

where J is the static nearest neighbor intermolecular interaction, $(n\alpha)$ labels a lattice site, $\hbar\,\omega_2$ is the energy of the molecular rotational phonon with dispersion effects neglected, γ denotes the coupling constant (in energy units) for the electron-libron interaction, and the term in brackets denotes the effects of the rotational vibrations in promoting electron transfer. Using linear response theory, Sumi (40) has shown that the nearest-neighbor interactions given by equation (31) the mobility perpendicular to the ab-plane can be written as

$$\mu_{c'c'} = (ea^2/\hbar)\eta\,(\overline{W}_2 + \varepsilon\,\overline{W}_1) \quad , \tag{32}$$

where $\eta = 8\pi\gamma/\Delta$, $\varepsilon\eta = 2/\pi\ (\Delta/S)\ (J/\hbar\,\omega_1)^2$; Δ is the conduction bandwidth for motion in the ab-plane, estimated to be about 700 cm^{-1} for naphthalene, S is the electron-phonon coupling energy for scattering within the ab-plane and ω_1 is the optical (or maximum acoustical) phonon energy. The quantities $\overline{W}_1$ and $\overline{W}_2$ are thermal averages of functions of the density of electron and phonon states whose explicit form need not concern us. Figure 12 illustrates for a particular set of parameters how $\overline{W}_2$ and $\overline{W}_1$

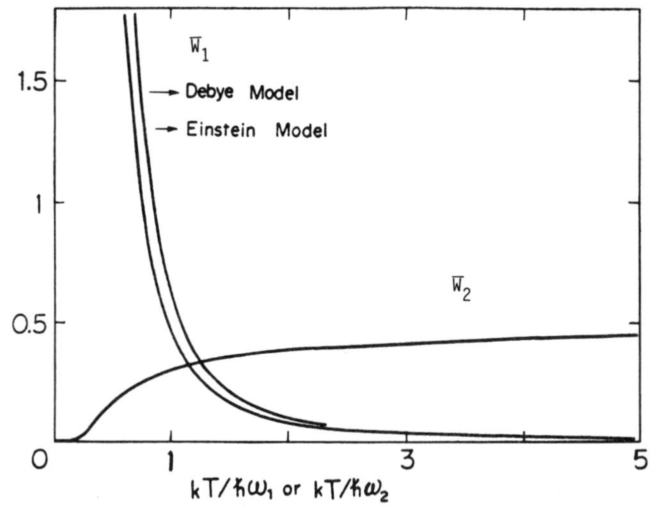

Fig. 12. The $kT/\hbar\omega_2$ dependence of $\overline{W}_2$, and $kT/\hbar\omega_1$ dependence of $\overline{W}_1$ calculated using either a Debye and Einstein model of the phonon spectrum. From Sumi (40).

contribute to the total electron mobility as a function of temperature. For high temperature, $T > 100K$, $\overline{W}_2 \sim 0.5$, and $\overline{W}_1 \ll 1$. Thus, if $\varepsilon < 1$, $\mu_{c'c'} \approx 4\pi a^2\gamma/h\Delta$, a quantity which is independent of temperature and therefore in agreement with experimental results. For $T < \hbar\omega_1/k \approx 100K$, $\overline{W}_2$ decreases to zero exponentially, implying that the mobility is determined primarily by $\overline{W}_1$. In this temperature regime

$$\overline{W}_1 \propto T^{-1} \exp\{\hbar\omega_2/kT\}. \qquad (33)$$

The dependence of $\mu_{c'c'}$ of temperature illustrated in Fig. 12 is quite similar to that of the mobility in naphthalene as shown in Fig. 10. Quantitative agreement with experimental results is afforded with coupling energies $S \approx 56\ cm^{-1}$ for electron-phonon interaction, $\gamma = 2.8\ cm^{-1}$ for electron-libron interaction and for a $\Delta \approx 700\ cm^{-1}$ for the electron bandwidth. The neglect of self-trapping tacitly assumed in Sumi's theory is justified since $S/\Delta \sim 0.1$ which is less than the critical value of 0.5 necessary for self-trapping effects (44). Sumi's treatment of carrier mobility, when applied to electron motion in the <u>ab</u>-plane, predicts band-type mobility with a temperature dependence of $T^{-1.5}$ – $T^{-2.0}$ provided $\hbar\omega_1 < kT < \Delta$. Generalizations of this theory to the case where transport along only the <u>a</u>-axis direction is coherent (43) do not appear necessary since the initial report (45) that the electron mobility along the <u>b</u>-axis is similar to that along the $\underline{c}'$-direction is unconfirmed (Schein, personal communication). Although Sumi's model of carrier transport is in accord with experimental data at low electric field strengths, the anomalous electric field behavior predicted for the $\underline{c}'$-direction electron mobility when the field F between neighboring $\underline{ab}$-planes of separation a) exceeds the librational energy $\hbar\,\omega_2$ ($\overline{F} \geqslant 2\text{x}10^5$V/cm) appears unconfirmed. In the context of Sumi's theory the drift velocity $V_{ii'}$ in an electric field F is given by the difference betwen the forward and backward velocity and can be shown to assume the form

$$V_{ii'} = a\,\{\,1 - \exp\,\{-eFa/kT\}\}\,\{J^2 f(F) + \gamma\,\hbar\,\omega_2 g(F)\,\} \quad (34)$$

For kT > eFa, the functions of f(F) and g(F), which depend upon electron and phonon density of states, are independent of F; thus $V_{ii'} \propto F$, i.e. $\mu_{ii'}$ is independent of field strength. However, when kT < eFa, Sumi (42) has argued that $f(F) \ll g(F)$ and that g(F) is independent of F, so that

$$V_{ii'} \propto \{\,1 - \exp\,\{\,-eFa/kT\,\}\;. \quad (35)$$

Saturation in the drift velocity at T = 140K for F > 1.8×10^5V cm^{-1} for anthracene in accord with equation (34) was reported by Nakano and Maruyama (46); unfortunately this experiment has not been confirmed by recent measurements (47). This lack of high field effects (up to 28V/ μm) for transport along the crystallographic axis which exhibits the hopping to band transition evident in Fig. 10 represents a perplexing problem. An additional difficulty of the Sumi treatment is the absence of predicted saturation of the electron mobility at high electric fields for T < 100K. Presumably an improved treatment of the density of electron and phonon state terms which appear in $\overline{W}_1$ and $\overline{W}_2$ (see equation (32)) and a different selection of coupling constants might allow these difficulties to be surmounted.

Although saturation in the electron velocity along $\underline{c}'$-direction for naphthalene has not been observed, field-dependent hole mobilities for F parallel to the $\underline{a}$-axis have been reported at low temperatures (48). This saturation behavior in the hole velocity is illustrated in Fig. 13. As noted previously, hole mobilities in both naphthalene and anthracene appear to be in accordance with band theory predictions. Presumably the electric field dependence at low temperature shown in Fig. 13 arises from hot carrier effects where the decrease in the phonon occupation at low temperature greatly enhances the carrier mean free path thereby allowing it to occupy high momentum, low velocity states near the edge of the Brillouin Zone. In particular, the data in Fig. 13 indicate that at T = 31K, hole velocity saturation occurs at $\sim 10^4$V cm^{-1}, corresponding to a potential drop of eFa $\sim 5\text{x}10^{-4}$ eV between neighboring molecules. Since hole bandwidth are

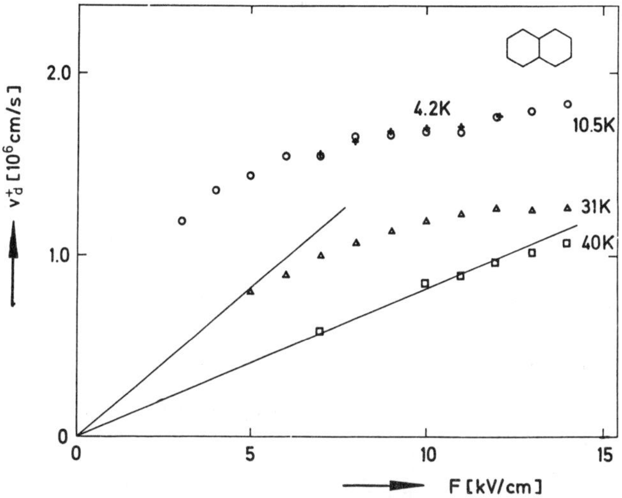

Fig. 13. Electric field dependence of the hole velocities in high quality naphthalene crystals obtained by the time-of-flight method with the electric field F parallel to a-axis (48). From Warta and Karl (48).

typically 0.1eV it follows that by traversing ∿ 500 lattice constants (∿ 2500Å) in the absence of scattering or phonon creation, the hole would acquire sufficient energy from the field to reach the edge of the Brillouin Zone. The observed limiting velocity is determined by the condition that the rate at which energy is absorbed from the electric field exactly equals the rate of loss due to phonon creation. A detailed analysis of the data in Fig. 13 should allow for a determination of the electron-phonon coupling constant.

It should be apparent from the above discussion that the current state of electrical conductivity in anthracene-like solids is still in need of further theoretical study.

REFERENCES

(1). See e.g., Rice, S.A. and Jortner, J, in: Physics and Chemistry of the Organic State, Vol. 3, eds. D. Fox, M.M. Labes and A. Weissberger (J. Wiley, New York, 1967, p. 201).

(2). Pope, M. and Swenberg, C.E., Electronic Processes in Organic Crystals, (Clarendon Press, Oxford University Press, New York, 1982).

(3). Jortner, J., Phys. Rev. Lett., 20, 244 (1968).

(4). Bounds, P.J. and Siebrand, W., Chem. Phys. Lett., 75, 414 (1980).

(5). Chance, R.R. and Braun, C.L., J. Chem. Phys., 59, 2269 (1973).

(6). Kato, K. and Braun, C.L., J. Chem. Phys., 72, 172 (1980).

(7). Onsager, L., Phys. Rev., 54, 554 (1938).

(8). Silinsh, E.A., Kolesnikov, V.A., Muzikante, I.J., and Balode, D.R., Stat. Sol.(b), 113, 379 (1982).

(9). Geacintov, N.E. and Pope, M., J. Chem. Phys., 47, 1194 (1967).

(10). Braun, C.L. and Hornig, J.F., 4th Mol. Cryst. Symp., Enschede, Holland (1968).

(11). Pope, M., Kallmann, H., and Giachino, J., J. Chem. Phys., 42, 2540 (1965).

(12). Vilesov, F.J., Zagrubskii, A.A., and Garbuzov, D.Z., Sov. Phys. Sol. Stat., 5, 1460 (1964).

(13). Kochi, M., Harada, Y., Hirooka, J., and Inokuchi, H., Bull. Chem. Soc. Jap., 43, 2690 (1970).

(14). Hernandez, J., and Choi, S., J. Chem. Phys., 50, 1524 (1969).

(15). Pope, M., and Burgos, J., Mol. Cryst., 3, 215 (1967).

(16). Tanaka, J., Bull. Chem. Soc. Jap., 36, 1237 (1963).

(17). Abbi, S.C., and Hanson, D.M., J. Chem. Phys., 60, 319 (1974).

(18). Bounds, P.J., and Siebrand, W., Chem. Phys. Lett., 85, 496 (1982).

(19). Bounds, P.J., Petelenz, P., and Siebrand, W., Chem. Phys., 63, 303 (1981).

(20). Bounds, P.J., Petelenz, P., and Siebrand, W., Chem. Phys., (in press).

(21). Sebastian, L., Weiser, G., and Bässler, H., Chem. Phys., 61, 125 (1981).

(22). Sebastian, L., Weiser, G., Peter, G., and Bässler, H., Chem. Phys., 75, 103 (1983).

(23). Bounds, P.J., Petelenz, P., and Siebrand, W., (submitted to Chem. Phys.).

(24). Siebrand, W., and Zgierski, M.Z., in Electronic Excitations and Interaction Processes in Organic Molecular Aggregates, eds.: P. Reineker, H. Haken and H.C. Wolf (Springer Series in Solid State Sciences, Berlin, 1983).

(25). Popovic, Z.D., and Sharp, J.H., J. Chem. Phys., 66, 5076 (1977).

(26). Batt, R.H., Braun, C.L., Hornig, J.F., J. Chem. Phys., 49, 1967 (1968).

(27). Batt, R.H., Braun, C.L., Hornig, J.F., J. App. Opt. Supp., 3, 20 (1969).

(28). Chance, R.R., and Braun, C.L., J. Chem. Phys., 64, 3573 (1976).

(29). Geacintov, N.E., and Pope, M., in Proc. 3rd. Photocond. Conf., 1969 (ed. E.M. Pell), Pergamon Press, Oxford (1971).

(30). Smith, R.A., "Semiconductors", Cambridge University Press, London, 1959.

(31). Gibbons, P.J., and Spear, W.E., J. Phys. Chem. Solids, 27, 1917 (1966).

(32). Schein, L.B., Duke, C.B., and McGhie, A.R., Phys. Rev. Lett., 40, 197 (1978).

(33). Berrehar, J., and Schott, M., Mol. Cryst. Liq. Cryst., 46, 223 (1978).

(34). Schein, L.B., Phys. Rev. B15, 1024 (1977).

(35). Karl, N., Ninth Molecular Crystal Symposium, Mittelberg-Kleinwalserthal, 1980.

(36). Efros, A.L., Soviet Physics Solid State, 9, 901 (1967).

(37). Schein, L.B., and McGhie, A.R., Chem. Phys. Lett., 62, 356 (1979).

(38). Schein, L.B., Chem. Phys. Lett., 48, 571 (1977).

(39). Schein, L.B., Warta, W., McGhie, A.R., and Karl, N., Chem. Phys. Lett., 75, 267 (1980).

(40). Sumi, H., Solid State Comm., 28, 309 (1978).

(41). Sumi, H., J. Chem. Phys., 71, 3403 (1979).

(42). Sumi, H., Solid State Comm., 29, 495 (1979).

(43). Sumi, H., J. Chem. Phys., 75, 2987 (1981).

(44). Sumi, H., J. Phys. Soc. Jap., 33, 327 (1972).

(45). Schein, L.B., and McGhie, A.R., Phys. Rev. B20, 1631 (1979).

(46). Nakano, S., and Maruyama, Y., Solid State. Comm., 35, 671 (1980).

(47). Schein, L.B., Narang, R.S., Anderson, R.W., Meyer, K.E., and McGhie, A.R., Chem. Phys. Lett. (in press).

(48). Warta, W., and Karl, N., Tenth Molecular Crystal Symposium, St. Jovette, Canada, 1982.

One of us (M.P.) acknowledges the support of the Department of Energy.

ELECTRICAL TRANSPORT PROPERTIES OF POLYACETYLENE AND RELATED COMPOUNDS

S. Roth, K. Ehinger and K. Menke

Max-Planck-Institut für Festkörperforschung,
7000 Stuttgart 80, FRG

ABSTRACT

A survey is given on electrical transport phenomena in polyacetylene. Data on the dependence of the conductivity on temperature frequency and doping concentration are reviewed and measurements of the magnetoresistance and of long-time aging effects under various storage conditions are presented. The experimental results are interpreted as superposition of intra-chain, inter-chain and inter-fiber conduction mechanisms, and resemble some aspects of the hopping transport in disordered semiconductors.

1. INTRODUCTION

Most polymers have a very low electrical conductivity. Fig. 1 shows a comparison of typical conductivities of various classes of solids /1/. The polymers range of the lower edge with σ 10^{-14} S/cm. The mechanisms ruling such low conductivities are by hopping between localized states /2/ and models developed for disordered semiconductors will also apply to polymers.

In Fig. 1 one polymer is included which is very close to the classical metals: $(CHI_{0.2})_n$, Iodine doped polyacetylene. In fact, polyacetylene belongs to a class of polymers whose conductivity can be changed by more than 14 orders of magnitude /3/, depending on the doping concentration and species (Fig. 2). These polymers

J. Ladik et al. (eds.), Quantum Chemistry of Polymers - Solid State Aspects, 165–190.

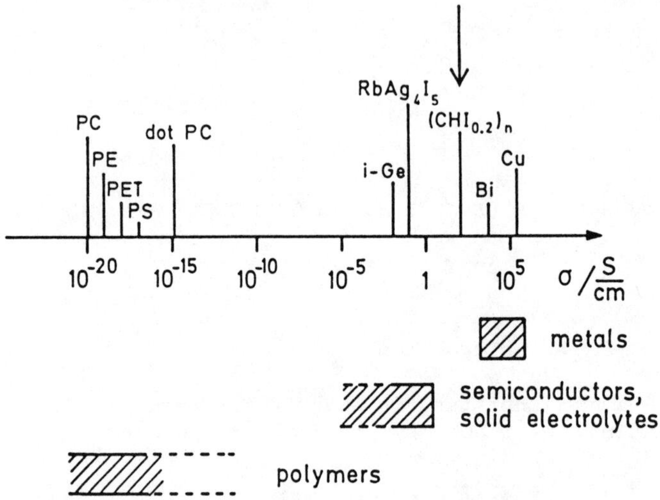
PC
PE
PET
PS
dot PC
i-Ge
RbAg4I5
(CHI0.2)n
Bi
Cu
10-20
10-15
10-10
10-5
1
105
σ / S/cm
metals
semiconductors, solid electrolytes
polymers

Fig. 1

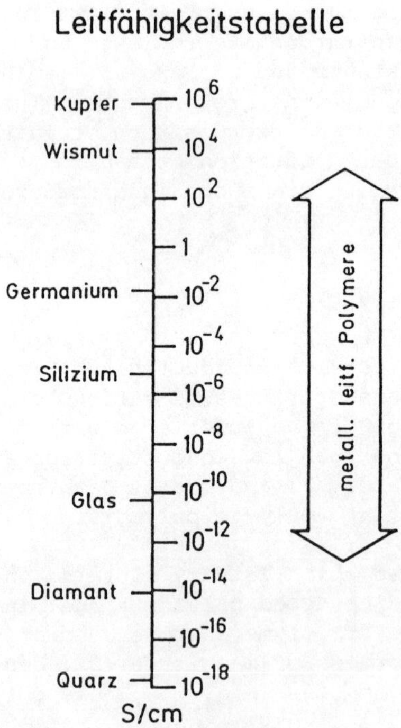
Leitfähigkeitstabelle
Kupfer
Wismut
Germanium
Silizium
Glas
Diamant
Quarz
10^6
10^4
10^2
1
10^-2
10^-4
10^-6
10^-8
10^-10
10^-12
10^-14
10^-16
10^-18
S/cm
metall. leitf. Polymere

Fig. 2

are characterized by extended systems of conjugated double bonds. The chemical structure of the most important representatives hereof is shown in Fig. 3.

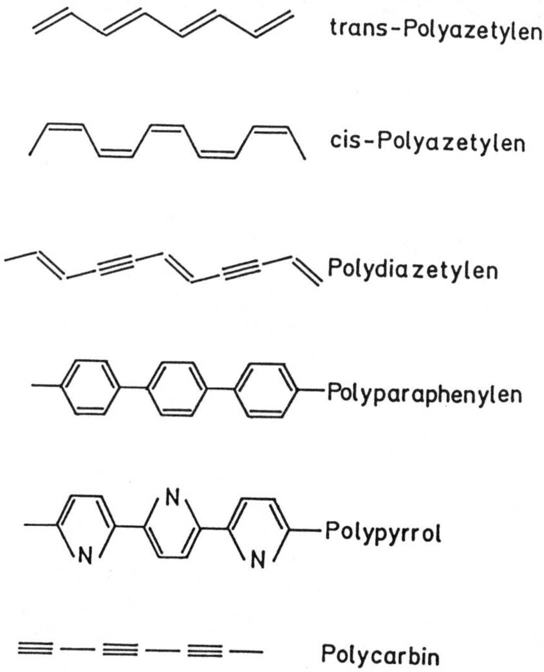

Fig. 3

We can expect that in these highly conducting polymers electrons are much less localized and that they will reflect certain features of inorganic metals. But the chain-like structure of the conjugated polymers will give rise to a very anisotropic behaviour and some aspects of one-dimensionality will show up, such as the metal-insulator transition (Peierls-transition /4/). The most exciting speculation is probably that on the existence of mobile conjugation defects, which can be described as non-linear excitations and are often referred to as solitons /5/ (Fig. 4). These defects share many properties with non-linear excitations in other fields of physics and offer an interdisciplinary connection all the way from chemistry to elementary particle physics and field theory /6/.

Many experts will agree that solitons would exist in very long isolated polyene chains, but there is some controversy on whether they can be observed in "real" polymers /7/. In the presently available polyacetylene films, for example, there exists a hierarchy of inhomogeneities, finite-length chains being bundeled

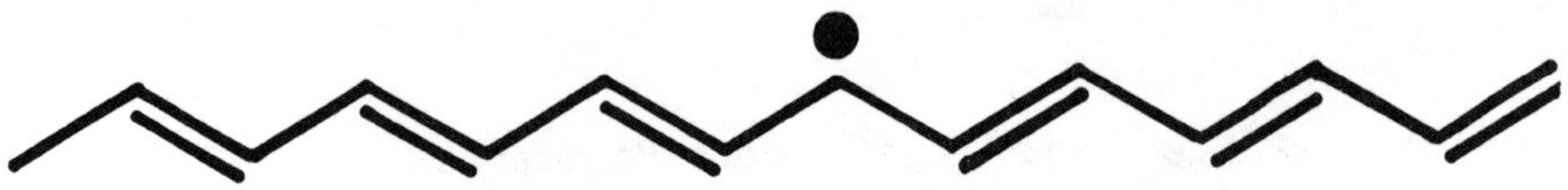

Fig. 4: "Soliton" in trans polyacetylene

to fibres and fibres forming a multiply-connected fleece. Therefore, any macroscopic electric conductivity will be governed by a superposition of intra-chain, inter-chain, and inter-fibre mechanisms (Fig. 5). Solitons would be only intra-chain and their effect might be very difficult to isolate.

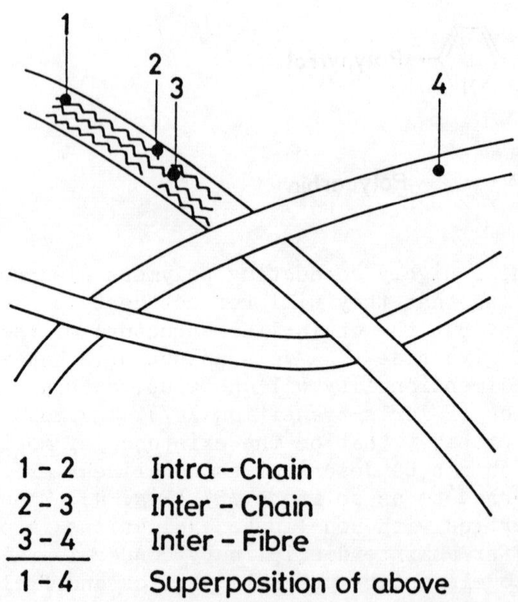

Fig. 5: Polyacetylene fleece schematically (typical fibre diameter 100-1000 A)

Even if the mechanism or mechanisms of the electrical conductivity in conjugated polymers is not solved, these substances might be useful for technical applications. In fact, many possible applications have been discussed and Fig. 6 presents a compilation of same relevant references.

- antistatic coatings, records etc.	
- electric conductors of low density	
- semiconducting devices (p-n junctions, Schottky barriers)	c, d, e, g, h;
- solar cells	a, b, d, e, g, h, i, p, q;
- electrochromens displays	
- light weight batteries	j, m, n;
- electrodes, membrans for electrochemistry, fuel cells	k, l;
- Image storage (xerography)	
- microwave shields	f;
- integrated circuits, molecular electronic devices ("bio-computer")	o, r;

----

a) Tani et al., Solid State Comm. 33, 499 (1980).
b) Chen et al., Appl. Phys. Lett. 36, 96 (1980).
c) Ozaki et al., Appl. Phys. Lett. 35, 83 (1979).
d) Weinberger et al.,Appl. Phys. Lett. 38, 555 (1981).
e) Grant et al., J. Appl. Phys. 52, 869 (1981).
f) Feldblum et al., J.Pol.Sci.; Pol.Phys.Ed. 19, 173 (1981).
g) Waldrop et al., Appl. Phys. Lett. 38, 53 (1981).
h) Tsukamoto et al., Jap.J.Appl.Phys. 20, 127 (1981).
i) Martin et al., Bericht, Centre Res. Macromol., Strasbourg (1982).
j) Nigrey et al., J.Elchem.Soc. 128, 1651 (1981).
k) Merz, A; Nachr.Chem.Tech.Lab. 30, 16 (1982).
l) Jahresrückblick, Nachr.Chem.Tech.Lab. 30, 118 (1982).
m) Shacklette et al., J.Chem.Soc.; Chem.Comm., im Druck.
n) Shacklette et al., J.Electrochem.Soc., im Druck.
o) De Rosnay, J., La Recherche 12, 870 (1981).
p) Weinberger et al., Synthetic Metals 4, 187 (1982).
q) Tsukamoto et al., Synthetic Metals 4, 177 (1982).
r) Carter, L.F., Naval Research Laboratory, Washington DC, NRL Memorandum Report 3940, 121 (1979), 4335, 35 (1980), 4717 (1982) und lecture at NATO-ARI, spring 1982 und "Molecular Electronics", Carter, F.L., Hgb. Marcel Dekker, 1982.

Fig. 6: Some applications of conducting polymers

Although conjugated polymers have been known and discussed for a long time by chemists they have been dicovered by physicists only very lately. In this respect they form the presently last bead in a sequence of one-dimensional substances, starting with KCP and running over TTF-TCNQ and $(SN)_x$ to $(CH)_x$. There are almost every year conferences specialized on this topic. Fig. 7 compiles the last few of these conferences and their proceedings. In Fig. 8 we present some review papers on "metally conducting polymers" and some general articles.

Saarbrücken (Sommerschule, DPG) "One Dimensional Conductors", Schuster, H.G., Hgb., Lecture Notes in Physics 34, Springer (1975).

Starnberg (NATO Advanced Study Institute) "Low Dimensional Cooperative Phenomena, the Possiblity of High-Temperature Superconductivity", Keller, H.J., Hgb. NATO Adv.Study Inst. 7B, Plenum Press, New York (1975).

Bozen (NATO Advanced Study Institute) "Chemistry and Physics of One-Dimensional Metals", Keller, H.J., Hgb., NATO Adv.Study Inst. 25B, Plenum Press, New York (1977).

Siofok (Hungarian Academy of Science) "Organic Conductors and Semiconductors", Pal,L., Grüner,G., Janossy,A. and Solyom, J., Hgb., Lecture Notes in Physics 65, Springer (1977).

New York "Synthesis and Properties of Low-Dimensional Materials", Miller, J.S. and Epstein, A.J., Hgb., Annals of the New York Academy of Science 313, New York (1978).

Dubrovnik "Quasi One-Dimensional Conductors" (Band I und II), Barisic,A., Bjelis,A., Cooper,J.R. and Leontic,B., Hgb., Lecture Notes in Physics 95, Springer (1979).

Les Arcs (NATO Special Program Panel on Material Science) "Molecular Metals", Hatfield, W.E., Hgb., NATO Conference Series VI, Material Science 1, Plenum Press, New York (1979).

Tomar (NATO Advanced Study Institute) "Chemistry and Physics of One-Dimensional Materials", Alcacer,L., Hgb., D. Reidel Publishing Company, Dordrecht (1979).

San José (Symposium organisiert von IBM Forschungslabor) "Structure and Properties of Highly Conducting Polymers and Graphite". The contribution have been published in Vol. 2 and 3 von Synthetic Metals 1 (1980).

Helsingør "Low Dimensional Synthetic Metals", Carneiro, K., Hgb., Chemica Scripta 17, No. 1-5 (1981).

Freiburg (Schweiz) "Physics in One-Dimension", Bernasconi,J., Schneider,T., Hgb., Serie in Solid State Sciences 23, Springer (1981).

Boulder (USA) "Low Dimensional Conductors" Epstein, A.J. and Conwell,E.M., Hgb., Mol.Cryst.Liq.Cryst. 77, 81, 83, 85 and 86 (1982).

Les Arcs "Conducteurs et Synthétiques à Basse Dimension", Veröffentlichung vorgesehen als Sonderbände von Journal de Physique (1983).

Fig. 7: Topical Conferences on One-Dimensional Metals Organic Conductors

Baeriswyl D., Harbeke G., Kiess H., and Meyer W., "Conducting Polymers: Polyacetylene" in "Electronic Properties of Polymers" Mort and Pfister, Eds., John Wiley, New York, in press

Etemad S., Heeger A.J., and Macdiarmid A.G., "Polyacetylene, $(CH)_x$: the prototype of Conducting Polymers", to be published Vol. 33 "Annual Reviews of Physical Chemistry"

Wegner G., Angew. Chem. 93, 352 (1981)

Hanack M. and Pawlowski M., Naturwissenschaften 69, 266 (1982)

Seeger K, Angew. Makromolekulare Chem. 109/110, 227 (1982)

S. Roth and K. Menke, Naturwissenschaften, in print

Fig. 8: Review and popular carticles on conducting polymers

2. EXPERIMENTAL DATA ON ELECTRONIC TRANSPORT IN POLYACETYLENE

In this section we present experimental data on the electric conductivity of polyacetylene, obtained in our laboratory. Polyacetylene is the prototype of conjugated polymers and the most intensively studied compound; in fact, since the first paper on doping of polyacetylene /3/ more than 500 publications have appeared on this substance.

Our samples have been polymerized by the Shirakawa technique at -80 °C /8/ and the films floating on the top of the catalyst solution have been used for the investigations. At -80 °C predominantly cis-polyacetylene is pbtained, for conversion into the trans modification the samples have been heated at 190 °C in an Argon atmosphere for 45 minutes. Some characteristic parameters of our samples are shown in Fig. 9.

Shirakawa Catalyst, Film floating on solution	
thickness:	20 to 200 μm
dask resistivity, dc:	$2\cdot10^{-6}$/Scm at RT (trans)
spin concentration (ESR):	400 ppm (trans)
ESR linewidth at RT:	1 Gauss (trans)
X-ray crystallinity:	80 %
elemental analysis:	CH 99.3 %
Raman-effective chain length:	20 % of chains with n > 20

Fig. 9: Some characteristics of typical samples of our laboratory

Doping was carried out by exposing the samples at room temperature to Iodine vapour or AsF_5 gas. In the case of AsF_5 several doping-pumping cycles were run to remove the side product AsF_3 and to ensure fairly homogenous doping ("Stuttgart procedure").

Fig. 10 shows the room temperature conductivity as function of the doping concentration for doping with Iodine, Bromine, and AsF_5 /9/. "Undoped" cis samples have a conductivity of 10^{-5} S/cm due to the casual doping, e. g. with Oxygen. By compensation with ammonia the resistivity can be increased by several order of magnitude.

In Fig. 11 we present data on the temperature dependence of the dc-conductivity for cis-polyacetylene doped with various concentrations of Iodine /9/. We obtain very similar conductivity values if we dope cis or trans $(CH)_x$. At high doping concentrations the conductivity is nearly temperature-independent, whereas it becomes more and more "thermally activated" at lower doping concentrations.

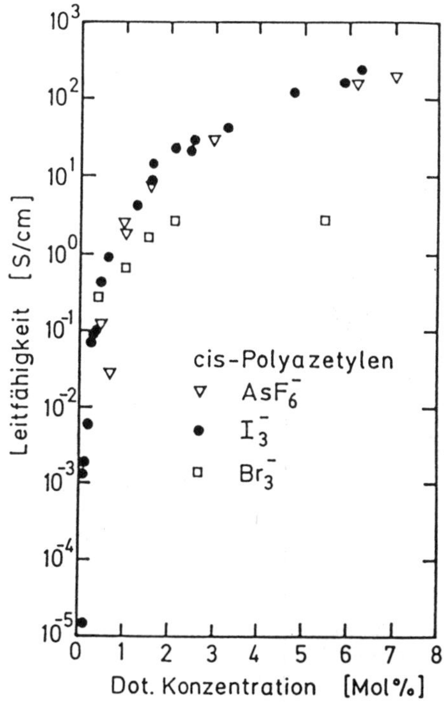

Fig. 10:

Room temperature conductivity after doping (Ref. 9)

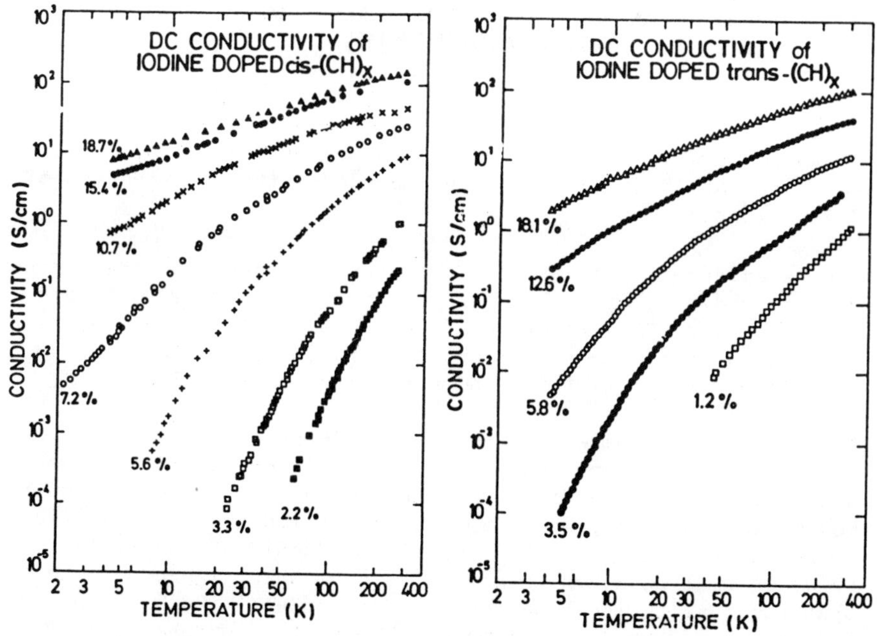

Fig. 11:

Temperature dependence of dc conductivity (Ref.9 and Les Arcs Proceedings)

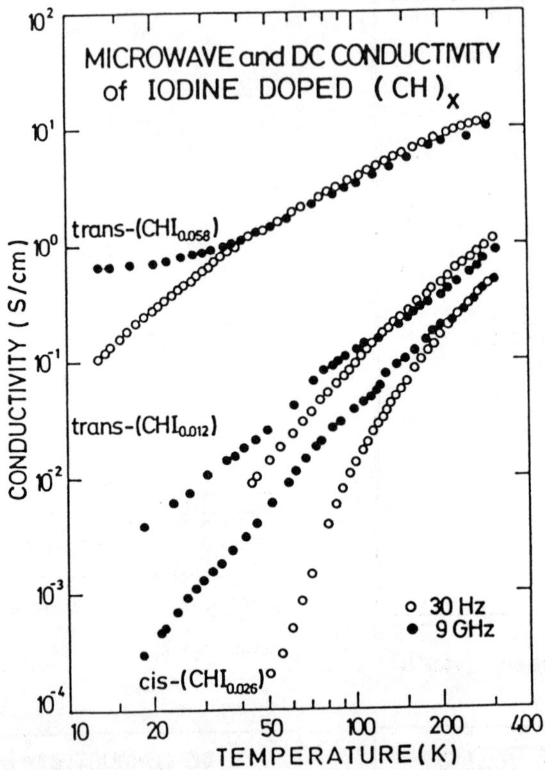

<u>Fig. 12:</u>

(Ref.9 and Les Arcs Proceedings)

In Fig. 12 we compare the dc and microwave (9 GHz) conductivity of Iodine-doped trans-polyacetylene in the temperature range from 10 to 300 K /9/. (dc measurements were carried out by a lock-in technique at 30 Hz to avoid ionic conductivity by migration of Iodine ions.)

A more extended study of the frequency dependence of the conductivity is shown in Fig. 13, where dc, microwave and IR data are combined /10/. The experimental curve (solid line) is tentatively decomposed into a free carrier contribution following a Drude behaviour $\sigma = \sigma_0/(1+\omega^2\tau^2)$, and a hopping contribution.

In Fig. 14 we compare the temperature dependence of the microwave absorption of cis and trans polyacetylene doped with AsF_5 and Iodine. The experiment was carried out at 50 GHz in a multimode-resonator /11/. The integral absorption shown in the figure corresponds roughly to the electric conductivity (the detailed relation will be given in Ref. 10). From this figure we conclude that the qualitative behaviour of the conductivity is much more determined by the doping species than by the conformation of the polymere (cis or trans). This result is important when the relevancy of solitons for electrical transport is discussed, since solitons should be mobile only in trans-polyacetylene.

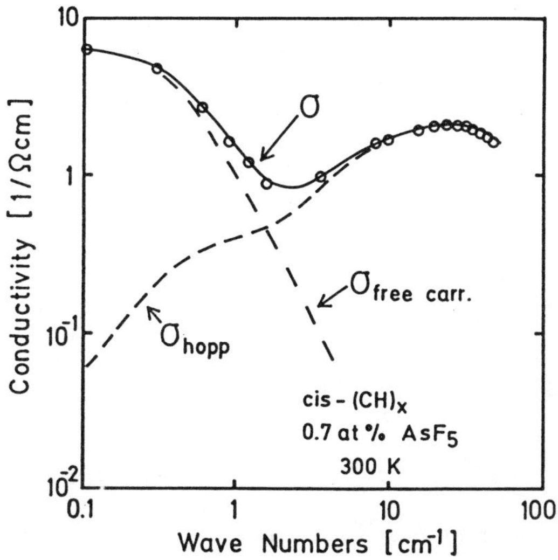

Fig. 13:

Frequence dependence of conductivity

(dc, microwave and IR data)

(Ref.10)

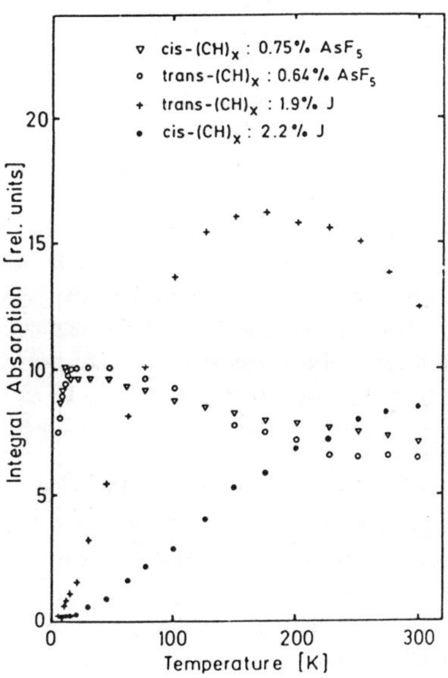

Fig. 14: Influence of dopant (AsF_5, Iodine) and of confirmation (cis, trans) on the temperature dependence of the microwave conductivity (Ref.10)

If one tries to vary additional parameters (like the electric or magnetic field) many anomalous effects occur in the electric conductivity of polyacetylene, but it is very difficult to draw conclusions relevant for the mechanism of conductivity from these effects. Fig. 15 shows the magnetoresistance of Iodine-doped samples /12/ as an example. It is interesting to note that for high doping concentrations a <u>negative</u> magnetoresistance is observed. This could be explained by special assumptions on the band structure or on electron localization .

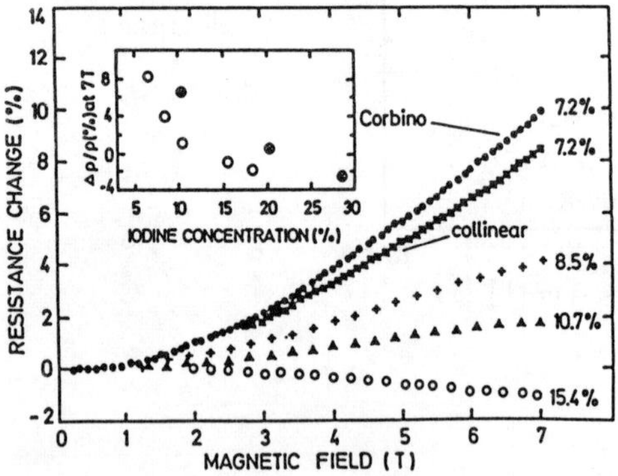

Fig. 15:

Magneto-resistance at 4 K

(Ref.14)

Conjugated polymers are very instable to oxygen, water vapour and other environmental influences. Fig. 16 shows an example of conductivity changes during long-time storage experiments. Highly doped samples are more stable but at low doping concentration the conductivity changes by several orders of magnitude if the samples are exposed to laboratory atmosphere for a month.

A compilation of publications concerning conductivity, photoconductivity, thermopower, Hall-effect and rectifying effects in polyacetylene is shown in Fig. 17, but by no means this list can claim completeness.

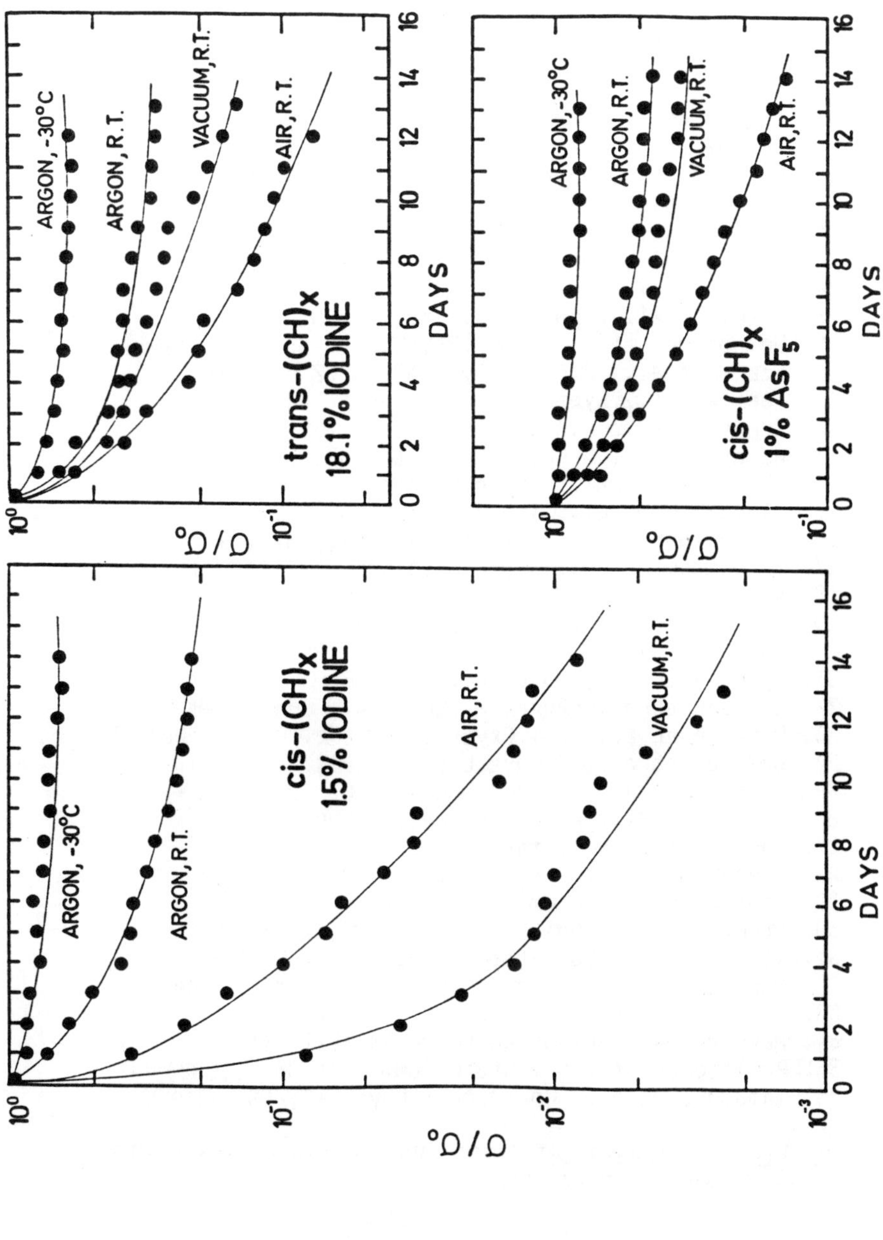

Fig. 16: Sample Degradation (Conductivity Loss) during long Time Storage (Ref. 9).

- Hall Effect, Magnetoresistance,	1, 2
- Thermopower	3 - 7
- Semiconductor Effects	8 - 18
- Frequency Dependence	19 - 25
- Photoconductivity	11, 26 - 31
- Non-linear Current-Voltage	2, 32 - 36
- Electrical Conductivity, General:	several hundred papers

1. Seeger et al., Sol.State Comm. 28, 873 (1978)
2. Seeger et al., Chemica Scripta 17, 129 (1981)
3. Kwak, Bull.Am.Phys.Soc. 23, 56 (1978)
4. Park et al., Sol.State Comm. 29, 747 (1979)
5. Kwak et al., Synth. Metals 1, 213 (1980)
6. Kwak et al., Sol.State Comm. 31, 355 (1979)
7. Moses et al., Sol.State Comm. 40, 1007 (1981)
8. Ozaki et al., Appl.Phys.Lett. 35, 83 (1979)
9. Weinberger et al., Appl.Phys.Lett. 38, 555 (1981)
10. Ozaki et al., J.Appl.Phys. 51, 4252 (1980)
11. Tani et al., Synth.Metals 1, 301 (1980)
12. Grant et al., J.Appl.Phys. 52/2, 869 (1981)
13. Waldrop et al., Appl.Phys.Lett. 38, 53 (1981)
14. Shirakawa et al., Synth.Metals 4, 43 (1981)
15. Tsukamoto et al., Jpn.J.Appl.Phys. 20, 127 (1981)
16. Weinberger et al., Synth.Metals 4, 187 (1982)
17. Tsukamoto et al., Synth.Metals 4, 177 (1982)
18. Kanicki et al., Mol.Cryst.Liq.Cryst. 83, 319 (1982)
19. Feldblum et al., J.Pol.Sci.; Pol.Phys.Ed.: 19, 173 (1981)
20. Grant et al., Sol.State Comm. 36, 291 (1980)
21. Epstein et al., Phys.Rev.Lett. 45, 1730 (1980)
22. Epstein et al., Phys.Rev.Lett. 47, 1549 (1981)
23. Epstein et al., Mol.Cryst.Liq.Cryst. 77, 81 (1981)
24. Gamoudi et al., J.Phys.Lett. (1981)
25. Epstein et al., Phys.Rev.Lett. 50, 1866 (1983)
26. Etemad et al., Phys.Rev.Lett. (1980)
27. Tani et al., Sol.State Comm. 33, 499 (1980)
28. Lauchlan et al., Phys.Rev. B24, 3701 (1981)
29. Etemad et al., Chem.Scr. 17, 159 (1981)
30. Etemad et al., Sol.State Comm. 40, 75 (1981)
31. Kiess et al., Sol.State Comm. 44, 1443 (1982)
32. Mortensen et al., Phys.Rev.Lett. 45, 490 (1980)
33. Epstein et al., Chem.Scr. 17, 135 (1981)
34. Mayr et al., Sol.State Comm. 43, 117 (1982)
35. Philipp et al., Sol.State Comm. 43, 857 (1982)
36. Yamamoto et al., Jpn.J.Appl.Phys. 19, 991 (1980)

Fig. 17: Compilation of literature on electrical transport in polyacetylene (not complete).

3. MECHANISMS FOR THE ELECTRIC CONDUCTIVITY IN POLYACETYLENE

The mechanism of the electric conductivity of polyacetylene is still highly controversal. From Fig. 5 one will have to conclude that there will not be one single mechanism governing the conductivity but rather a superposition of several mechanisms. It is reasonable to assume that the measured resistivity is the sum of three contributions

$$R = R_1 + R_2 + R_3$$

where R_1 refers to charge motion along a chain, R_2 to transport from one chain to an other and R_3 to transport from one fibre to the other (a typical fibre diameter in polyacetylene is 100 to 1000 Å).

It seems plausible that the inter-fibre resistance behaves similarly to that between metallic grains in evaporated thin films or metallic particles in insulting polymer matrices. For this case a model of fluctuation-induced tunneling has been successfully applied /13/. It also fits perfectly to the temperature dependence of highly doped polyacetylene /14,9/ (Fig. 18). We assume

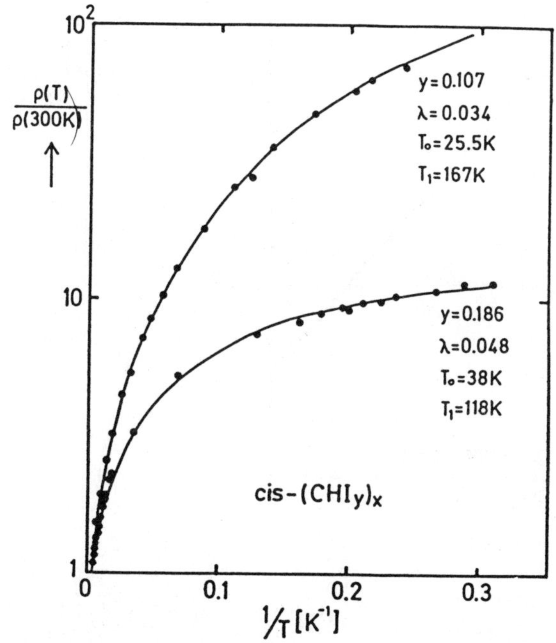

Fig. 18
Fluctuation-induced tunneling model for highly doped polyacetylene (Ref. 9)

that in highly doped polyacetylene R_1 und R_2 are neglegible compared to R_3.

At lower doping concentrations $R_1 + R_2 >> R_3$. A very elegant theory for R_1 is the propagation of charged solitons. This will be dealt with in some more detail in the next section. Due to the short chain length in today available polyacetylene R_2 will be more important than R_1 and the electrical conductivity of moderately and lightly doped polyacetylene will be mainly due to interchain hopping.

In the dc and low frequency case this process can be approximated by variable range hopping leading to $\sigma = A/\sqrt{T}\,\exp(T_o/T)^{1/4}$. The power 1/4 is characterisitc for three-dimensional variable range hopping and fits well to conductivity data obtained from polyacetylene samples doped with a few at% of Iodine (Fig. 19).

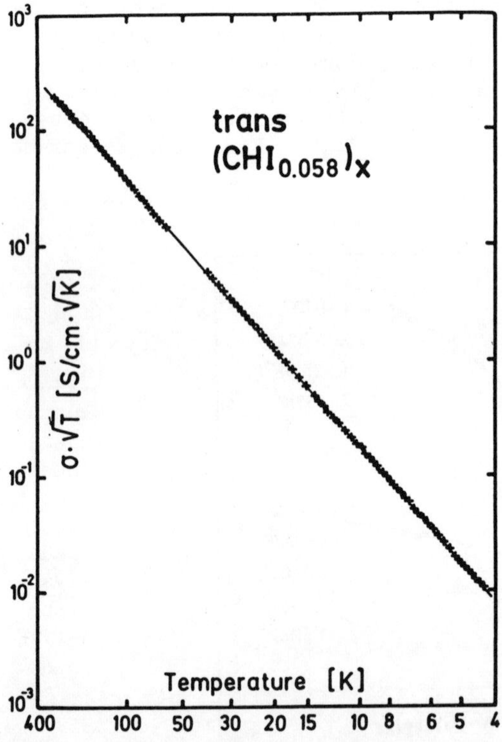

Fig. 19
Variable range hopping model for moderately doped polyacetylene (Ref. 9)

Although the hopping model, originally developed for amorphous semiconductors, is quite sucessful in reproducing the temperature dependence of the dc-conductivity of polyacetylene, it does not well describe the frequency dependence. This deficiency probably can be corrected for by taking into account an anisotropic hopping probability or hops between interrupted chains. Such a model has been proposed by Alexander et al. /15/ and fits well to the disordered charge transfer salt Qn(TCNQ)$_2$ and the ionic conductor holandite, investigations of the applicability of this model to polyacetylene are in progress.

For undoped or very lightly doped polyacetylene Kivelson /16/ has proposed a mechanism of "soliton assisted hopping" or intersoliton charge exchange. This mechanism makes use of trapped charged solitons and mobile neutral solitons. Whenever a neutral soliton passes near a charged soliton there is a chance of charge exchange. This model fits well to experimental data of Epstein et al. /17/. Starting from the equivalent network theory of Miller and Abrahams recently Chroboczek and Summerfield /18/ have developed a generalized hopping theory, not specifying the hopping sites or the hopping species. A comparison of this theory with experimental data is shown in Fig. 20.

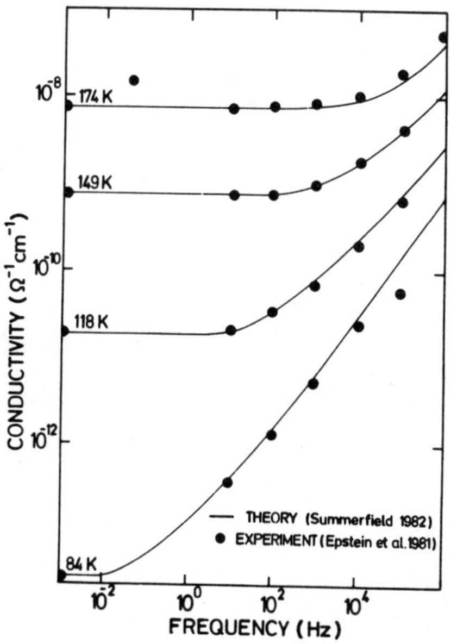

Fig. 20 General hopping theory fit to temperature and frequency dependence of conductivity in undoped trans polyacetylene (Ref. 7)

4. SOLITONS

For charge transport along chains with conjugated double bonds the mechanism of soliton motion has been proposed (this would be a theory for R_1 in the terminology of the previous section). If the conjugation defect of Fig. 4 is called "soliton", this term is used to stress two features of this defect: it does not disperse while it moves (just as solitary water waves do not disperse. This is where the name comes from) and it has certain symmetry properties. If most polymer chains in today's polyacetylene films are very short, solitons will not be able to move very far since they are confined to the polyene chains. Therefore the non-dispersivity will be hard to test. In addition solitons will not be important for electrical conductivity under these circumstances.

On the other hand the symmetry properties of solitons will persist even in short chains. They are a direct consequence of bond alternation. One such property is, for example, the fact, that solitons can only be created pairwise, as soliton plus anti-soliton, never alone (in the chemical bond picture this is rather trivial: breaking a bond results in two nonbonding electrons). Another consequence is the midgap state: because of the bond alternation there is a gap between π and π^* states. Where the alternation is interrupted (conjugation defect), the gap is locally suppressed and there is a state at midgap. Fig. 21 shows the creation of midgap states by breaking of double bonds through chemical oxidation (doping) /19/.

The soliton in Fig. 4 is electrically neutral: the charge of the dangling electron is compensated by the ionic charge sitting on the carbon nucleus. If the electron is removed by oxidation (doping), the soliton does not vanish: the conjugation defect remains (in first approximation the intact double bonds are not effected by this oxidation). Oxidation transforms the neutral soliton into a positively charged soliton (resembling positive holes in semiconductors), which (in some theories) leads to electrical conductivity in polyacetylene and related compounds. There exists an interesting spin-charge inversion for solitons in polyacetylene: neutral solitons carry spin (dangling electrons), charged solitons are spinless (there are only paired electrons). This spin-charge inversion is schematically shown in Fig. 22.

Fig. 23 shows ESR measurements of the decrease of the spin concentration during doping /20/. In undoped polyacetylene there are usually several hundred ppm of neutral solitons. Light doping converts them into charged and spinless solitons, hence the spin concentration decreases on doping.

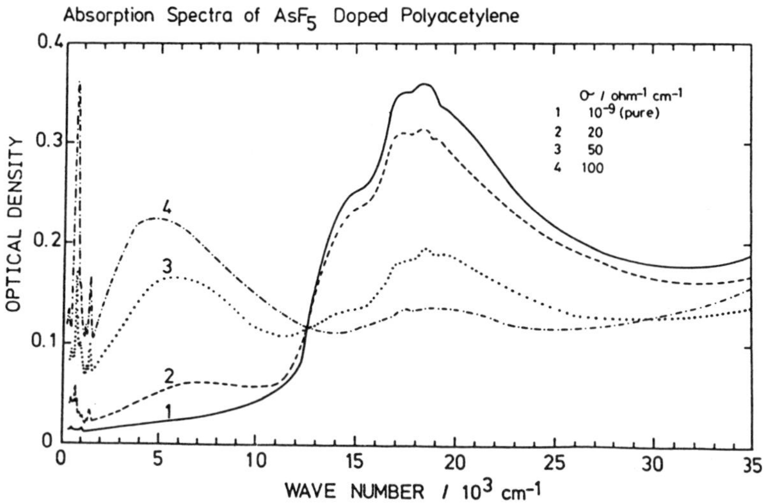

Fig. 21 Midgap state in polyacetylene
(Data from Fujimoto, Tanaka and Tanaka)

In addition to charging neutral solitons doping will break double bonds and create pairs of new solitons. If the soliton concentration is high enough the conjugation breaks down completely and the electrons rearrange to a gapless structure with a partially filled band. This is expected to occur when the Pauli susceptibility sets in at a critical doping concentration (Fig. 24). A similar step is observed in the NMR lineshift and it is tempting to interpret this as Knight-shift and relate it to the Pauli susceptibility. On the other hand, at the same critical concentration there is an abrupt change in the Raman spectra and all these steps might be due to structural rearrangements rather than overlap of soliton wave functions.

Double bonds can also be broken by excitation with light and thus solitons can be photogenerated. Photoexcitation is more likely to create charged solitons than neutral solitons and photoconductivity experiments might be a good way to test this concept. Several such experiments have been carried out (Fig.17). They show that photoconductivity is more pronounced in trans than in cis polyacetylene. In cis-polyacetylene solitons are unstable because of symmetry reasons (lattice forces will drive solitons to the chain ends (Fig. 25)) and so these experiments seem to support the soliton theory.

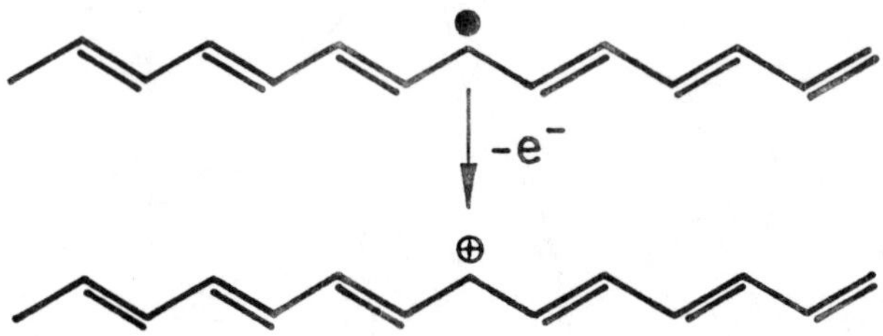

Fig. 22 Transformation of spin-carrying neutral soliton into spinless charged soliton by electron removal (oxydative doping)

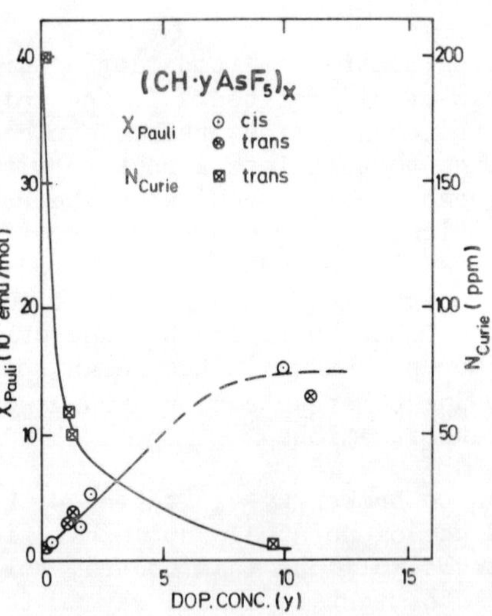

Fig. 23 Decrease of spin concentration (Curie constant) on doping (Data from Davidov at al.)

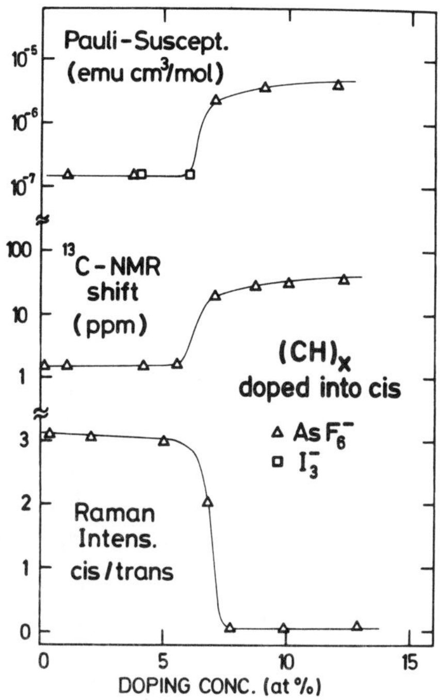

Fig. 24 Changes in cis-polyacetylene at critical doping concentration (Ref. 7)

Fig. 25 Soliton in cis-polyacetylene. The soliton moves to the right hand side "eating up" the high-energy trans-cisoid structure and leaving behind low-energy cis-transoid

In a recent experiment /21/ on the Schubweg of photogenerated carriers it has been shown however, that these carriers do not drift very far in an electric field. Apparently they get trapped soon after their creation - within a few picoseconds or less - and from then on they migrate by hopping (repeated trapping and detrapping).

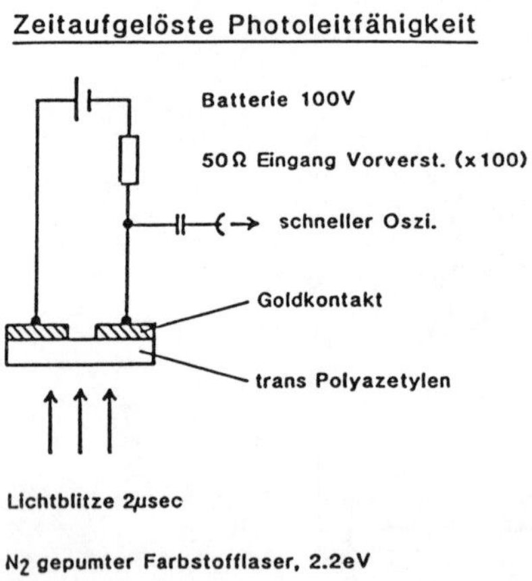

Fig.26a Experimental setup for Schubweg measurement in polyacetylene

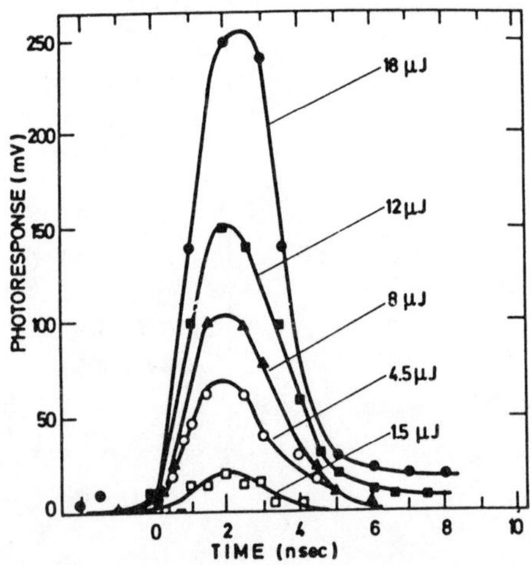

Fig.26b Photosignal after light-pulses (Ref. 21)

In view of these results it is very difficult to decide whether motion of charged solitons along polyene chains is a significant process in electric conduction and whether solitons are important for transport phenomena at all. Nevertheless, photoinduced excitation of polyene chains will lead to many interesting phenomena and are studied intensively (Fig. 27).

The experimental setup and the photosignal of the Schubweg investigations are shown in Fig.26a and 26b,respectively. The Schubweg is calculated from the integrated intensity of the photosignal. The soliton free drift here is estimated using "reasonable" values for the soliton mobility to be $\tau = 8\times10^{-14}$ sec!

The magnetic moment of neutral solitons can be observed in ESR experiments. The resonance of trans polyacetylene has a Lorentzian

- Salaneck et al., Phys. Rev. Lett. 49, 801 (1982)
- Orenstein et al., Phys. Rev. Lett. 49, 1043 (1982)
- Vardeny et al., Phys. Rev. Lett. 49, 1657 (1982)
- Vardeny et al., Journal de Physique: in print (Les Arcs Proceeding)
- Vardeny et al., Phys. Rev. Lett. 50, 2032 (1983)
- Blanchet et al., Phys. Rev. Lett. 50, 1938 (1983)
- Weinberger, Phys. Rev. Lett. 50, 1693 (1983)
- Shank et al., Phys. Rev. Lett. 49, 1660 (1982)

Fig. 27 Some recent literature on photoinduced excitations in polyacetylene

shape and shows considerable narrowing (about 1 G or less at room temperature, whereas a linewidth of 23 G is estimated from the hyperfine interaction). This has been interpreted as motional narrowing and a very high mobility of neutral solitons has been deduced. From the frequency dependence of the relaxation time it has been concluded that this motion is mainly one-dimensional /22/.

Recently it has been proposed to use solitons for data storage and data handling on a molecular level /23/. A moving soliton has the property of innerchanging single and double bond in trans-polyacetylene. If there were a possibility to find out whether there is a single or a double bond between two particular carbon neighbours, solitons could be used in a molecular computer. Such read-out devices can be constructed by building special side groups to the polymer chain. The molecule of Fig. 28 could work as a read-out unit: in the ground state there is a double bond between the two carbons. Absorption of light leads to an excited state, where a dipole moment is created and the double bond moves outwards to the nitrogen atoms. If this molecule is built into a polyene -chain (Fig. 29) it can absorb light only if all polyene double bonds are as in Fig. 29). Passage of a soliton along the

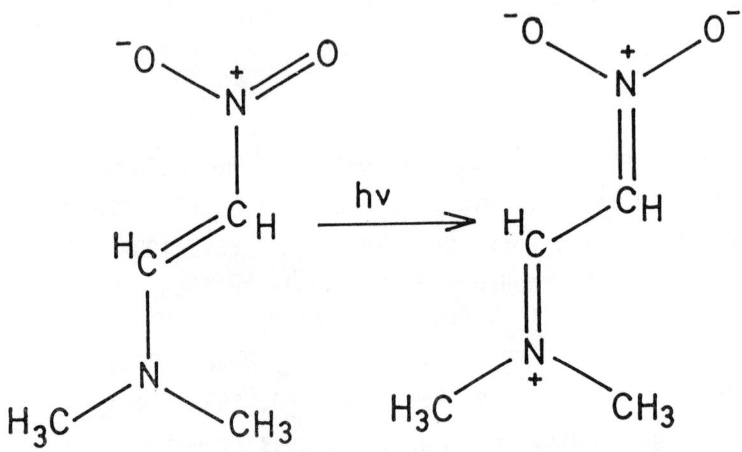

Fig. 28 Molecular read-out unit for soliton switching

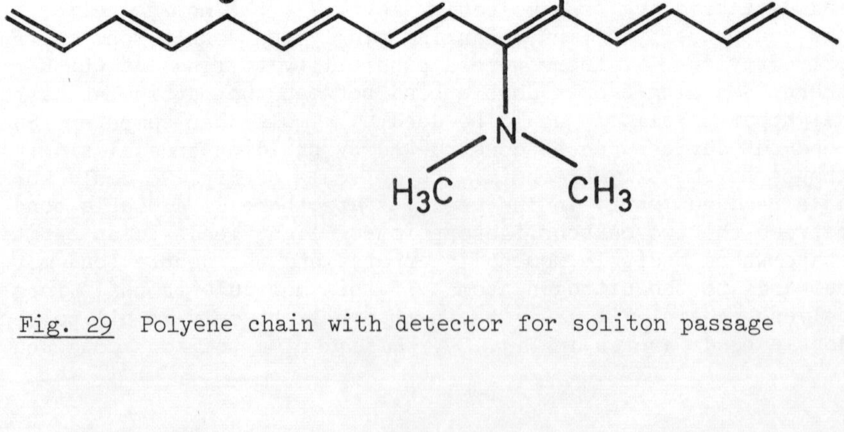

Fig. 29 Polyene chain with detector for soliton passage

chain would inhibit absorption. Although solitons might turn out to be of little importance for charge transport in macroscopic samples, they might be helpful in discussing how the two ends of a finite-length polyene (e. g. carotene) communicate with one another.

5. CONCLUSION

Polymers with conjugated double bonds appear to be very interesting systems for experimental and theoretical investigations. By doping their electrical conductivity can be changed by many orders of magnitude from insulating via semiconducting to metallic behaviour. Therefore several applications as synthetic metals or semiconductors are proposed. In the discussion of electronic excitations in these systems the concept of solitons is very fruitful. Even if the relevancy of solitons for experiments which can reproducibly be carried out in today-available polymers is often overestimated,"ideal" polyenes are fascinating model substances,the investigation of which leads to elegant insights into a large class of phenomena in many fields of physics. Fig. 30 brings the result of a literature survey with the key words "solitons" and "polyacetylene".

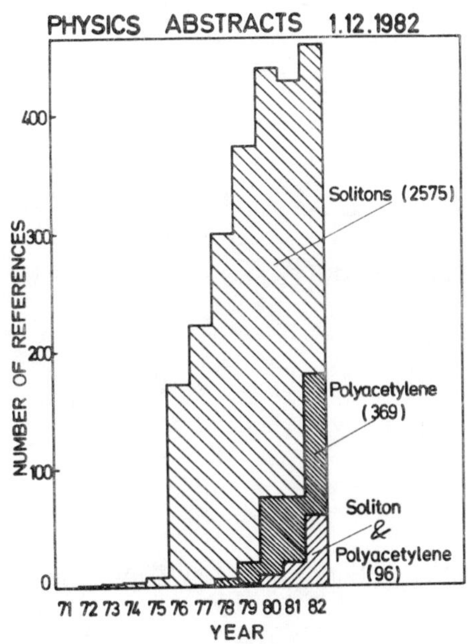

Fig. 30 Literature survey

REFERENCES

1. H. Kiess and W. Rehwald: Colloid & Polymer Sci. 258, 241 (1980)
2. N.F. Mott: Phil. Mag. 19, 835 (1969) and Festkörperprobleme 9, 22 (1969)
3. C.K. Chiang, C.R. Fincher, Y.W. Park, A.J. Heeger, H. Shirakawa, E.J. Louis , S.C. Gau, A.G. MacDiarmid: Phys. Rev. Lett. 39 1089 (1977)
4. R.H. Friend and D. Jérome: J. Phys. C.: Solid State Phys. 12, 1441 (1979)
5. M.J. Rice: Phys. Lett. A71, 152 (1979) W.P. Su, J.R. Schrieffer, and A.J. Heeger: Phys. Rev. Lett. 42, 1698 (1979)
6. D.K. Campbell and A.R. Bishop: Phys. Rev. B24, 4859 (1981)
7. S. Roth, K. Ehinger, K. Menke, M. Peo, and R.J. Schweizer: Journal de Physique, to be published (Les Arcs Proceedings)
8. H. Shirakawa and I. Ikeda: Polym. J. 2, 231 (1971)
9. K. Ehinger, Thesis, University of Konstanz
10. L. Genzel, F. Kremer, A. Poglitsch, G. Bechthold, K. Menke, and S. Roth: to be published
11. F. Kremer and L. Genzel: 5th Int. Conf. on Infrared and Millimeter Waves, Miami Beach, 1981 (IEEE Cath. No 81CH 1645-1MTT)
12. W. Röss, A. Philipp, K. Seeger, K. Ehinger, K. Menke, and S. Roth: Solid State Comm. 45, 933 (1983)
13. B. Abeles: Appl. Sol. State Sci. 6, 1 (1976) P. Sheng: Phys. Rev. B 21, 2180 (1980)
14. A. Philipp, W. Mayr and K. Seeger: Solid State Comm. 43, 857 (1982)
15. S. Alexander, J. Bernasconi, W.R. Schneider, R. Biller, W.G. Clark, G. Grüner, R. Orbach, and A. Zettl: Phys. Rev. B24, 7474 (1981)
16. S. Kivelson: Phys. Rev. Lett. 46, 1344 (1981)
17. A.J. Epstein, H. Rommelman, . Abkowitz, and H.W. Gibson: Mol. Cryst. Liqu. Cryst. 77, 81 (1982)
18. J.A. Chroboczek and S. Summerfield: Journal de Physique, to be published (Les Arcs Proceedings)
19. N. Suzuki, M. Ozaki, S. Etemad, A.J. Heeger, A.C. Mac Diarmid: Phys. Rev. Lett. 45, 1209 (1980)
20. B.R. Weinberger, J. Kaufer, A.J. Heeger, A. Pròn, and A.G. MacDiarmid: Phys. Rev. B20, 223 (1979) D. Davidov, S. Roth, W. Neumann, and H. Sixl: Z.Phys. B51, 145 (1983)
21. Y. Yacoby, S. Roth, K. Menke, F. Keilmann, and J. Kuhl: Solid State Commun. (in print)
22. M. Nechstein, F. Devreux, R.L. Greene, T.C. Clarke, and G.B. Street: Phys. Rev. Lett. 44, 356 (1980)
23. F.C. Carter, Ed., "Molecular Electronic Devices", Marcel Dekker, 1982

THE PREPARATION AND PROPERTIES OF ORDERED AND DISORDERED DIACETYLENE POLYMERS

D. Bloor

Department of Physics, Queen Mary College,
Mile End Road, London E1 4NS.

INTRODUCTION

The current depth of understanding of the physics of non-polymeric solids has been made possible by the availability of large, highly perfect single crystals for experimental study. The approach to a similar level of understanding for polymers has been hindered by the structural complexity of the majority of polymeric materials. A notable exception to this is a subset of the polymers based on disubstituted diacetylenes, which are available in the form of macroscopic single crystals. The fact that these polymers have conjugated backbones adds considerably to their value as model materials, particularly in view of the interest in other less perfect conjugated polymers, such as polyacetylene, which can be doped to obtain high electrical conductivity.

The single crystal nature of the product of the solid-state polymerization of certain diacetylenes has been known for over a decade (1). During this time there has been a considerable amount of research into the preparation and properties of the resulting polymers. A number of reviews of both these areas have appeared (2-8) and it is not possible, in the space available, to present all the extant material in detail. The emphasis here will be on recent experimental results and their interpretation.

Experimental studies of polydiacetylenes give us more information about the fundamental properties of the conjugated polymer chain than similar measurements made on less perfect polymers. In the latter case the fundamental, anisotropic

J. Ladik et al. (eds.), Quantum Chemistry of Polymers - Solid State Aspects, 191–220.

properties must be deduced indirectly from the isotropic properties of the amorphous or partially micro-crystalline bulk samples or the anisotropy of partially aligned samples. Despite the absence of the complications due to such complex morphologies neither the experiments nor their interpretation have always been straightforward even for polydiacetylenes. These problems will be considered below. A cause of some difficulties is the fact that polydiacetylenes are a class of polymers, not a single compound. Thus, though all polydiacetylenes contain essentially the same conjugated backbone, they have sidegroups which have a wide variety of chemical structures. The sidegroups will affect the properties of the backbone so that care must be exercised when making generalisations, particularly if they are based on measurements on a small selection of polydiacetylenes.

Despite these problems a coherent picture of the properties of polydiacetylene crystals is emerging. It is, therefore, possible to use this as a basis for studies and interpretation of the properties of disordered samples. Again the polydiacetylenes provide model systems since the morphology of disordered samples can be controlled and characterized. Thus, in principle, the effect of particular types of disorder can be studied. This work is clearly of value in understanding the behaviour of polymers for which only disordered samples are available. Less effort has been expended on such studies but it is an area of growing interest.

The preparation of polydiacetylene single crystals will be outlined. Studies of oligomeric intermediates observed during polymerization will be described since they are of theoretical interest and provide information on the properties of short polymer chains which cannot be obtained in any other way. The electronic properties of these crystals, as revealed by studies of their spectra and electrical conductivity, will be discussed together with models used to interpret them. Finally studies of disordered systems will be described, in particular the recent observations of carrier recombination in damaged and disordered samples. The outstanding experimental and theoretical problems will be emphasised.

PREPARATION OF POLYDIACETYLENES

The starting point for the preparation of polydiacetylene crystals (hereafter referred to as PDAs) is the synthesis of disubstituted diacetylenes. The synthetic routes to these compounds are well established (9) and will not be discussed here; typical examples are given in Figure 1. The abbreviations given in the figure will be used to refer to particular materials below. The solid-state polymerization of the monomers proceeds by a 1,4 - reaction of the acetylenic carbons, which may be initiated chemically, thermally, by ionizing radiation (UV, X and γ-rays) or by deformation of the crystal to give the conjugated polymer chain shown in Figure 1.

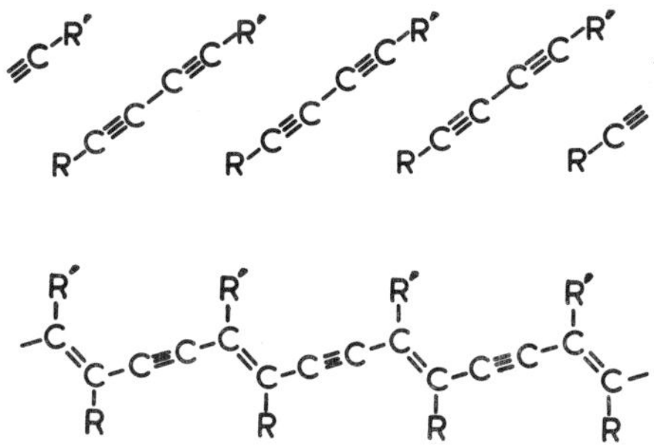

1OH	R= $-CH_3$, R'= $-CH_2OH$
TS	R=R'= $-CH_2OSO_2C_6H_4CH_3$
TS12	R=R'= $-(CH_2)_4OSO_2C_6H_4CH_3$
MS	R=R'= $-CH_2OSO_2C_6H_2(CH_3)_3$
ETU	R=R'= $-CH_2OCONHC_2H_5$
TCDU	R=R'= $-(CH_2)_4OCONHC_6H_5$
nBCMU	R=R'= $-(CH_2)_nOCONHCOOC_4H_9$
B9	R=R'= $-(CH_2)_9OCOCH_2C_6H_5$
BPG	R/R'= $-C_6H_4OCO(CH_2)_3OCOC_6H_4-$

Figure 1. Schematic representation of the solid-state polymerization of disubstituted diacetylenes, with examples of substituent groups.

Solid-state chemistry is less widely used than solution chemistry but the particular molecular configuration and interaction imposed by the crystal lattice leads to a well defined reaction product. The very specific nature of the lattice packing required for reaction, however, makes the solid-state route less predictable than solution chemistry. Solid-state polymerization has been observed to occur by bulk, topotactic and topochemical reactions. In bulk reactions the product is amorphous because the crystal lattice is destroyed during the reaction. Topotactic reactions produce a product phase which has a small number of orientations, possibly only one, with respect to the initial crystal. Topochemical reactions are those in which the reaction path is determined by the crystal structure and may lead to a topotactic product.

For the product of reaction to retain the order of the monomer crystal, the reaction must be topochemical with a unique topotactic product. During the reaction any strains due to mismatch of monomer and product polymer must not lead to segregation of the two phases (10). An adequate criterion for the occurence of a solid-state reaction in diacetylenes has been shown (8) to be that originally proposed by Schmidt (11). This requires that the reacting atoms are separated by less than 0.4 nm. The role of the monomer endgroups (polymer sidegroups) is crucial since they are the major factor in determining lattice packing. In addition to bringing the molecules close enough together for a reaction to be possible they must define a unique polymerization direction and mediate the interfacial strain between monomer and polymer. The effect on lattice packing and reactivity of the substituent groups is illustrated in Figure 2. In this the monomer separation and angle of inclination (d and γ) with respect to the direction of closest approach of the monomers, which is the polymer chain axis in reactive crystals, are plotted using available crystallographic data. This shows (a) the adequacy of Schmidt's criterion for reactivity and (b) the very large mismatch that can occur between monomer separation and the polymer repeat distance (0.49 nm).

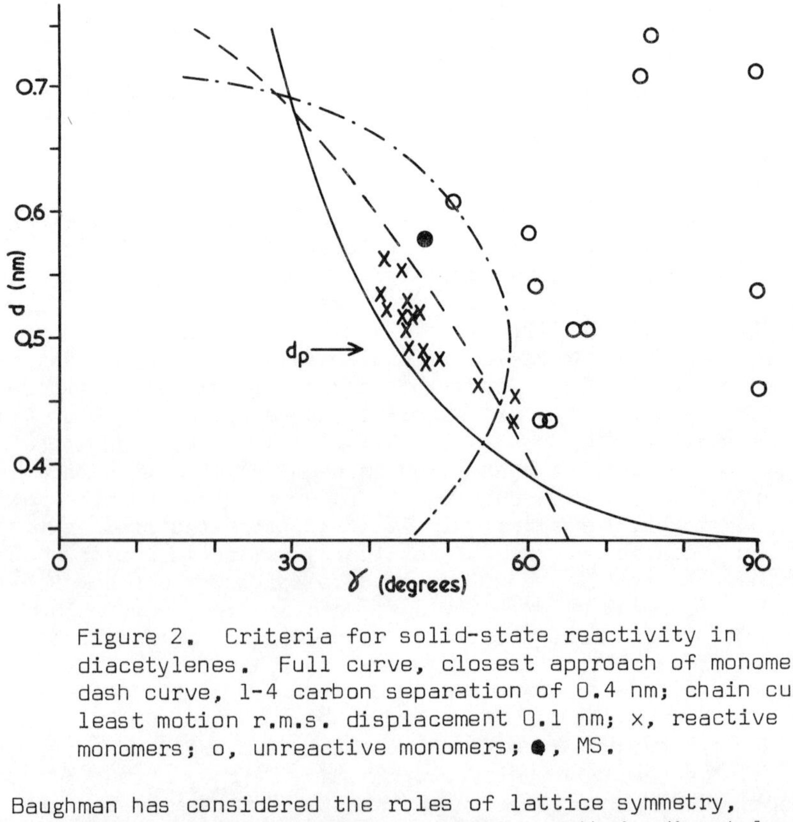

Figure 2. Criteria for solid-state reactivity in diacetylenes. Full curve, closest approach of monomers; dash curve, 1-4 carbon separation of 0.4 nm; chain curve, least motion r.m.s. displacement 0.1 nm; x, reactive monomers; o, unreactive monomers; ●, MS.

Baughman has considered the roles of lattice symmetry, molecular and lattice energies and reaction path in diacetylene polymerization (3, 12). He considered in plane motion of rigid rod molecules and calculated r.m.s. molecular displacements. This model does not seem to be a realistic one since the curves for constant r.m.s. displacement do not form a boundary between reactive and unreactive monomers. In addition, the rate of reaction does not correlate with the size of the r.m.s. displacement calculated for this model. In this context we have recently determined the crystal structure of MS (see Figures 2 and 3) which has a small r.m.s. displacement but is unreactive since the 1-4 carbon separation is more than 0.4 nm (13). Unfortunately

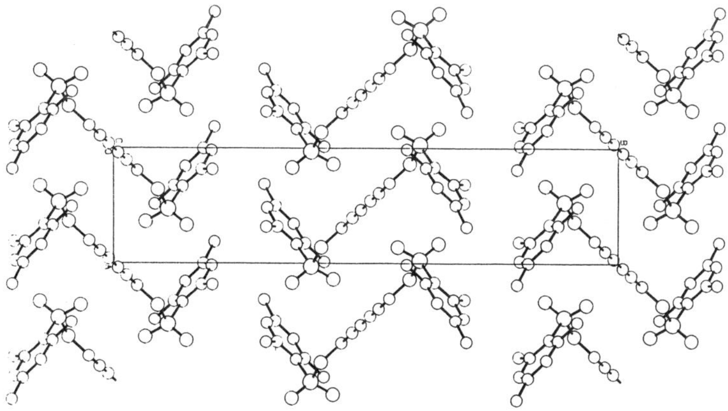

Figure 3. Crystal structure of MS monomer projected on to the ab-plane.

because of the large size of the disubstituted diacetylene molecules accurate calculations of lattice structures are not possible (14). Thus, the goal of the molecular design of reactive monomers has not yet been reached and will require the development of more powerful and accurate methods of predicting crystal structures for large, non-rigid molecules.

Four distinct behaviours of the crystal lattice have been observed during polymerization. First, the polymer occurs as a solid-solution in the monomer lattice so that the physical properties change gradually as polymerization proceeds. The best example of this is TS, for which the lattice parameters, and other properties, change continuously during polymerization, see (15 - 17) and Figure 4. Secondly, a phase transition is induced from a less to a more reactive phase; the transition is collective and the polymerization proceeds in both phases by the solid-solution mechanism. This has been reported for DCH during γ-ray polymerization (18); the lattice parameters changing discontinuously at about 30% conversion to polymer. Thirdly, a phase transition to a more reactive phase is induced locally so that the final polymer crystal is a mixture of two phases. The monomer ETU polymerizes in this manner for thermal polymerization (19). Finally, the crystal segregates into a polymer and a monomer phase leading to a fibrous micro-crystalline or amorphous product. This behaviour is common, e.g. for the thermal polymerization of DCH (20).

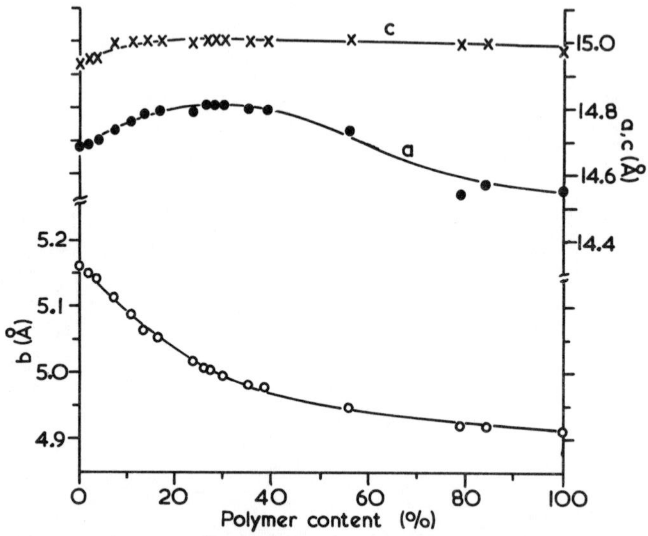

Figure 4. Variation in lattice parameters, measured at room temperature, during the polymerization of TS.

Despite these differences in microscopic behaviour the kinetics for thermal polymerization are similar in the first three cases with an initially slow rate, followed by an 'autocatalytic' acceleration and a final slow approach to complete conversion. A theoretical model for polymerization via solid-solutions, in which the kinetic chain length is controlled by lattice strain, has been discussed (15, 21) and subjected to criticism (22, 23). There has been an extensive body of work into the polymerization of TS and BCMU monomers, see for example (24 - 26) but this will not be discussed further here.

Such studies did, however, suggest that the initial step in the polymerization reaction is a diradical (26), see Figure 5, apparently contrary to the evidence of the existence of carbene radicals in thermally polymerizing crystals (27). This apparent conflict has been resolved by an extensive series of studies of the photo- and thermal polymerization of diacetylenes at low temperatures conducted at the Universities of Stuttgart and Bayreuth. A review of this work is in the course of publication and references to the extensive literature will be found there (28 - 30). The most thorough studies have been made of the monomer TS, similar but less detailed results have been obtained for a number of other diacetylene monomers (31). In most of the experiments monomer crystals at 4 K are irradiated with ultraviolet radiation and studied by optical, EPR and ENDOR spectroscopy. At 4 K broad band irradiation produces stable oligomeric intermediates since polymerization cannot proceed by thermal activation.

The evolution of further intermediates can then be followed by selective irradiation or by heating the sample. Intermediates produced by such thermal annealing can be frozen in by cooling the sample to 4 K.

(a)
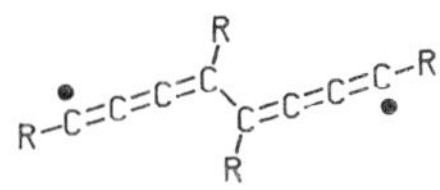

(b)
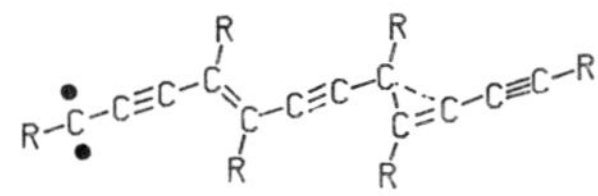

(c)
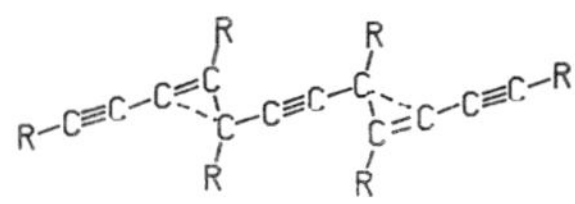

(d)
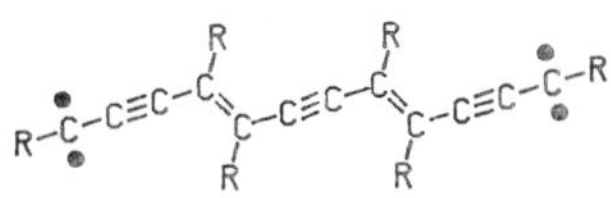

(e)
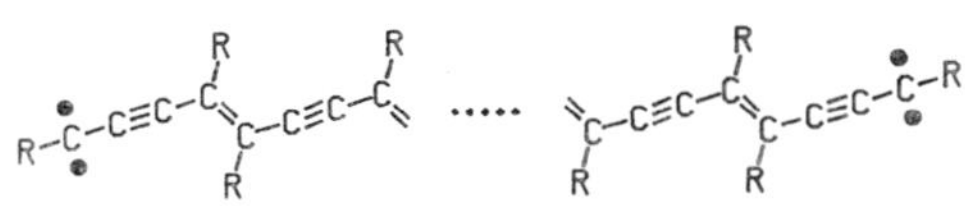

(f)
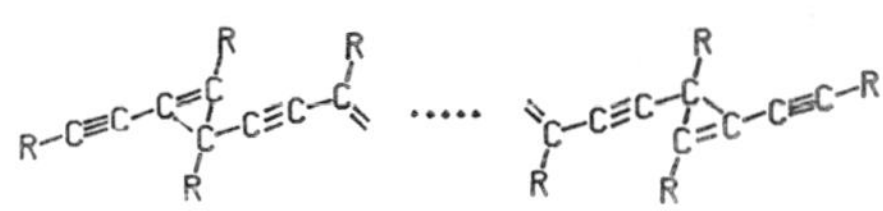

Figure 5. Species observed during the photo-reaction of PDAS at low temperature, (a) diradicals, (b) asymmetric carbenes, (c) stable oligomers, (d) and (e) dicarbenes and (f) stable polymer, from (30).

The results of these studies are summarised in Figure 5. For TS, the first intermediate observed is a dimer diradical, which can be converted thermally or by selective irradiation into a trimer, tetramer, etc.. Irradiation, however, also produces an asymmetric carbene trimer, which in turn can be converted into a series of oligomers. There is then a further side reaction which converts the asymmetric carbene trimer into a stable oligomer. Conversion of a diradical into an asymmetric carbene and a carbene into a stable oligomer occur only by the addition of a further monomer unit to the intermediate. These intermediates appear as the direct products of photo-polymerization below 80 K. Above 100 K dicarbene radicals with singlet ground states and excited triplet and quintet states are observed. Some of these are sufficiently long for the interaction of the carbene radicals to be negligible. A small population of dicarbenes has been observed as photoproducts at 4 K in perdeuterated TS (32). Above 150 K much longer stable polymer chains are formed which are distinguishable from polymer produced by thermal initiation at room temperature by a shifted optical absorption spectrum.

Diradicals longer than 6 monomer units have not been observed in TS though asymmetric carbenes and stable oligomers have been reported with ten or more repeat units. Thus, it seems that in TS the diradicals are stable only for short chain lengths and at room temperature these are rapidly converted into carbene species (33). Although similar results have been observed for some diacetylenes, these show differences in detail, in particular the annealing processes at higher temperatures vary markedly from compound to compound.

The optical spectra of these intermediates are the only significant source of information concerning the relationship of chain length and absorption energy for diacetylene oligomers. The results for TS are shown in Figure 6. Similar results have been reported for TS12, which exhibits longer diradical species but apparently no stable oligomers (34). For TS the results show a much less rapid variation for the longer oligomers than application of a modified free-electron model (35) suggests.

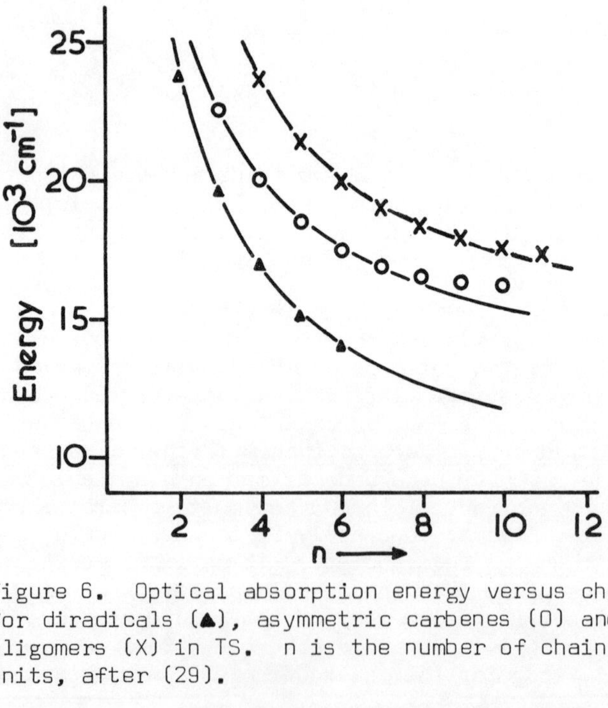

Figure 6. Optical absorption energy versus chain length for diradicals (▲), asymmetric carbenes (O) and stable oligomers (X) in TS. n is the number of chain repeat units, after (29).

The dicarbene radicals have attracted attention since although there is extensive data and theory for triplet states in organic crystals there has been little discussion of coupled triplets other than as a mechanism of exciton annihilation (36). Angular dependent ESR spectra (32, 37) and ENDOR (38) have been measured for the dicarbenes in TS. These can be interpreted using the fine structure parameters determined for the isolated carbene triplets by prior studies (27). The case of two carbene states coupled by electrostatic interactions has been discussed by Benk and Sixl and coworkers (39, 40) and the coupling of a doublet and triplet has also been considered (41). The coupled carbenes have a singlet ground state and a low lying quintet state well separated from the triplet state. The singlet-quintet splitting is sufficiently small that for TS the states are mixed. The theory predicts quintet state fine structure parameters that are in excellent agreement with the experimental values.

An outstanding question in this area is the nature of the polymerization initiation site. It has often been assumed that molecules near lattice defects act as exciton traps and also as initiation sites. Recent X-ray topographic studies show that good quality TS crystals contain a very small number of line defects (42, 43). Molecular defects, e.g. solvent and oxygen, exist in TS crystals but their presence seems to have very little effect on the polymerization process. In addition, there is relatively little information available about the triplet states of diacetylene monomers (44, 45). Analysis of the low temperature photo-reactions indicates that initiation is effected by the interaction of a triplet exciton on one monomer and a vibrational excitation of the neighbouring monomer (46). At high temperatures this is a facile interaction but does not shed any light on the process of thermal polymerization other than to suggest that an initial photo-reaction may be necessary. In certain cases specific initiation reactions have been identified. For example, the initiation reaction in phenazine-diacetylene complexes at low temperature has been shown to be an electron transfer between an excited phenazine molecule and a diacetylene monomer (47).

PHYSICAL PROPERTIES OF POLYDIACETYLENE CRYSTALS

PDA crystals display striking mechanical, optical and electronic properties. The emphasis in this discussion will be on the latter two areas. The former area is worth a brief discussion since, because of the ordered packing of the polymer chains, it is possible to make a connection between the macroscopic elastic modulus along the polymer-chain axis and the microscopic force constants of the polymer backbone.

The Young's modulus for PDA crystals is not very high since, because of the bulky sidegroups, each polymer chain has a large cross-sectional area. Typical values are 45 GPa for a urethane sidegroup polymer (48), 44 GPa (49) and 43.3 GPa (50) for TS and 45 GPa for DCH (51). For ETU polymer, which has somewhat smaller sidegroups, a value of 60 GPa has been measured (52). These macroscopic values can be compared with the microscopic force constants using the method of Treloar (53). The microscopic force constants are obtained from an analysis of Raman spectra. The Raman spectra contain a small number of lines since the backbone vibrations are resonantly enhanced. This simplifies the analysis but limits the set of force constants that can be deduced (54); this set is, however, adequate for the calculation of the Young's modulus. For TS the calculated Young's modulus is 50 GPa for a force constant set optimised to fit the Raman spectra recorded in an unstrained crystal (53). A value of 45 GPa is found for a set optimised to fit the shift in vibrational frequencies with applied strain (53). Similar results have been obtained for ETU polymer (55) and a calculated modulus of 74 GPa, somewhat higher than the experimental value, has been obtained. The observations of frequency shifts with applied strain allow the bond-anharmonicity to be evaluated directly. This information is valuable since it enables the force constant set to be transferred to other PDAs (54) and allows the evaluation of bond lengths in situations where they cannot be determined directly (56 - 58).

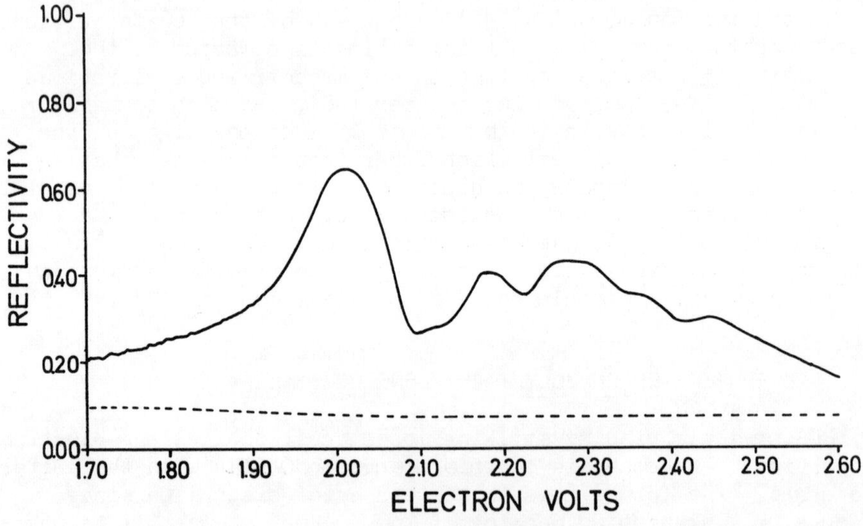

Figure 7. Reflection spectrum of TS polymer at room temperature, full curve, light polarized parallel to the polymer chains, dash curve, perpendicular polarization.

The electronic excited states of PDAs have been studied by both spectroscopic methods and by measurements of electrical conductivity. The most striking feature of PDA crystals is their metallic lustre which is caused by a relatively narrow peak in the reflectivity for light polarized parallel to the polymer chains, see Figure 7. The absorption is highly dichroic with an absorption coefficient for light polarized parallel to the chains several orders of magnitude larger than the values for perpendicular polarization (59). Reflection spectra have been reported for a number of PDAs (7). These are similar to that of TS. On cooling, a splitting was observed in the TS spectrum which was attributed either to a Davydov splitting or to inter-chain coupling before the real cause, a phase transition producing chains with two different structures, was discovered (60). This phase transition has attracted considerable attention, see for example (61) and references therein, but will not be discussed further here.

The asymmetry of the main reflection peak, and more particularly the asymmetry in the ε_2 peak determined by Kramers-Kronig analysis, was initially adduced as a signal for a van Hove singularity (62). The ε_2 profile for TS was a good fit to the square-root divergence expected of the van Hove singularity for an interband transition in 1D. Similar shapes were observed for BPG (63) but a careful analysis of the TS spectra cast doubts on this interpretation (64) and showed that the model of Toyazawa (65), which was valid for excitons, gave a better fit. Since that time evidence has been slowly accumulating and shows that the intense absorption in PDAs is due to an excitonic state. Exciton surface polaritons have been observed for TS crystals (66). Photo-conduction has been shown either to pass through a minimum coincident with the absorption peak, e.g. for TS (67-69) or to be zero until the photon energy is somewhat greater than that at the absorption peak (20, 70). In either case the absorption must be due to an uncharged species, i.e. an exciton, a model originally proposed in 1974 (71). Recently the reflection spectrum of TS observed using a synchrotron radiation source has been interpreted in terms of an exciton model (72). This assignment has been confirmed by studies of Raman scattering from TS crystals (58) and by electric field modulated reflection spectroscopy (73 - 75).

In principle the absorption profile for an exciton can be written in the Franck-Condon approximation as

$$A(E) = k \sum_{\tilde{n}} |\langle \tilde{n}_e \, I \, 0_g \rangle|^2 \, E_{\tilde{n}} \, L_{\tilde{n}} (E) \qquad (1)$$

where k is a constant proportional to the transition dipole moment, $L_{\tilde{n}}(E)$ is a lineshape function, $E_{\tilde{n}}$ is the energy of the transition from the ground state to the $\tilde{n}$th vibrational level of the excited state and the matrix element is a multidimensional Franck-Condon overlap integral of the general form

$$<\tilde{m}_g \, I \, \tilde{n}_e> = \prod_{a=1}^{\ell} <m_{ag} \, I \, n_{ae}> \qquad (2)$$

where ℓ is the number of normal modes for which the displacement of the potential minima in the ground- and excited-states along the normal co-ordinate is significantly greater than zero. The complexity of these overlap integrals prevents a direct analysis of the absorption data. At first the expression for the resonant Raman scattering cross-section looks more complicated; see for example (76). For PDAs, however, this can be significantly simplified since (a) there are only four important normal modes, (b) the line width is less than the normal mode frequencies so that cross-terms can be neglected and (c) if the displacement of ground and excited states is small the products of overlap integrals contribute a constant exponential factor. More detailed discussion of these simplifications are given elsewhere (77). The final expression for the simultaneous excitation of a phonon in modes 1 and 2 is

$$|\alpha_{zz}|^2 = K \sum_{n_1} \sum_{n_2} \frac{\left[<1_{1g} I n_{1e}><n_{1e} I 0_{1g}><1_{2g} I n_{2e}><n_{2e} I 0_{2g}>\right]^2}{\left[E_0 + n_1 hF_1 + n_2 hF_2 - E_L\right]^2 + \Gamma_e^2}$$

where F_1 and F_2 are the frequencies of modes 1 and 2 in the excited state, E_L is the laser energy and Γ_e is the exciton line-width. Corresponding but simpler expressions can be written for single mode excitation. These expressions may be readily evaluated (58) using the Franck-Condon factors derived by Keil (78).

Raw resonant Raman data must be corrected for sample reflection, absorption and refraction. Applying these corrections to the raw data for TS gives unrealistic excitation profiles. This has been attributed to surface roughness invalidating the refraction correction when the penetration depth of light into the sample is small (58). Neglecting this correction the data can be analysed to give the parameters listed in Table 1; typical corrected data and calculated excitation profiles are shown in Figure 8. A check is provided on these results by the reasonable fit of the experimental absorption spectrum and that calculated using the parameters of Table 1. In addition the excited state vibrational frequencies can be used to calculate the excited state bond lengths using the known bond anharmonicity. The values obtained are 0.140, 0.121 and 0.140 nm respectively for the single, double and triple bonds which have lengths of 0.143, 0.119 and 0.136 nm in the ground state. These bond length changes are consistent with the charge-transfer character of the exciton first discussed by Philpott (79) and calculated for bipolaron states by Cade and Movaghar (80).

Table 1

Electronic and vibrational properties of TS deduced from an analysis of resonant Raman scattering (58).

Phonon[a] (mode)	f(ground) (cm^{-1})	F(excited) (cm^{-1})	D(nm)	dE/dQ[b] (eV/nm)
1	2086	2040	2.1×10^{-3}	40
2	1485	1420	3.4×10^{-3}	33
3	1203	1170	1.4×10^{-3}	9.2
4	952	940	2.4×10^{-3}	9.6

a: The vibrational modes are 1- triple bond stretch; 2- double bond stretch; 3- single bond stretch/double bond bend and 4- triple bond bend (54).
b: dE/dQ is given for a quadratic potential function by $-4\pi^2 mf^2 D$.

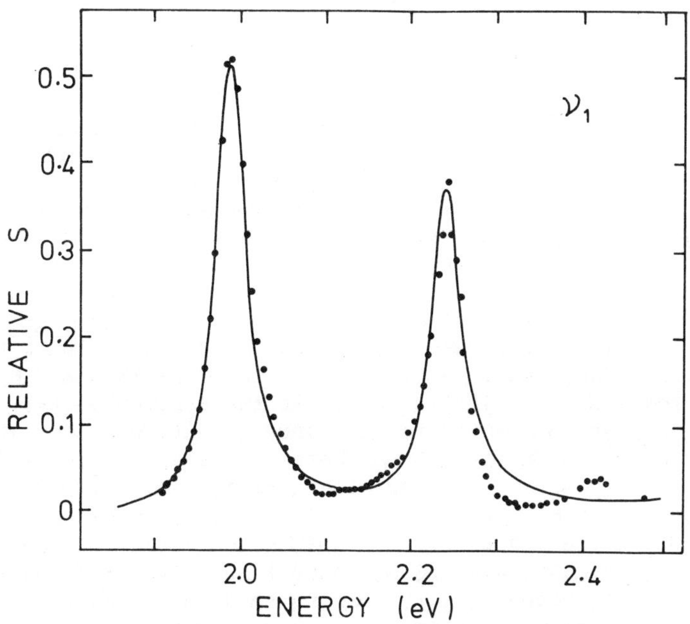

Figure 8. Resonant Raman excitation profile for triple bond stretching mode in TS polymer. Experimental data is represented by points, the calculated profile by the full curve, S in the corrected scattered intensity, from (58).

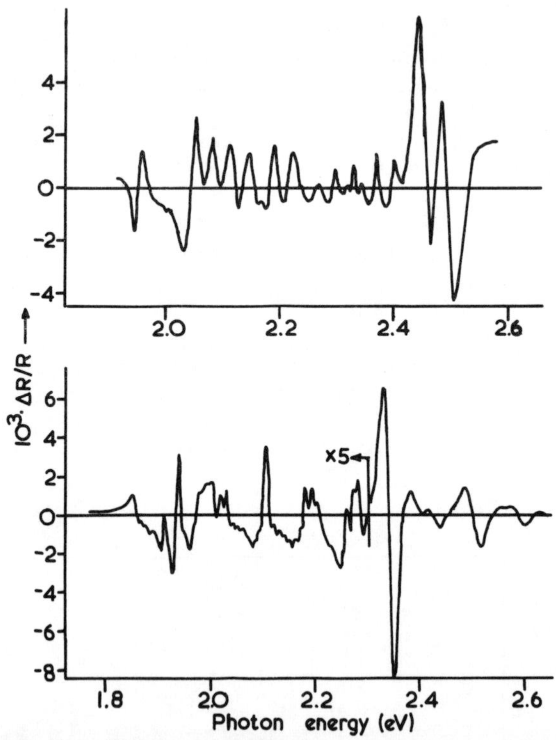

Figure 9. Electro-reflectance spectra of TS polymer (upper curve) and DCH polymer (lower curve) recorded at 2 K with applied fields of about 23 kV/cm, after (74).

Electro-reflectance spectra show unambiguously that the inter-band-transition for PDAs lies about 0.5 eV above the excitonic state (73-75). The electro-reflectance spectra of TS and DCH at low temperatures are shown in Figure 9. At the exciton energy, about 2 eV, a feature appears which is a second derivative of the absorption profile, characteristic of a charge-transfer exciton (81). At 2.5 eV a larger signal is observed with a first derivative profile, characteristic of a Frenkel exciton. In the absence of an observable absorption feature at 2.5 eV these signals must be assigned to inter-band transitions. Two peaks occur for exciton and inter-band transitions of TS because of the two non-equivalent polymer chains present in the low temperature phase. The data deduced from the electro-reflectance spectra of four PDAs are listed in Table 2. The data for DCH have been interpreted to indicate a broad conduction band leading to a low effective electron mass of only 0.05 m_e and a carrier mobility of 2,800 cm^2/Vs (73).

A complete picture of the energy levels in TS has been provided by careful photoemission studies (82). These give an ionization potential of 5.5±0.1 eV and hence the bottom of the conduction band lies 3.1±0.1 eV below the vacuum level. Very similar values have been found for DCH, 5.8 and 3.4 eV respectively, by studies of electron and hole injection using a range of electrodes with different work functions (83). Thus, the nature and energies of the electronic excitations in PDAs are well established.

Table 2.

Exciton and inter-band transitions for four PDAs as observed by reflection and electro-reflection spectroscopy (Sebastian and Weiser (75)).

	DCH	TS	TCDU(1)[a]	TCDU(2)[a]	BPG
E_{Ex}(eV)	1.858	1.937	1.940	2.321	2.128
ε_2(max)	40	56	10.4	5.4	4.9
q/e (CT)	0.38	0.19	0.19	-	0.16
E_g(eV)	2.333	2.447	2.516	2.983	2.624
E_B(exciton)	0.475	0.510	0.576	0.662	0.496

a: TCDU(1) is the as polymerized phase; TCDU(2) is the form induced by applied stress (118).

The interpretation of measurements of dark- and photo-conductivity in PDAs is still a matter of debate. One reason for this is the fact that the behaviour of different PDAs shows much larger variations when conductivity is measured than when spectra are measured. Thus, while electron and hole injection has been observed for DCH, similar electrodes form blocking contacts on TS crystals (83). In addition there appear to be no deep traps in DCH (84) while photoconduction in TS is dominated by a level lying 0.8 eV below the conduction band (69 85). In much of the early literature on electrical conductivity in TS a low mobility was deduced on the assumption that carriers were injected to give space charge limited currents (85). There was, however, a consensus that both dark- and photo-currents, and mobilities, were about 10^3 larger for carrier motion parallel to the polymer chains than perpendicular (69, 85, 80). In addition, photo-carrier generation was shown to be well described by Onsager's theory for geminate recombination in one dimension (69, 87, 88).

The absence of charge injection in TS was first noted by Spannring and Baessler (89) and Donovan and Wilson (90). The former still, however, assumed that photo-currents were limited by recombination. This assumption was not made by the latter workers who deduced a high mobility. The absence of recombination centres was subsequently shown in an experiment in which the polymer chain lengths were reduced by laser ablation of the crystal surface (91). The distance over which electrons moved was shown to be equal to the distance between the ablated regions. The absence of recombination centres and application of the Onsager model leads to a carrier drift velocity of about 2×10^3 m/s, which is saturated down to very low fields and implies a very high intrinsic mobility, of the order 2×10^5 cm^2/Vs. To support these experimental findings Wilson has published a series of papers discussing carrier motion in one-dimensional systems (92 - 95).

There is a growing body of other evidence for high electron mobility in PDAs. The electro-reflectance data gives such a result by an independent route (73) and a high value is indicated by the injection studies for DCH (83). In addition, studies of recombination in dosordered samples, described in the next section, indicate that the density of such centres is low even in highly imperfect samples (96, 97). Recent studies of the field dependence of photo-carrier generation have led to a value for the microscopic anisotropy of carrier mobility in excess of 10^4 (98). At very low fields a deviation from the 1D Onsager model has been reported, with carrier generation efficiency becoming approximately isotropic (99). The theoretical consequences of this result have not yet been investigated.

It is somewhat surprising in the light of this evidence that attempts to observe carrier transit have been unsuccessful (100 - 102). After a short light pulse the photocurrent decays monotonically with no features attributable to carrier transit. Such a response, of the form $t^{-\alpha}$, is in fact typical for amorphous materials; an apparent paradox in view of the quality of the PDA crystals used. This problem has been resolved by recent theoretical and experimental work (103, 104). These show that the apparent anomaly is due to a breakdown in linear response theory. Carrier motion in 1D in the presence of strong scattering centres is shown to be extremely sensitive to the density of scattering centres. An increase in the fraction of centres from 10^{-5} to 10^{-4} increases the transit time in a crystal from 10^{-2} to 10^5 seconds. This is consistent with the long time decays seen in imperfect samples (96, 97). This theory also indicates that the intrinsic carrier mobility is larger than 10^3 cm^2/Vs. The failure of linear response theory in a range of 1D transport problems is of considerable general interest.

The availability of detailed information about the electronic states of PDAs makes them ideal systems to test molecular quantum mechanical theories. The earliest calculation for a model PDA chain with simple sidegroups gave rather poor values for the band-gap, see (7). In most of these calculations Coulomb correlations were neglected so that only band structures were deduced. Further work along these lines has included the use of an ab initio crystal orbital method (105), studies of the ground state geometries (106), a priori Hartree Fock crystal orbital calculations (107) and a non-empirical effective Hamiltonian technique (108). These show that the acetylene structure of figure 1 is the lowest energy chain structure. Suhai obtains reasonable values for the ionization potential and electron affinity, this is also true for the ionization potential determined by the effective Hamiltonian technique, though accurate calculation of the band-gap remains difficult.

Calculations including Coulomb interactions were first reported by Yarkony (109) and more general models were considered by Gasser and Boerner (110) and Cade and Young (111). Yarkony and co-workers compared band and exciton models and cyclic and linear chains. Ukrainskii (112) used the experimental exciton binding energy to evaluate the nearest neighbour electron Coulomb repulsion in PDAs. He also concluded that the electron repulsion accounted for about 60% of the observed band gap with only about 40% due to the bond-alternation. PDAs have also been treated using molecular M.O. methods by a number of authors (79, 113, 114).

Recently the theoretical models developed for polyacetylene have been applied to PDAs (80, 115, 116). A degenerate ground state does not occur for PDA so that this discussion is limited to polarons. The potential surface for the ground state structures deduced by this method is very similar to that calculated by the ab initio method (80, 106) indicating that the alternative butatriene structure is metastable. A further conclusion is that the band gap is primarily due to the bond alternation produced when the π-electrons are added to the bare PDA backbone. These calculations are currently being extended to make realistic exciton calculations for PDAs (117). The occurrence of acoustic-mode polarons and exciton-polarons in PDAs has also been discussed recently (95, 145).

Many of the theoretical papers have involved calculations of the idealised acetylenic and butatrienic structures. The case for the occurrence of the latter structure has not been discussed here in detail since there is now a consensus that shifts in optical and vibrational excitation energies can be accounted for by deformations of the acetylenic structure (57, 108, 118). Extremely large shifts in exciton energy are observed as a function of temperature for partially polymerized diacetylenes, see Figure 10, since the monomer lattices can have large thermal expansion coefficients which dramatically affect the environments of the polymer chains. A particularly striking example is DCH where a 6% compression along the chain axis at the 142 K monomer phase transition is accompanied by a 0.4 eV increase in exciton energy (119). The molecular deformation produced by this compression has been shown to be principally changes in bond-angles. It would be interesting to see theoretical predictions of the electronic states for such highly distorted structures.

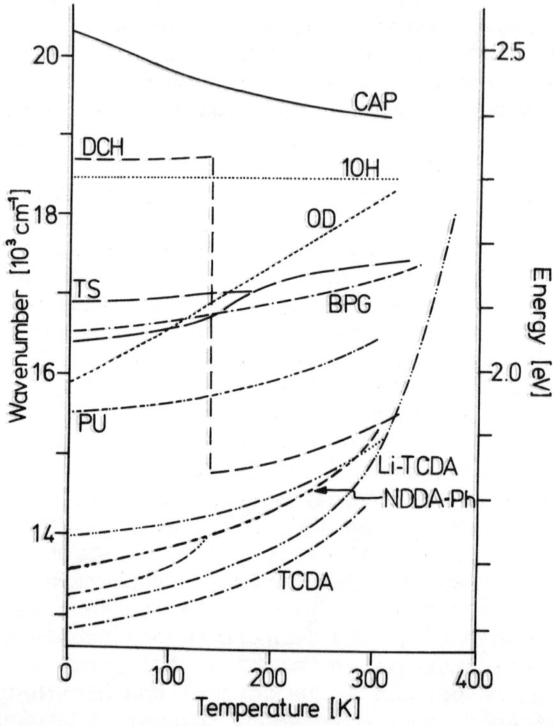

Figure 10. Temperature dependence of the optical absorption in partially polymerized diacetylene monomer crystals; the level of conversion was generally less than 1%, after (129).

The recent theoretical studies of PDAs have made greater use of the available experimental data and provide realistic microscopic models. The current state of the art is such that one can look forward to theory being used increasingly in a predictive role rather than an interpretive one.

PROPERTIES OF DISORDERED POLYDIACETYLENES

Disordered samples of polydiacetylenes can be prepared by a number of routes. First, the less perfectly polymerizing crystals can be obtained in a fibrilar form due to disruption of the crystal lattice by internal strain, e.g. (120). Secondly, similar samples can be obtained by the removel of unreacted monomer from partially polymerized crystals. Thirdly, single crystals can be deformed mechanically. Fourthly precipitates in the form of films or powders can be obtained for soluble diacetylenes and finally the solutions of such soluble materials can be studied directly.

Samples prepared in this way allow a range of morphologies from microcrystalline to amorphous to be obtained. In this review it will be possible to cover only part of the literature describing measurements on such systems. The discussion will, therefore, be limited to three topics, a) the characterization of highly disordered polymers in solution and solid phases, b) photo-currents in slightly disordered solids and c) doping of PDAs to enhance conductivity.

The prinicpal effect of increasing disorder is to smear out the well defined energy levels of the extended PDA chain. The excitonic state will be broadened inhomogeneously, due to static perturbations of the backbone, and homogeneously, due to the increased chain motion possible in disordered materials. It is difficult to distinguish these effects from the absorption spectrum alone and Raman spectroscopy has proved vital in characterising disordered samples. Three different situations have been observed, a) a continuous distribution of unperturbed chain sequences, b) a small number of distinct chain structures and c) a single average structure.

The first case is observed for true solutions of PDAs. A number of readily soluble PDAs are now known; the most widely studied are the nBCMU and related polymers developed at Allied Chemicals (118, 121, 122 and references therein). Other examples are TS-12 (123) and the derivatives of the docosadiyne diol (124, 125). These solutions have been characterized by measurements of viscosity, osmotic pressure, light scattering and ^{13}C-NMR with comparable results for all the materials. The mean polymer chain lengths are in the range 1000 to 2000 repeat units with a polydispersity of between 3 and 5 (123 - 126). The molecules adopt a random coil form despite the rigidity of the backbone, whether this is due to gradual deformation or sharp kinks has not been established. The NMR studies show that the acetylenic backbone of Figure 1 is retained in solution (121, 125). Colour changes occur on the addition of a non-solvent; this has been attributed to a random-coil to rigid-rod transformation (80, 126, 127). This result has been questioned because of neglect of polydispersity (123) and recent results in our laboratories suggest the formation of micro-crystals on addition of a non-solvent.

The optical spectra of solutions are all similar and have the form shown in Figure 11. The absorption maximum lies at about 2.6 eV in comparison with the 1.8 to 2.3 eV observed for crystals, see Table 1. This reflects a general trend of the absorption maxima towards higher energy as disorder is introduced (80, 128, 129). The absorption has been interpreted in terms of conjugation lengths using either a free electron (8, 112, 130) or a Hückel model (131). The fit to the experimental data is, however, just as good as that for a single heavily broadened absorption band (132, 133). Raman spectroscopy enables these two possibilities to be distinguished. The vibrational frequencies are observed to shift continuously as the excitation frequency is varied as expected for a distribution of conjugation lengths, see Figure 12 (130, 132). The same effect is also observed for the amorphous surface layer produced by the heavy deformation of TS single crystals (57).

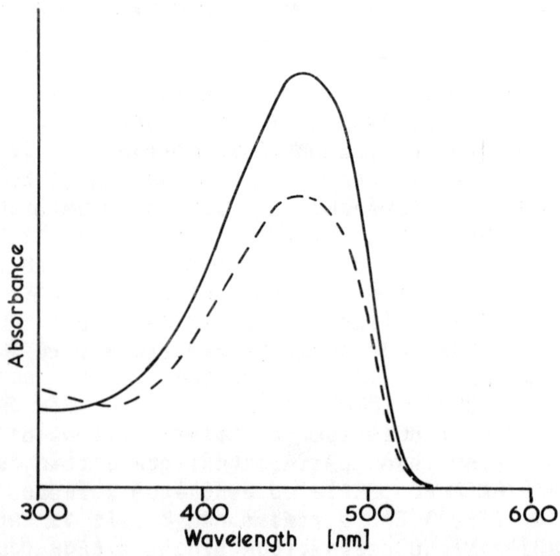

Figure 11. Absorption sepctra of 4BCMU (solid curve) and B9 (dash curve) polymers dissolved in chloroform at room temperature.

An intermediate case is observed in Langmuir-Blodgett films of diynoic acids. The absorption profile has slightly more structure than that of a solution but Raman spectroscopy shows that this is produced by the superposition of the spectra of just three different types of PDA chain, see Figure 13 (134). TS-colloids, precipitated from solution of low molecular weight TS in chloroform, show in turn slightly more structure in their spectra and Raman spectroscopy shows only one principal set of vibrations (133). This result is somewhat surprising since the zero-phonon absorption peak appears at 2.5 eV. This led to the suggestion that such a species could contribute to the 'solution' spectrum (133). Recently studies of glasses produced by quenching solutions to liquid nitrogen temperatures have revealed exciton spectra at and above 2.5 eV with linewidths comparable to those of crystals (135).

Thus, we can conclude that PDAs in solution in general adopt a random coil conformation and similar disruption of conjugation occurs in highly deformed solids. There are also states of disorder where the molecules aggregate or adopt a molecular conformation to produce a well defined average structure having

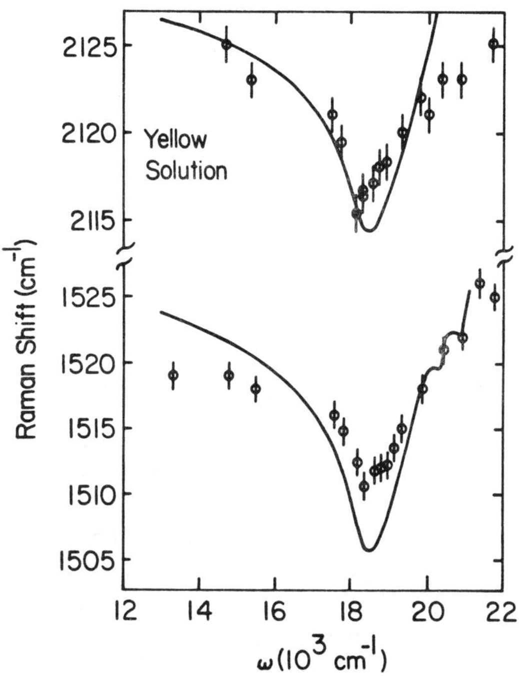

Figure 12. Calculated (full curve) and observed (circles) dependence of the principal Raman bands on excitation frequency for 3BMCU solution, from (132).

a high exciton energy. The exact nature of these structures is not known but possibilities include a) chain folded lamelar crysals with a short, regular fold length, b) quasicrystalline particles, c) backbone kinks and d) grossly deformed backbone structures. In the latter context it is worth noting that changes of bond angles are more likely than changes in bond length since the angular force constants are weaker.

A phenomenon that has received little attention is the appearance of fluorescent emission for disordered PDAs. PDA crystals show negligible fluorescence but it is observed from solutions and other disordered samples (133, 136). The quantum efficiency is low (135) suggesting that disorder introduces a radiant decay channel which competes with the non-radiative mechanism operative in extended chain PDAs. A possible mechanism for this has been suggested (117). The bipolaron, and also the exciton, in PDAs has a rather flat potential curve, see Figure 14. Excited excitons will, therefore, be rather extended and emission of a phonon destroying the initial state, i.e. non-radiative decay, is likely. If the exciton is confined by defects the emission of phonons is less likely and fluorescence can occur.

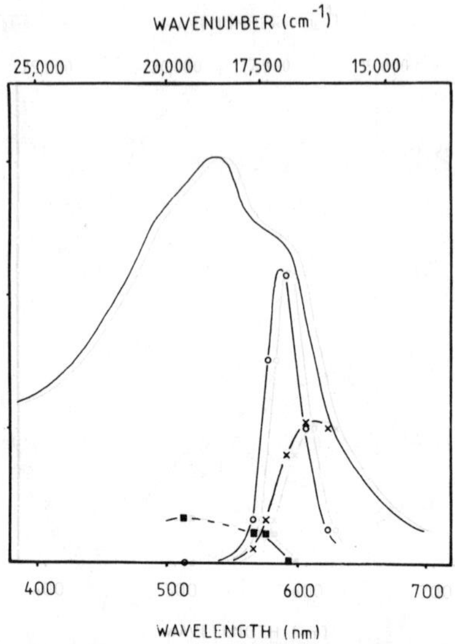

Figure 13. Absorption profile, full curve, and resonance Raman excitation profiles (uncorrected) for the double bond stretching modes at 1448 cm^{-1} (X), 1486 cm^{-1} (O) and 1514 cm^{-1} (■) in a PDA Langmuir-Blodgett film, from (134).

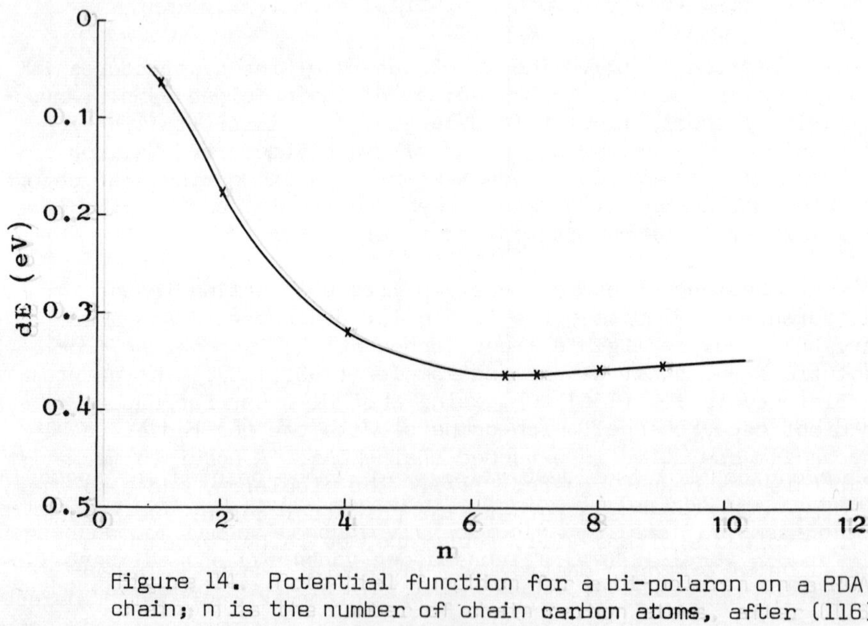

Figure 14. Potential function for a bi-polaron on a PDA chain; n is the number of chain carbon atoms, after (116).

This model can also explain carrier generation effects observed for excited exciton states and provides a more realistic model for interpreting absorption profiles than those used previously.

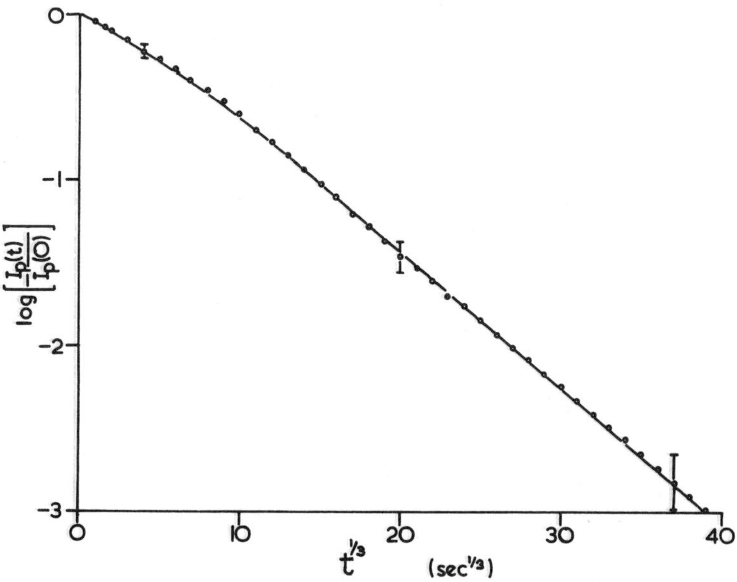

Figure 15. Decay of photo-current in an orientated fibrous film of 10H polymer at room temperature for a field of 5 kV/cm showing $\exp(-t^{1/3})$ dependence.

The band-states of the PDAs will be smeared into the band-gap with increasing disorder and a mobility gap will appear with the lowest states becoming localised. Introducing disorder has a dramatic effect on the time scale for the decay of photo-currents. In TS single crystals the decay has a $t^{-\alpha}$ dependence, the photo-current falls by an order of magnitude in 100 μs, but a weak tail persists for about 1s (104). For TS crystals damaged by He-ion bombardment the decay at intermediate fields and times goes at $\exp(-t^{1/3})$ becoming exponential at times of the order 10^4s (96). An $\exp(-t^{1/3})$ decay law has been observed over times up to 2×10^4s in films of 10H polymer, see Figure 15 (97). This unusual law is that expected theoretically for diffusive motion in one dimension (137). In addition, the theory predicts a change to an exponential law depending on defect concentration, time, applied electric field and temperature. The predicted behaviour with time and field have been observed for the damaged TS samples (96). Our recent data has shown the theoretical temperature dependence for 10H. It should be noted that these dramatic increases in decay time do not require very high defect densities since the time for an electron to transit the sample depends on the defect density, x, to the power $(1-\alpha)^{-1}$ (104). Experimentally α=0.85 which leads to a power law in x of approximately x^7. These results show that even in imperfect PDA samples the electron motion is highly anisotropic; a lower limit for the microscopic anisotropy is 10^4, comparable with that reported for polyacetylene.

The main focus of recent interest in conjugated polymers has been the metallic conductivity achieved when strong donors or acceptors interact with them. The apparent absence of comparable effects for PDAs has been somewhat surprising since the relevant electronic properties of the PDA backbone are similar to those of other dopable polymers (138). This has been rationalised in terms of the presence of the large potentially reactive sidegroups, which are required to ensure solid-state polymerization, and the extremely small penetration of large dopant molecules into the closely packed polymer crystal lattices. The modification of conductivity by adsorbed surface species has been known for some time (87). These facts suggest that doping will be most likely for fibrous samples with simple sidegroups (139). Even for fibrous samples of 10H polymer, however, the final conductivity is low (10^{-6}S/cm) despite the uptake of large amounts of I_2.

Dopant inclusion by co-crystallisation and vapour phase doping during polymerization have also been investigated (140). Conductivities as high as 10^{-4}S/cm have been observed, but despite the apparently unperturbed crystal quality the conductivities of the doped samples were essentially isotropic. It should be noted that TS crystals grown in air contain considerable quantities of O_2, this does not seem to affect their conductivity but is capable of oxidation of the polymer chain immediately after photo-excitation of charge transfer between the polymer backbone and the O_2 (141).

Recently doping of DCH polymer has been reported first by the electrolytic field driven injection of ions into microscopic channels parallel to the polymer chains (142) and secondly by the action of SbF_5 on thermally polymerized DCH (143). In the latter case the crystal fibrilates on polymerization. Consideration of the crystal structure, however, indicates that the charge transfer must be mediated by the sidegroups since direct approach of dopants to the polymer chain is not possible. DCH polymer is stable against photo-oxidation in consequence (141). Conductivities up to 10^{-3}S/cm are reported for doped DCH.

We have observed lower conductivity values for radiation polymerized DCH crystals, which are more perfect than thermally polymerized samples (144). A small reversible increase in conductivity is found on cycling the doped samples from vacuum to a normal atmosphere and back. In the case of 10H a smaller increase has been observed on SbF_5 doping with a much larger increase on admitting air. The final conductivity observed was 2S/cm which decayed over a period of a few days. The results with 10H are irreproducible due to variations in fibril dimensions and packing and the sensitivity of this polymer to photo-oxidation. It appears, however, that the enhancement of conductivity in PDAs can be attributed to the surface adsorption of dopants.

CONCLUSIONS

Polydiacetylenes remain the only conjugated polymers available in an ideal chain-extended form. As such, their continued use as model systems, which allow direct investigations of fundamental physical properties, is assured. Even for PDAs, it has taken some time to gain a proper understanding of these properties. The interplay of experiment and theory that has been necessary to gain this understanding will continue and will shed more light on the physics of quasi-one-dimensional systems.

It has not been possible to review all the areas of current interest in PDAs; this reflects the growth of interest in these unique materials. In particular the non-linear optical properties of PDAs appear likely to lead to applications in the near future. A number of other technological uses are emerging and it seems probable that PDAs will be of increasing interest to applied scientists in the future.

ACKNOWLEDGEMENTS

I would like to thank Drs. N. Cade, B. Movaghar, E.G. Wilson, R.J. Young and Profs. H. Sixl, H. Baessler and G. Weiser for communication of results and manuscripts prior to publication. I would like to thank my colleagues in the Polymer Research Group at Queen Mary College and Dr. B. Movaghar for useful discussions. Some of the results reported were obtained with the support of the Science and Engineering Research Council.

REFERENCES

1. Wegner, G., 1969, Z. Naturforsch., 24b, p. 824.
2. Baughman, R.H. and Chance, R.R., 1978, Ann. NY. Acad.Sci., 313, p. 705.
3. Baughman, R.H. and Yee, K.C., 1978, J.Polym.Sci. Macromol. Revs., 13, p. 219.
4. Wegner, G., 1979, Molecular Metals (Plenum Press), p. 209.
5. Wegner, G., 1980, Faraday Discussion No. 68, p. 494.
6. Enkelmann, V., 1980, Quantum Theory of Polymers (Springer Lecture Notes in Physics), 113, p. 1.
7. Bloor, D., 1980, Quantum Theory of Polymers (Springer Lecture Notes in Physics), 113, p. 14.
8. Bloor, D., 1982, Developments in Crystalline Polymers - 1. (Applied Science Publ.), p. 151.

9. Cadiot, P. and Chodkiewicz, W., 1969, Chemistry of Acetylenes (Marcel Dekker), p. 597.
10. Kaiser, J., Wegner, G. and Fischer, E.W., 1972, Israel J.Chem., 10, p. 157.
11. Schmidt, G.M.J., 1967, Reactivity of the Photo-excited Organic Molecule (Wiley), p. 227.
12. Baughman, R.H., 1974, J.Polym.Sci.Polym.Phys. Ed., 12, p. 1511.
13. Milburn, H., Werninck, A., Blair, E., Ando, D.J., Bloor, D., Motevalli, M. and Hursthouse, M.B. - to be published.
14. Williams, R.L., 1981, Thesis, University of London.
15. Baughman, R.H., 1978, J.Chem.Phys., 68, p. 3110.
16. Enkelmann, V., Leyrer, R.J., Schleier, G. and Wegner, G., 1980, Makromol.Chem., 180, p. 1787.
17. Day, R. and Bloor, D. - to be published.
18. Enkelmann, V., Leyrer, R.J., Schleier, G. and Wegner, G., 1980, J.Mater.Sci., 15, p. 168.
19. Galiotis, C., Young, R.J., Ando, D.J. and Bloor, D., 1983, Makromol.Chem., 184, p. 1083.
20. Yee, K.C. and Chance, R.R., 1978, J.Polym.Sci.Polym.Phys.Ed., 16, p. 431.
21. Baughman, R.H. and Chance, R.R., 1980, J.Chem.Phys., 73, p. 4113.
22. Krohnke, C., Enkelmann, V. and Wegner, G., 1980, Chem.Phys. Lett., 71, p. 38.
23. Bloor, D., 1983, Mol.Cryst.Liq.Cryst. - in press.
24. Chance, R.R. and Patel, G.N., 1978, J.Polym.Sci.Polym.Phys.Ed., 16, p. 859.
25. Prock, A., Shand, M.L. and Chance, R.R., 1982, Macromolecules, 15, p. 238.
26. Patel, G.N., Chance, R.R., Turi, E.A. and Khanna, Y.P., 1978, J.Amer.Chem.Soc., 100, p. 6644.
27. Eichele, H., Schwoerer, M., Huber, R. and Bloor, D., 1976, Chem.Phys.Lett., 42, p. 342.
28. Sixl, H., Gross, H. and Neumann, N., 1983, Studies in Inorganic Chemistry (Elsevier), 3, p. 493.
29. Gross, H. and Sixl, H., 1983, Mol.Cryst.Liq.Cryst. - in press.
30. Sixl, H., Adv. in Polym.Sci. - in press.
31. Gross, H. and Sixl, H., 1982, Chem.Phys.Lett., 91, p. 262.
32. Huber, R.A. and Schwoerer, M., 1980, Chem.Phys.Lett., 72, p. 10.
33. Niederwald, H., 1982, Thesis, University of Bayreuth.
34. Siegel, D., Sixl, H., Enkelmann, V. and Wenz, G., 1982, Chem.Phys., 72, p. 201.
35. Baughman, R.H. and Chance, R.R., 1976, J.Polym.Sci.Polym.Phys. Ed., 14, p. 2037.
36. Swenberg, C.E. and Geacintov, N.E., 1973, Organic Molecular Photophysics (Wiley), 1, Chap. 10.
37. Huber, R.A., Schwoerer, M., Benk, H. and Sixl, H., 1981, Chem.Phys.Lett., 78, p. 416.
38. Hartl, W. and Schwoerer, M., 1982, Chem.Phys. 69, p. 443.

39. Benk, H. and Sixl, H., 1981, Mol.Phys., 42, p. 779.
40. Kollmar, C., Sixl, H., Benk, H., Denner, V. and Mahler, G., 1983, Chem.Phys.Lett. - in press.
41. Kollmar, C. and Sixl, H., 1983, Chem.Phys. - in press.
42. Dudley, M., Sherwood, J.N., Bloor, D. and Ando, D., 1982, J.Mat.Sci.Lett., 1, p. 479.
43. Dudley, M., Sherwood, J.N., Ando, D.J. and Bloor, D., 1983, Mol.Cryst.Liq.Cryst. - in press.
44. Bertault, M., Fare, J.L. and Schott, M., 1979, Chem.Phys.Lett., 62, p. 161.
45. Bubeck, C., Sixl, H., Bloor, D. and Wegner, G., 1979, Chem. Phys.Lett., 63, p. 574.
46. Gross, H., Neumann, W. and Sixl, H., 1983, Chem.Phys.Lett. - in press.
47. Bubeck, C., Nguyen Xuan, T.H., Sixl, H., Tieke, B. and Bloor, D., 1983, Ber.Bunsenges Phys.Chem. - in press.
48. Baughman, R.H., Gleiter, H. and Sendfeld, N., 1975, J.Polym. Sci. Polym.Phys. Ed., 13, p. 1871.
49. Leyrer, R.J., Wegner, G. and Wettling, W., 1978, Ber.Bunsenges Phys.Chem., 82, p. 697.
50. Rehwald, W., Vonlanthen, A. and Meyer, W., - to be published.
51. Galiotis, C., Young, R.J. and Batchelder, D.N., 1983, J.Mat. Sci.Lett., 2, p. 263.
52. Galiotis, C. and Young, R.J., 1983, Polymer - in press.
53. Batchelder, D.N. and Bloor, D., 1979, J.Polym.Sci. Polym.Phys. Ed., 17, p. 569.
54. Lewis, W.F. and Batchelder, D.N., 1979, Chem.Phys.Lett., 60. p. 232.
55. Galiotis, C., Young, R.J. and Batchelder, D.N., J.Polym.Sci. Polym.Phys. Ed. - in press.
56. Cottle, A.C., Lewis, W.F., Batchelder, D.N., 1978, J.Phys.C. Sol.St.Phys., 11, p. 605.
57. Batchelder, D.N., Kennedy, R.J., Bloor, D. and Young, R.J., 1981, J.Polym.Sci. Polym.Phys. Ed. 19, p. 677.
58. Batchelder, D.N. and Bloor, D., 1982, J.Phys.C. Sol.St.Phys., 15, p. 3005.
59. Bloor, D. and Preston, F.H., 1976, Phys.Stat.Solidi a, 37, p. 427.
60. Enkelmann, V., 1977, Acta Cryst., B33, p. 2842.
61. Bertault, M., Collet, A. and Schott, M., 1981, J. de Phys.Lett., 42, p. L-131.
62. Reimer, B., Baessler, H., Hesse, J. and Weiser, G., 1976, Phys.Stat.Solidi b, 73, p. 709.
63. Bloor, D., 1976, Chem.Phys.Lett., 42, p. 342.
64. Bloor, D. and Preston, F.H., 1977, Phys.Stat.Solidi a, 39, p. 607.
65. Toyazawa, Y., 1958, Progr.Theor.Phys. (Kyoto), 20, p. 53.
66. Brillante, A., Pockrand, I., Philpott, M.R. and Swalen, J.D., 1978, Chem.Phys.Lett., 57, p. 395.
67. Reimer, B. and Baessler, H., 1975, Phys.Stat.Solidi a, 32, p. 435.

68. Chance, R.R. and Baughman, R.H., 1976, J.Chem.Phys. 64, p. 3889.
69. Siddiqui, A.S., 1980, J.Phys.C. Sol.St.Phys. 13, p. 2147.
70. Lochner, K., Baessler, H., Tieke, B. and Wegner, G., 1978, Phys.Stat.Solidi b, 88, p. 653.
71. Bloor, D., Ando, D.J., Preston, F.M. and Stevens, G.C., 1974, Chem.Phys.Lett., 24, p. 407.
72. Tokura, Y., Mitani, T. and Koda, T., 1980, Chem.Phys.Lett., 75, p. 324.
73. Sebastian, L. and Weiser, G., 1981, Phys.Rev.Lett., 46, p. 1156.
74. Sebastian, L. and Weiser, G., 1981, Chem.Phys., 62, p. 447.
75. Sebastian, L. and Weiser, G. - to be published.
76. Clark, R.J.H. and Stewart, B., 1979, Structure and Bonding, 36, p. 1.
77. Batchelder, D.N. and Bloor, D., 1983, Advances in Infrared and Raman Spectroscopy (Wiley) - in press.
78. Keil, T.H., 1965, Phys.Rev., 140, p. A601.
79. Philpott, M.R., 1977, Chem.Phys.Lett., 50, p. 18.
80. Cade, N.A. and Movaghar, B., 1983, J.Phys.C. Sol.State Phys., 16, p. 539.
81. Sebastian, L., Weiser, G. and Baessler, H., 1981, Chem.Phys., 61, p. 125.
82. Murashov, A.A., Silinsh, E.A. and Baessler, H., 1982, Chem. Phys.Lett., 93, p. 148.
83. Spannring, W. and Baessler, H., 1981, Chem.Phys.Lett., 84, p. 54.
84. Lochner, K., Baessler, H., Sebastian, L., Weiser, G., Wegner, G. and Enkelmann, V., 1981, Chem.Phys. Lett., 78, p. 366.
85. Siddiqui, A.S. and Wilson, E.G., 1979, J.Phys.C. Sol.St.Phys., 12, p. 4237.
86. Lochner, K., Reimer, B. and Baessler, H., 1976, Chem.Phys.Lett., 41, p. 388.
87. Lochner, K., Reimer, B. and Baessler, H., 1976, Phys.Stat. Solidi b, 76, p. 533.
88. Donovan, K.J. and Wilson, E.G., 1981, Phil.Mag., 44, p. 31.
89. Spannring, W. and Baessler, H., 1979, Ber.Bunsenges Phys.Chem., 83, p. 433.
90. Donovan, K.J. and Wilson, E.G., 1979, J.Phys.C. Sol.St.Phys., 12, p. 4857.
91. Donovan, K.J. and Wilson, E.G., 1981, Phil.Mag., 44, p. 9.
92. Wilson, E.G., 1980, J.Phys.C. Sol.St.Phys., 13, p. 2885.
93. Wilson, E.G., 1982, J.Phys.C. Sol.St.Phys., 15, p. 3733.
94. Wilson, E.G., 1982, Chem.Phys.Lett., 90, p. 221.
95. Wilson, E.G., 1983, J.Phys.C. Sol.St.Phys., 16, p. 1039.
96. Seiferheld, U., Baessler, H. and Movaghar, B. - to be published.
97. Hunt, I.G., Bloor, D. and Movaghar, B., 1983, J.Phys.C. Sol.St.Phys. - in press.
98. Seiferheld, U., Ries, B. and Baessler, H., 1983, J.Phys.C. Sol.St.Phys. - in press.
99. Siddiqui, A.S. - to be published.
100. Chance, R.R., Baughmann, R.H., Reucroft, P.J. and Takahashi, K., 1976, Chem.Phys., 13, p. 181.

101. Reimer, B. and Baessler, H., 1976, Chem.Phys.Lett., 43, p. 81.
102. Reimer, B. and Baessler, H., 1978, Phys.Stat.Solidi b, 85, p. 145.
103. Movaghar, B. - to be published.
104. Movaghar, B., Murray, D., Donovan, K. and Wilson, E.G., 1983, J.Phys.C.Lett. - in press.
105. Kertesz, M., Koller, J. and Azman, A., 1978, Chem.Phys., 27, p. 273.
106. Karpfen, A., 1980, J.Phys.C. Sol.St.Phys., 13, p. 5673.
107. Suhai, S., 1980, Chem.Phys., 54, p. 91.
108. Bredas, J.L., Chance, R.R., Silbey, R., Nicolas, G. and Durand, Ph., 1981, J.Chem.Phys., 75, p. 255.
109. Yarkony, D.R., 1978, Chem.Phys., 33, p. 171.
110. Gasser, W. and Boerner, M., 1977, Acta.Phys.Polon., A51, p. 61.
111. Cade, N.A. and Young, W., 1979, J.Phys.C. Sol.St.Phys., 12, p. 819.
112. Ukrainskii, I.I., 1981, Phys.Stat.Solidi b, 106, p. 55.
113. Boudreaux, D.S. and Chance, R.R., 1977, Chem.Phys.Lett., 51, p. 273.
114. Dinur, U. and Karplus, M., 1982, Chem.Phys.Lett., 88, p. 171.
115. Bishop, A.R., 1980, Sol.St.Commun., 33, p. 955.
116. Cade, N.A. and Movaghar, B., 1982, J.Phys.C. Sol.St.Phys., 16, p. L807.
117. Cade, N.A. - to be published.
118. Chance, R.R., 1980, Macromolecules, 13, p. 396.
119. Kennedy, R.J., Chalmers, I.F. and Bloor, D., 1980, Makromol. Chem.Rapid.Commun., 1, p. 357.
120. Tieke, B., Bloor, D. and Young, R.J., 1982, J.Mater.Sci., 17, p. 1156.
121. Babbitt, G.E. and Patel, G.N., 1981, Macromolecules 14, p. 554.
122. Chance, R.R., Patel, G.N. and Witt, J.D., 1979, J.Chem.Phys., 71, p. 206.
123. Wenz, G. and Wegner, G., 1982, Makromol.Chem.Rapid.Commun., 3, p. 231.
124. Plachetta, C., Rau, N.O., Hauck, A. and Schultz, R.C., 1982, Makromol.Chem.Rapid.Commun., 3, p. 249.
125. Plachetta, C. and Schultz, R.C., 1982, Makromol.Chem.Rapid. Commun., 3, p. 815.
126. Patel, G.N. and Walsh, E.K., 1979, J.Polym.Sci.Lett.Ed., 17, p. 203.
127. Lim, K.C., Fincher Jnr., G.R. and Heeger, A.J., 1983, Phys. Rev.Lett., 50, p. 1934.
128. Patel, G.N. and Miller, G.G., 1981, J. Macromol.Sci.Phys., B20, p. 111.
129. Bloor, D., 1981, A.C.S. Symposium Series, 162, p. 81.
130. Chance, R.R., Shand, M.L., Le Postollec and Schott, M., 1981, J.Polym.Sci.Polym.Lett. Ed., 19, p. 529.
131. Chance, R.R., Shand, M.L., Hogg, C. and Silbey, R., 1980, Phys.Rev.B, 22, p. 3540.
132. Shand, M.L., Chance, R.R., Le Postelloc, M. and Schott, M., 1982, Phys.Rev.B., 25, p. 4431.

133. Bloor, D., Batchelder, D.N., Ando, D.J., Read, R.T. and Young, R.J., 1981, J.Polym.Sci. Polym.Phys.Ed., 19, p. 321.
134. Tieke, B. and Bloor, D., 1979, Makromol.Chem., 180, p. 2275.
135. Rughooputh, S., Philips, D. and Bloor, D. - to be published.
136. Bhattecharjee, H.R., Preziosi, A.F. and Patel, G.N., 1980, J.Chem.Phys., 73, p. 1478.
137. Movaghar, B., Sauer, G.W. and Wurtz, D., 1981, Sol.St.Commun., 39, p. 1179.
138. Baughman, R.H., Bredas, J.L., Chance, R.R., Elsenbaumer, R.L. and Shacklette, L.W., 1982, Chem.Revs., 82, p. 209.
139. Bloor, D., Hubble, C.L. and Ando, D.J., 1979, NATO Conf.Ser. Ser. VI, 1, p. 243.
140. Nakanishi, H., Hasumi, K., Mizutani, F., Kato, M., Ishimura, K. and Fujishige, S., 1981, Seni's Kob. Zair. Kenk. Kenk.Hap. Shirjo., 120, p. 92.
141. Batchelder, D.N., Poole, N.J. and Bloor, D., 1981, Chem.Phys. Lett., 81, p. 560.
142. Seiferheld, U. and Baessler, H., 1983, Sol.St.Commun. - in press.
143. Sandman, D.J. - to be published.
144. Ferrer-Anglada, N., Hercliffe, R.J., Chalmers, I.F. and Bloor, D. - to be published.
145. Wilson, E.G. - to be published.

DOPED CONJUGATED POLYMERS: THEORY AND EXPERIMENT

R.R. Chance, D.S. Boudreaux, H. Eckhardt,
R.L. Elsenbaumer, and J.E. Frommer

Corporate Research Center, Allied Corporation
Morristown, New Jersey 07960

J.L. Brédas*

Laboratoire de Chimie Théorique Appliquée
Facultés Universitaires Notre-Dame de la Paix
rue de Bruxelles, 61 B-5000 Namur (Belgium)

R. Silbey

Department of Chemistry and Center for Materials
Science and Engineering, Massachusetts Institute
of Technology Cambridge, Massachusetts 02139

ABSTRACT

Recent theoretical and experimental work in the conducting polymers area is reviewed. The specific topics to be discussed are: 1) General experimental aspects of the synthesis and doping of conjugated polymers, 2) Electronic properties of doped conjugated polymers, 3) Conducting polymer solutions, 4) Polyacetylene spectroscopy, 5) Theoretical prediction of electronic and electrochemical properties of conjugated polymers and 6) Charged defect formation and transport in doped polymers.

*Cherchewe qualifié of the Belgian National Science Foundation (FNRS)

J. Ladik et al. (eds.), Quantum Chemistry of Polymers - Solid State Aspects, 221–248.

I. INTRODUCTION

The conducting polymers area has experienced phenomenal growth over the past few years. Interest began with the discovery(1) that polyacetylene, PA, could be "doped" with electron acceptors and electron donors to conductivity levels (~1000 S/cm) approaching those of some metals and comparable to those obtained with organic charge transfer crystals (2). The latter is especially important since it was obvious that doped PA is highly disordered, compared to the single-crystal charge transfer complexes, and would probably provide a tortuous path for carrier transport. Interest, both theoretical and experimental, expanded rapidly. The list of effective dopants for polyacetylene grew(3), drawing from the rather extensive work on dopants in the intercalated graphite area(4). Electrochemical doping of polyacetylene was demonstrated by Nigrey et al.(5) in 1979; this work lead to interest in the application of conducting polymers in rechargeable batteries(6,7). Theoretical work on polyacetylene grew in concert with the heavy experimental effort(8-10). This work benifited from the extensive theoretical work on polyenes dating back over several decades. Particularly notable in this regard is the 1962 work of Pople and Walmsley(11), who anticipated many of the important and interesting aspects of neutral and charged defect formation in polyacetylene.

A significant breakthrough occurred in 1979 with the discovery(12) that poly(p-phenylene), PPP, could be doped to conductivity levels quite comparable to those obtained in the polyacetylene, PA, system. This discovery was important in that it demonstrated the nonuniqueness of the PA system and paved the way for the discovery of a number of new polyaromatic-based conducting polymer systems. These polyaromatics now include poly(p-phenylene sulfide)(13,14), polypyrrole(15), polythiophene (16), and polyquinoline(17). The PPP system was also important theoretically, since the generally similar behavior experimentally(18) of PPP and PA cast doubt on theories of doping and transport which were highly specific to PA, i.e., the soliton theory of Su, Shrieffer and Heeger(8) and later others(19-24) as applied to polyacetylene. A coherent theory of doping and transport applicable to a broad range of conducting polymers is now emerging(25,26), and will be discussed herein.

The understanding of the electronic structure of the parent (undoped) polymers is now at a fairly high level. Many different theoretical techniques(9,10,27-30) have been applied to the various conjugated polymers, with varying degrees of success in predicting, for example, ionization potentials, bandgaps, bandwidths, and photoemission spectra. The ionization potential is especially important, since it determines whether or not a particular electron acceptor is capable of ionizing the polymer chain--the first step in obtaining a conducting polymer. Iodine, for example, will ionize PA to produce a conducting complex, but has little effect on the conductivity of PPP(18). This result is due to the fact that iodine is a relatively weak electron acceptor and PPP has a higher ionization potential than PA. The bandwidth is important, at least in a qualitative sense, in that it gives some indication of the extent of electron delocalization in the system and suggests how mobile the carriers will be once produced by the ionization process. Only a qualitative correlation can be expected, since large geometric changes will occur on ionizing any organic material. The most successful technique developed to date--in terms of accuracy, computational cost, and general applicability--is the Valence Effective Hamiltonian (VEH) technique(31,32). The VEH technique has been broadly applied to hydrocarbon(33-35), nitrogen-containing(36), and sulfur-containing polymers(35,37) with remarkable success. For example, oxidation and reduction potentials(38,39) are now being obtained with the VEH technique which are in very good agreement with experiment; this is important given the current high level of interest in application of conducting polymers in batteries(6,7).

In this review, we will summarize recent experimental and theoretical developments in the conducting polymers area, with emphasis placed on work carried out in the Allied Corporation laboratories. In the next four sections we discuss experimental work, including synthesis and doping of the polymers, conducting polymer solutions, and spectroscopy of neutral and doped polymers. The last two sections deal with theory as applied to understanding the undoped polymer precursors and the possible charged defect structures that result from doping. The latter includes a discussion of a model(26) for "spinless" electrical transport based on bipolarons (dications or dianions). This model is shown to yield a coherent picture of doping and carrier transport of general applicability to the wide range of doped polymer systems.

II. SYNTHESIS AND DOPING OF CONJUGATED POLYMERS

A key development in the evolution of the conducting polymers area is the discovery by Ito et al.(40) of a synthetic route to free-standing, high-quality films of polyacetylene, PA. The synthetic routes for PA and other conducting polymer precursors are illustrated in Figure 1. The synthesis of PA involves a Ziegler-Natta catalyst which is coated onto the walls of the reaction vessel. This coating is exposed to acetylene gas usually at -78C and the reaction proceeds to produce a coherent PA film which coats the catalyst-exposed surfaces. The films, typically a few hundred microns in thickness, can then be peeled off the interior surface of the reaction vessel. The catalyst is usually washed out of the films by extraction with organic

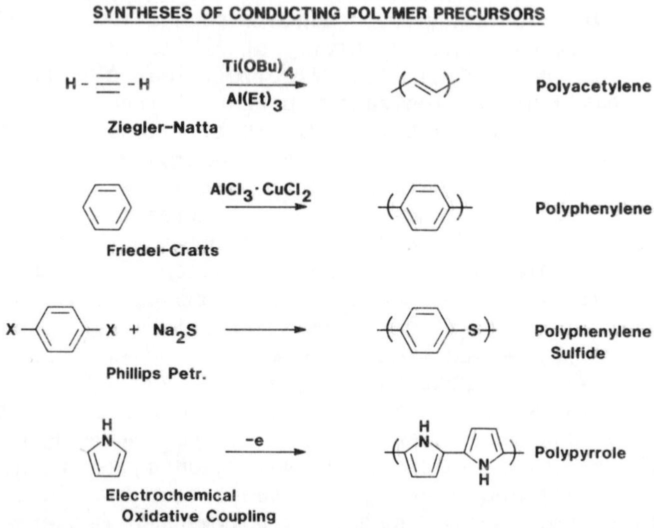

Figure 1. Illustration of synthetic routes to conducting polymer precursors.

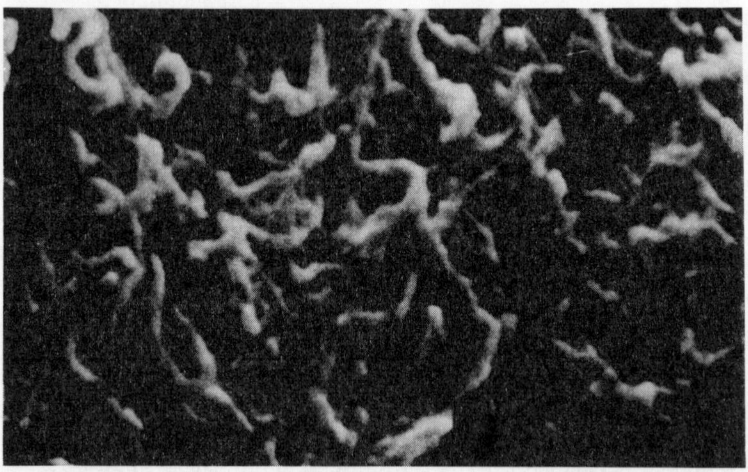

Figure 2. Electron micrograph of polyacetylene film. The average fiber diameter in these films is about 800 A (44).

solvents. PA produced in this mannner consists mainly of the cis conformation. Isomerization to the trans conformation proceeds slowly at room temperature; near complete conversion to the trans isomer is achieved by heating to typically 170C for twenty

trans cis

minutes. The isomerization process is activated by about 15 kcal/mole of CH units (at low conversions)(41); the trans isomer is more stable than the cis by about 10 kcal/mole(10). Doping of the cis isomer with electron acceptors or electron donors also results in isomerization to the trans isomer. The isomerization process, as achieved by heating at least, results in additional disorder in the system probably due to radical initiated cross-linking. This crosslinking produces interchain single bonds which serve to interupt conjugation at the crosslink on each of the chains involved. The resulting conjugation length dispersion has important effects on the electronic properties as evidenced by UV-visible spectroscopy and Raman spectroscopy(42,43). This point will be discussed in Section IV. Disorder is expected to strongly influence both the optical and the electrical properties of the doped compositions.

Figure 2 shows a scanning electron micrograph of a PA film(44). The film is composed of a bundle of fibers, the average fiber diameter being around 800 A. The films are about 70% voids, so that the overall density of the film is about 0.4 gm/cm3 compared to a theoretical density of about 1.2 gm/cm3. This fibular morphology is particularily important in its effect on electrochemical properties, since the resulting high surface area(~60 m^2/gm) aids in the rapid oxidation or reduction of the PA films. In conducting polymer batteries, this effect leads to very high power densities(6,7). The PA films show a relatively high degree of crystallinity--sufficient for a fairly high quality determination of the molecular structure via X-ray diffraction(45). The structural results, and recent NMR results(46), indicate a bond length alternation (difference between double and single bond lengths) in the range 0.08 to 0.10 A for trans PA. Theoretical predictions of the bond length alternation are around 0.1 A according to ab initio results(10) and recent MNDO results(47).

A typical doping curve for PA is shown in Figure 3. These data are taken from the work of Park et al.(48) The results are characteristic of the doping of PA in that they show a S-curve shape with a precipitous rise in conductivity (semiconductor-to-metal transition?) at a doping level of about 2 to 4%. (More

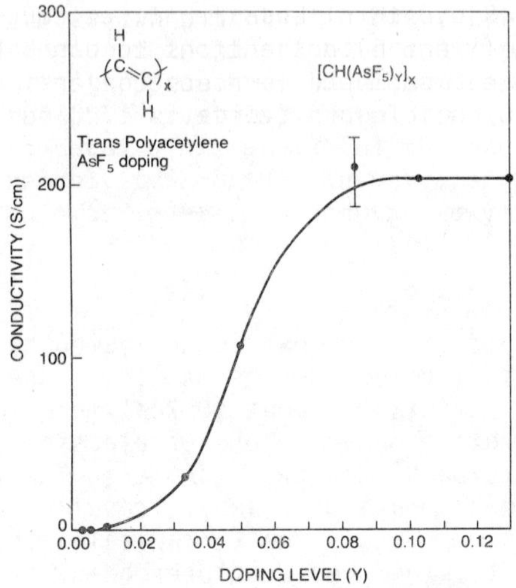

Figure 3. Conductivity versus doping level for the AsF_5 doping of polyacetylene(48).

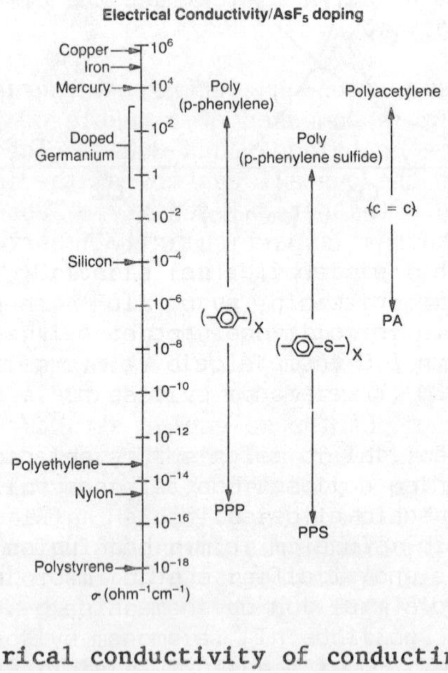

Figure 4. Electrical conductivity of conducting polymers obtained with AsF_5 doping, compared to conductivities of various metallic, semiconducting, and insulating materials.

recent data(49) show an earlier rise in conductivity, a result which is attributed to improvements in the uniformity of the doping.) The final conductivity achieved with AsF_5 doping in this example is about 200 S/cm. More recent work, including work on oriented PA samples, has yielded values in excess of 1000 S/cm (49). The identity of the dopant ion is still not fully resolved, but is probably AsF_6^- with some quantity of AsF_3 incorporated as a solvating agent. The solvating role of AsF_3 will become more apparent in Section IV where we discuss conducting polymer solutions. On doping, AsF_5 probably undergoes the reaction:

$$3\ AsF_5 + 2e \longrightarrow 2\ AsF_6^- + AsF_3$$

Evidence for this reaction is provided by identification of the AsF_6^- species in infrared spectra of the doped polymer and the observation of AsF_3 evolution in the gas phase during doping.

Poly(p-phenylene), PPP, reacts with AsF_5 to produce a composition(18) with conductivity comparable to that of doped PA. This discovery was first announced by Ivory et al.(12) in 1979. PPP is prepared by Friedel Crafts coupling of benzene (Figure 1). The preparative method, extensively studied by Kovacic and coworkers(50), involves low reaction temperatures with an aluminum chloride/cupric chloride catalyst/oxident system and yields almost exclusively the desired para-linked product. Chain lengths are only about 15 to 20 phenyl units, though the chain length is probably increased somewhat on doping with AsF_5(18,51). With these relatively short chains, interchain transport of charge carriers must be very important and perhaps rate-limiting in the conduction process. The same statement is likely to be true for other conducting polymer systems, though chain lengths are generally much longer(52).

On exposure to AsF_5 the conductivity of a PPP powder pressed into a pellet rises rapidly to about 1 S/cm and then more slowly over a period of hours to a final conductivity of about 500 S/cm (18). The kinetics of doping is limited by diffusion of the dopant into the interior of the PPP pellet. Conductivity versus doping level curves have not been obtained for PPP, and would be of limited value given the obvious inhomogeneity of the process. From the data that are available, there is no indication of a semiconductor-to-metal transition and it is expected that the conductivity rises proportionately with doping level.

Poly(p-phenylene sulfide), PPS, is produced by the coupling reaction(53) illustrated in Figure 1. This process is practiced commerically by Phillips Petroleum Company (X=Cl). PPS is a thermally processible, high molecular weight thermoplastic. The discovery of conducting compositions derived from PPS(13,14) was particularly exciting because of its processibility and commercial availability. PPS produces a conducting composition on exposure to AsF_5 in a manner that differs in important ways from the two preceding examples. Conductivity levels are generally lower--at least with conventional techniques (but see Section IV). Typically, conductivities in the range 1 to 10 S/cm are achieved on exposure to AsF_5 vapor. PPS is also distinctly different in that significant chemical modification(54) of the precursor polymer occurs during the doping process. This chemical modification primarily envolves intramolecular bridging to produce planar dibenzothiophene structures. The limit of this process is poly(benzothiophene), which has a substantially lower calculated ionization potential than the parent PPS(37). It is believed that this chemical modification is required to reach the upper limit of conductivity of doped PPS. The inhomogeneity of The inhomogeneity of the doping process is well documented with PPS and will be discussed briefly in Section IV.

Polypyrrole, PPY, has been extensively studied by researchers at IBM(16,55). It is produced by an electrochemical oxidative coupling process which yields an acceptor-doped conducting complex directly, such as $PPY^+BF_4^-$ (Figure 1). The conductivity of this complex is typically 100 S/cm and surprisingly shows a high degree of air stability. (Conducting compositions of PA, PPP, and PPS show poor air stability because of reactions with water vapor.) The ionization potential of PPY is very low--about 0.7 eV lower than PA. This too is surprising since VEH(36,38) and other calculations(9) show that the highest occupied valence band is almost identical to that of cis PA with no contribution from the nitrogen atoms. Nevertheless, the VEH computational technique (as discussed in Section VI) reproduces the low ionization potential of PPY with remarkable accuracy(38).

Conductivities for these four polymers and various donor and acceptor dopants are compiled in Table I. Note that for polymer/dopant systems that achieve reasonable conductivity levels, there is little sensitivity to the actual identity of the dopant. For example, I_2 and AsF_5 doped PA yield essentially the same conductivities. The same is true in donor doping with Li, Na, and K yielding similar conductivity values for PPP and PA (not indicated in the table). Conductivities obtained with donor dopants are generally lower than those with acceptor dopants; for PA, defect calculations(47) suggest that the anion is intrinsically more localized which might lead to a lower mobility for negatively charged carriers. The smaller size of the donor dopant counterions might also tend to localize carriers. Note that I_2 does not ionize PPP and PPS; this is due to the higher ionization potentials(34) of these polymers compared to PA (see Section VI). Alkali metals such as K in Table I fail to dope both PPS and PPY, probably because of the low electron affinities of these polymers(34,38).

Conductivities comparable to those displayed in Table I can also be achieved in many cases by electrochemical doping. In this process the polymer is placed in an electrolyte, $Li^+AsF_6^-$ in propylene carbonate for example, with a counter electrode such as lithium metal(5-7). For acceptor doping, a voltage is applied such that the polymer is made positive with respect to the lithium electrode; at sufficient applied voltage electrochemical oxidation of the polymer occurs. As the polymer is oxidized, the anion, $AsF6^-$, diffuses into the polymer and the cation, Li^+, is reduced to $Li^\bullet$ and plated out on the lithium electrode. This simple doping process will stop when the polymer becomes sufficiently oxidized so that competing processes, such as solvent oxidation or chemical degradation of the polymer, become more favorable energetically than further oxidation of the polymer. For practical conditions, electrochemical doping yields doping levels similar to those displayed in Table I. The open circuit voltage, which measures the difference in chemical potential between the two electrodes,

Table I. Conductivities of polyacetylene[PA], poly(p-phenylene) [PPP], poly(p-phenylene sulfide)[PPS], and Polypyrrole [PPY] with various dopants

Polymer	Dopant Ion (Conc.)[a]	Conductivity(S/cm)	Ref. and Notes
PA	I_3^- (0.10)	550	3
PA	AsF_6^-(0.10)	1100	3
PA	BF_4^-(0.09)	100	3, b
PA	K^+(0.16)	50	56, c, d
PPP	I_3^-	$<10^{-5}$	18
PPP	AsF_6^-(0.4)	500	18
PPP	BF_4^-(0.2)	70	57, b
PPP	K^+(0.6)	20	18
PPS	I_3^-	$<10^{-5}$	54
PPS	AsF_6^-(0.7)	5	54
PPS	BF_4^-	----	
PPS	K^+	$<10^{-5}$	54, b
PPY	I_3^-(--)	600	55
PPY	AsF_6^-(0.3)	100	55
PPY	BF_4^-(0.3)	100	55
PPY	K^+	$<10^{-5}$	55

[a]Probable dopant cation or anion is given along with the molar concentration of the dopant on the basis of moles of that ion to moles of the minimum chemical repeat unit of the polymer,

e.g., CH for PA, C_6H_4 for PPP, C_6H_4S for PPS, and C_4H_3N for PPY.

[b]Doping with BF_4^- achieved by exposure to $NO_2^+BF_4^-$. In PPP the high doping level shown was obtained by further doping with electrochemical techniques(57).

[c]Doping with potassium (and other alkali metals) is usually achieved by exposure to potassium naphthalide solution.

[d]Recent work involving the annealing of donor-doped PA has yielded conductivites up to about 500 S/cm(58).

---

offers fundamental information on the energetics of the doping process and the relative energetics of doping the various polymers. This is particularly true at low doping levels as will become evident when we discuss in Section VI the theoretical prediction of redox potentials of these polymers.

The electrochemical doping process described above provides the basis for the application of conducting polymers in rechargable batteries. The doping process in the above example provides the energy storage mechanism and is therefore the battery charging process. If a load is placed in the external circuit of the charged polymer battery, current will flow and work will be done as the battery discharges i.e., as the polymer "dedopes" (undergoes reduction back to the neutral state) and lithium metal is oxidized. In the battery application the lithium can be replaced by lithium doped polymer to make the so-called "double polymer battery". A third, and lowest voltage, configuration is lithium doped polymer versus lithium, wherein the polymer is doped during discharge and undoped during charge. The principal problems challenging battery applications are solvent stability, self discharge, and polymer stability.

III. ELECTRONIC PROPERTIES OF DOPED CONJUGATED POLYMERS

One of the main driving forces in research in the conducting polymers area is the hope of combining the electronic properties of metals with the materials properties (especially processibility) of plastics. A number of applications can be envisioned involving the displacement of metals, electromagnetic shielding being one of the most attractive. One of the promising features of doped polymers is that the conductivity can be varied in a controlled manner over many orders of magnitude, as illustrated in Figure 4. Two problems hamper the advance toward application: 1) poor environmental stability and 2) loss of desirable mechanical properties on doping. Only in the PPS case is processibility maintained through the discovery(59) of conducting polymer solutions (Section IV). In this section, we will briefly summarize the electronic properties, beginning with some general comments on the doping process followed by a discussion of spectrocopic changes on doping.

Figure 5 illustrates the doping and compensation process and the usual configuration for conductivity measurement. The precursor polymer (an electrical insulator) is mounted with four probes typically in a van der Pauw arrangement(60). The voltage V is measured between two adjacent probes while a current i is passed between the other two. The conductivity σ is determined as $(\ln 2)i/\pi dV$, where d is the sample thickness. In the insulating state σ is very low and strongly temperature activated. For example, in undoped trans PA σ is about 10^{-6} S/cm with an activation energy of about 0.3 eV(56). In undoped PPP, σ is much lower (Figure 4) and the activation energy is higher,~0.7 eV(18). On exposure to an electron acceptor the conductivity increases to, in favorable cases, values in excess of 100 S/cm. The activation energy for σ decreases substantially on doping to near zero for PA and PPP, an indication of metal-like behavior. With acceptor doping, Hall-coefficient measurements(18) clearly show that the conduction is p-type, i.e., positively charged carriers. This is an important point in that it demonstrates that the majority carriers move through the organic material and not through the dopant array. This question arises in the doped polymers because of the relatively high dopant concentrations, much higher than those encountered typically in the doping of inorganic semiconductors. The acceptor-doped polymer can be "compensated" (neutralized) at this point by addition of an electron donating agent, as illustrated in Figure 5. Optimally, this process returns the polymeric material to its starting composition. We have already mentioned that with PPS the doping process results in significant chemical modification(54) of the polymer. The processes illustrated in Figure 5 can be written analagously for donor doping.

The increase in conductivity on doping is accompanied by dramatic changes in optical properties. As expected for such highly conducting materials, substantial free-electron like absorption is observed in the infrared region of spectra of doped polymers. In most cases this absorption effectively masks the infrared spectrum of the polymer. There are also dramatic changes in optical spectra at higher energies, as illustrated in Figure 6(56,51,13,55). The intrinsic absorption of the precursor polymer decreases as new lower energy near-infrared absorptions appear. At high doping levels the bandgap absorption of the polymer [1.5 eV for PA(43,56), 3.4 ev for PPP(18), 3.6 ev for PPS(13), and 3.0 eV for PPY(55)] completely disappears. The spectrum for PA shown in Figure 6 is taken at an intermediate doping level; the new peak at about 0.7 eV appears immediately on

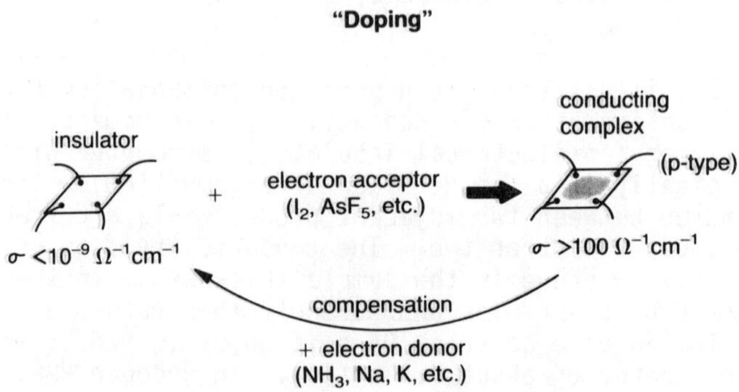

Figure 5. Illustration of the doping and compensation process.

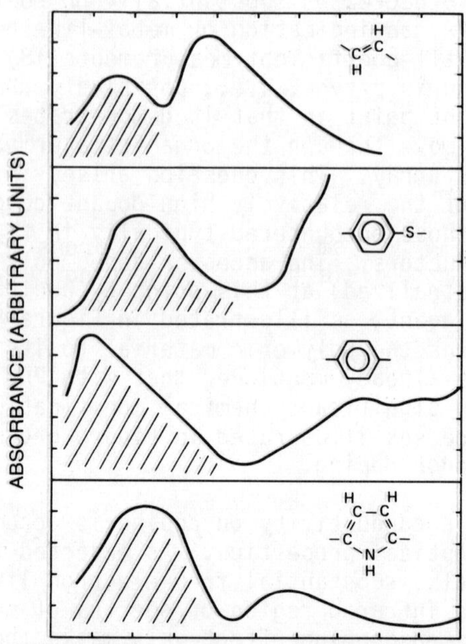

Figure 6. Spectroscopic changes on doping of polacetylene, poly(p-pheneylene sulfide), poly(p-phenylene), and polypyrrole. Data are taken from references 56,51 13, and 55, respectively. The shaded areas show the new transitions which appear on doping.

doping and grows in intensity as the doping level is increased. This peak has been assigned to a "soliton", i.e., to an electronic transition from the valence band of PA to an empty (cation; acceptor case) or filled (anion; donor case) midgap level(8). We will discuss solitons in Section VII. The soliton doping and transport mechanism has been promoted by Su, et al, (8) to explain the observation of spinless transport in doped PA(49,56) using the optical absorption data of Figure 6 as primary evidence for solitons. Bredas et al.(25) have shown this explanation of the optical data to be nonunique, since radical cation or radical anion (polaron) absorption explains not only the PA absorption data, but also those for other polymers to which the soliton theory does not apply.

One attractive aspect of the soliton theory of charge transport is that the carriers (cations or anions) carry no spin, i.e., the conducting compositions do not contain unpaired electrons. ESR experiments on the doping of PA(49) show that in certain intermediate doping regimes the spin concentration is much lower than expected from the observed conductivity values; this phenomenon is referred to as spinless conductivity. If the conduction involved a normal process of defect-induced hole or electron transport, there would be a direct correlation between ESR determined spin concentration and conductivity. The same conclusion of spinless conduction is obtained from ESR experiments on doped PPP(61); however the soliton theory is not applicable to the PPP system(25). In Section VII, we present an alternate transport mechanism based on bipolarons (dications or dianions) which is applicable to all conducting polymer systems(26).

IV. CONDUCTING POLYMER SOLUTIONS

Poly(p-phenylene sulfide), PPS, is the only commericially available processible precursor to a conducting polymer complex. Gas phase doping of PPS films with AsF_5 leads to substantial chemical and physical alteration of the polymer and a substantial loss of the mechanical properties of the parent polymer. In this section we will describe a recently discovered method which allows the retention of processibility in the PPS system.

Frommer et al.(62) have recently shown that the addition of AsF_3 vapor during the doping process enhances the AsF_5 doping rate in the PPS system by a factor of 1000. AsF_3 can act as a weak Lewis acid (AsF_5 is a strong Lewis acid) but alone has no effect on the conductivity of PPS. When AsF_3 is combined with AsF_5, a synergistic effect on the doping rate is obtained as shown in Figure 7. AsF_3 appears to act as a solvating agent for the doped polymer, lowering the energy of the doped complex and allowing enhanced dopant diffusion into the interior of the PPS

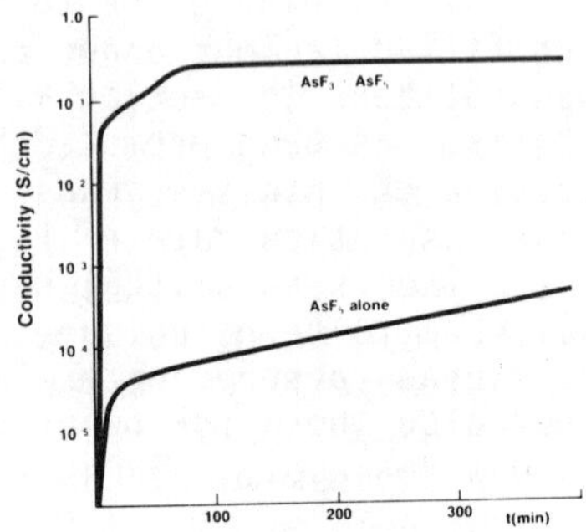

Figure 7. Conductivity versus time in the AsF_5-doping of 0.025 cm thick films of poly(p-phenylene sulfide) illustrating effect of added AsF_3 on doping rate(62).

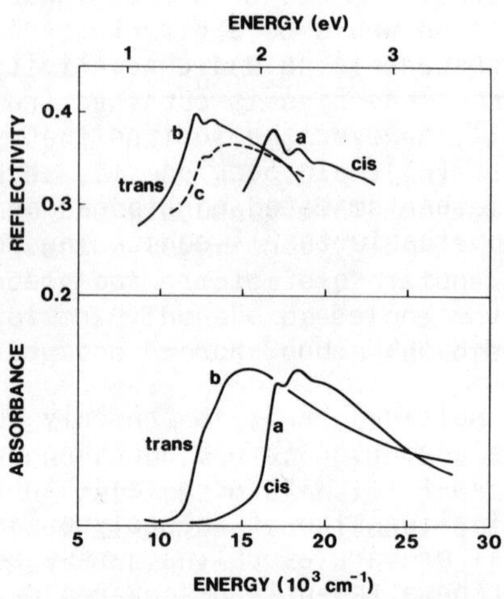

Figure 8. Specular reflection (upper) and absorption (lower) spectra of cis and trans polyacetylene films. The reflection data are taken on thick, freestanding films. Absorption data are taken on thin films (~1000 A). Isomerization conditions are: 170 C, 30 min for curve b (upper), 170 C, 180 min for curve c, and 110 C, 24 hr for curve b (lower). All data are taken from Eckhardt(43).

film. Films doped in this manner show improved mechanical properties compared to those prepared with AsF_5 alone.

In extending the use of AsF_3 from the gas phase to the solution phase, Frommmer et al.(59) discovered the first example of a conducting polymer solution: AsF_5-doped PPS in liquid AsF_3. High quality conducting polymer films can be cast from this solution. These films show electrical properties which are substantially better than previously obtained with conductivities ranging up to 200 S/cm. This conductivity is 10 to 100 times better than that found with gas phase doping. Furthermore, the films show good strength and flexibility compared to the brittle, inhomogeneously doped films obtained on doping melt-molded PPS films in the absence of AsF_3.

The solutions are bright blue in appearance and spectroscopically they are quite similar to the AsF_5-doped PPS films (Figure 6). One might expect that the effective conjugation length for the ionized PPS chain in solution would be substantially shorter than in the solid state because of the more highly disordered chain configurations in the solution phase. This does not seem to be the case--at least to the extent that these shorter conjugation lengths would be reflected in the spectroscopic properties. The delocalization length for a charge or for an excited state on a PPS chain can be estimated to be about 3 to 4 repeat units by comparison to oligomer data. Thus the spectroscopic properties of the conducting polymer solutions suggest that local order on this scale is maintained in the solution phase. This observation further suggests the possibility that the ionized PPS chains in solution are capable of intra-chain conduction. ESR experiments(59,63) on the solution are consistent with this idea in that they show a strong Dysonian lineshape indicative of the transport of conduction electrons in the solution. The basic similarities between optical spectra (both electronic and vibrational) obtained on conventionally doped PPS and on doped PPS solutions confirm the description of the AsF_3 solution as a "conducting polymer solution".

V. SPECTROSCOPY OF POLYACETYLENE

The optical properties of PA have been extensively studied over the past few years(56,64-66). One of the most emphasized results is the contrasting behavior of cis and trans PA. The visible spectrum, reflection or transmission, for cis PA shows a peak at 2.1 eV corresponding to the 0-0 π-π^* transition, followed by a vibrational progression. In contrast, the trans spectrum is broad and unstructured with a peak at about 1.8 eV. In this section we will present new results obtained by Eckhardt (43), which demonstrate that vibrational structure in the reflection spectrum of trans PA can be retained after isomerization from cis PA--if isomerization conditions are carefully controlled. This is an important finding, since it establishes that electron-phonon coupling in the trans PA excited state is much stronger than previously thought, and probably not that different from the cis case. Furthermore, it strongly suggests that the lowest energy optical peak in trans PA is excitonic in origin, a result already supported by photoconduction experiments(66) and electron energy loss spectroscopy(65).

The spectroscopic results obtained by Eckhardt(43) are shown in Figure 8. Reflection spectra are given for thick films of cis and trans PA, the latter being obtained via a careful isomerization at 170C for 30 minutes. It is clear that vibrational structure is maintained on isomerization to trans PA; in fact, the lowest energy peak at 1.5 eV is clearly the most intense, which suggests a 0-0 assignment for this peak. The spectrum which results from excess exposure to high temperatures during isomerization is also shown. The intensity of the 0-0 band is decreased, and the spectrum is evolving towards the broad structureless spectra previously reported for trans PA(56,64,66).

In the bottom portion of Figure 8, transmission spectra for thin films of PA are shown. These are in good agreement with results in the literature(56,64,66). The 0-0 transition evident in the trans PA reflection spectrum appears only as a shoulder (at 1.5 eV) in the transmission spectrum--a result which suggests more disorder in the thin films than in thick films. Etemad et al.(67) have assigned this shoulder as a polaron absorption; that assignment is ruled out by these data.

These results clearly point to the important role played by disorder in trans PA. In most samples studied to date the conjugation lengths are highly dispersed, with an average length on the order of 25 to 50 double bonds(42,43). This conjugation length dispersion is supported not only by the data of Figure 8 (stong absorption at high energies), both also very strongly by resonance Raman spectroscopic studies(42). The defect which disrupts conjugation length is likely to be a carbon-carbon single bond crosslink formed during the radical initiated isomerization process. It is unlikely that charge carriers, which are repelled by chain ends(47), can get by these defects without hopping to the next chain. Thus charge transport over even microscopic distances must importantly involve interchain hopping-probably as the rate-limiting step in conductivity.

VI. THEORETICAL STUDIES OF CONJUGATED POLYMERS

In this section we will briefly summarize the application of the Valence Effective Hamiltonian (VEH) technique to polymers of interest to the conducting polymers area. In a series of recent papers(33-39) we have demonstrated the applicability of the VEH technique in predicting ionization potentials, bandgaps, bandwidths, and X-ray photoelectron emission spectra. Most recently we have extended this work to include oligomers and to calculate the redox properties of oligomers and polymers. We will discuss this recent work in some detail here, since it serves to demonstrate the remarkable utility of the VEH method to the conducting polymers area.

The VEH technique employs atomic potentials derived from double zeta ab initio calculations on small molecules to compute the valence band electronic structure of large molecules(31,32). It is important to emphasize that the VEH technique is completely theoretical. No experimental information has been used to derive the VEH atomic potentials. Nevertheless, the technique yields ab initio double zeta quality results with negligible computer time, since only one electron integrals are evaluated and SCF iterative cycles are completely avoided. The only experimental information necessary is the geometry of the oligomer or polymer being considered. Often in our work this is unknown. For this reason, we employ the MNDO (Modified Neglect of Differential Overlap) semi-empirical procedure(68) to compute geometries; the MNDO geometry then serves as input to the VEH program.

Some of our results are given in Table II for oligomers and polymers in four systems which serve as precursors to conducting polymers: PA, PPP, PT (polythiophene), and PPY. The theoretical results are summarized as gas phase ionization potentials (IP) and bandgaps (E_g) and compared where possible to experimental determinations. Agreement with experiment is quite good, especially with regard to the IP values.

Previous work(33,34,37) with the VEH technique has demonstrated that IP values for the polymers, after a 1.9 eV downward correction for polarization energy, are in good agreement with experimental estimates. We have also considered in some detail the diphenylpolyene series and have reproduced(38,39) to within a 0.1 eV or less the gas phase IP values obtained by Hudson et al. (74) for diphenylpolyenes with up to 8 polyene double bonds. This is remarkably good agreement considering the fact that these results are obtained with a technique that is completely theoretical. We are also able to compute oxidation and reduction potentials for the diphenylpolyenes which are in excellent agreement with experiment(38,39). For the calculation of redox potentials, a scale factor is necessary to adjust from the gas phase IP scale to the Saturated Calomel Electrode (SCE) scale for the electrochemical measurements. We find for the scaling: $IP-6.3 = E_{ox}$ and $IP-E_g-6.3 = E_{red}$, where E_{ox} is the oxidation potential and E_{red}

Table II. Gas Phase Ionization Potentials(IP) and Bandgaps(Eg). All energies are given in eV. Experimental IP values are taken from gas phase experiments; experimental E_g values are taken from absorption peaks in solution spectra.

System	Chain Length(n)	VEH IP	VEH E_g	Experiment IP(ref)	Experiment E_g(ref)
PA	1	10.16	7.85	10.51(69)	~7.5(70)
	2	8.78	5.29	9.06(69)	5.7(70)
	3	8.12	4.09		4.6(70)
	∞	6.67	1.45		1.5(43)
PPP	1	9.30	6.69	9.24(71)	5.9(72)
	2	8.34	4.88	8.32(71)	4.9(72)
	3	7.97	4.18	8.20(71)	4.4(72)
	∞	7.45	3.23		3.6(18)
PT	1	9.29	5.51	8.95(69)	5.4(72)
	2	8.14	3.65		4.1(72)
	3	7.71	2.93		3.5(72)
	∞	7.01	1.71		
PPY	1	8.13	7.01	8.23(73)	6.0(72)
	2	6.89	5.24		4.4(72)
	3	6.43	4.35		3.6(72)
	∞	5.68	2.99		3.0(15)

is the reduction potential. The 6.3 eV correction factor is in good agreement with previous comparisons of experimental IP values with E_{ox} values (75). Using this same scale factor we also find good agreement between VEH predictions and observed redox potentials for the other oligomers considered in Table II.

Having established the reliablity of the VEH predictions for the redox properties of oligomers, we may now consider the important problem of predicting the redox properties of the polymers. This is important because these redox potentials will determine the voltage characteristics of battery cells constructed with the polymers, i.e. the open circuit voltage (V_{OC}) of the battery. For example, V_{OC} for a battery with acceptor-doped PA for one electrode and Li for the other should be roughly the difference between E_{ox} for PA and E_{red} for Li (E_{red} same as E_{ox} for Li). Similarly, V_{OC} for the double polymer battery mentioned earlier should be roughly E_{ox}-E_{red} for polymer, which is just E_g by construction; experiments show that this is approximately true for both PA and PPP(5-7).

Table III shows a comparison of VEH theory and experiment for the oxidation and reduction potentials of the four polymers considered in this section. It should be pointed out that the calculations refer to the first oxidation or reduction potentials of the polymers (initial removal or addition of one electron). Therefore, E_{ox} or E_{red} would correspond to onset values in terms of the observed oxidation or reduction of the polymers. The experimental values shown in Table III are derived accordingly (38,39).

Table III. Comparison of VEH predictions of oxidation and reduction potentials (volts versus SCE) to those derived from experimentally observed oxidation and reduction onsets.

Polymer	Oxidation Potential		Reduction Potential	
	VEH Theory	Exp.(ref)	VEH Theory	Exp.(ref)
PA	0.4	0.2(57,76)	-1.1	-1.3(57,76)
PPP	1.2	0.9(57)	-2.1	-2.1(57)
PTP	0.7	0.6(71)	-1.0	---
PPY	-0.6	-0.4(55)	-3.6	->2.9(77)

The agreement between theory and experiment shown in Table III is quite good; in fact the agreement is probably within the uncertainity of the experimental determination of the redox onsets. It is especially important to point out that the surprisingly low oxidation potential of PPY is predicted by VEH theory with good accuracy.

In summary the VEH technique has proven to be a powerful tool in the calculation of the electronic properties of conjugated polymers. Though intended only for application to the valence band related properties such as IP, we also find that in many cases the lowest energy optical transition is computed with good reliability.

VII. NEUTRAL AND CHARGED DEFECTS IN CONJUGATED POLYMERS

Most of the theoretical work in the conducting polymers area has been concerned with neutral and charged defects on the polymer chains. We will not attempt to review this work completely, but instead will discuss some of the general aspects of the problem and recent results from our laboratory, including an emerging coherent picture of doping and carrier transport applicable to all conducting polymer systems. The discussion will concentrate on PA and PPP.

Figure 9 shows the various charged defects that could be formed on ionization of PA or PPP by an electron acceptor, A: polarons (radical cations), bipolarons (dications), and charged solitons (cations). Most of the work in the literature has been concerned with solitons, both neutral (radical) and charged. This work dates back to Pople and Walmsley(11), who recognized that a radical on a PA backbone would be able to migrate isoenergetically along the PA chain because the structures on either side of the defect have the same energy. This idea was extended and promoted by Su et al.(8) in the language of field theory as an explanation of much of the interesting physics of the PA system. Of special interest was the possibility of charge transport via charged solitons. This idea was appealing because of the experimental observation of spinless transport, i.e. the ESR measurement(49) of a spin concentration too low to account for the observed level of conductivity. However, as we have already pointed out the same experimental observation is made for PPP(61). PPP cannot support solitons since the structures on either side of the charged defect would not be of the same energy. This is illustrated in Figure 9 for the cation defect on PPP; the benzoid structure to the left of the defect is significantly lower in energy than the quinoid structure to the right. Thus, the soliton theory of doping and transport is unappealing because of its lack of generality to the numerous conducting polymer systems other than PA.

We began our work in this area by asking the question: What defect structure do we expect on removal of one electron from a conjugated polymer, such as PPP or PA? Previous work(8) would imply that a one electron ionization of PA would yield an unbound charged soliton/neutral antisoliton pair. This picture was obviously not valid for PPP, since the defect pair must be bound due to the nondegenerate ground state structure of PPP. In other words, the ionization event in PPP should produce a polaron with a structure like that shown in Figure 9. In fact, we have been able to show(24,25) that the same is true for PA, and that the defect pair is strongly bound in forming a polaron. Our calculations, which use Huckel theory and a construction very similar to that employed by Su et al.(8), predict that ionization of either PA or PPP yields a polaron. We are also able to show(24,25) that in both PA and PPP two polarons will interact to produce a bipolaron (Figure 9), but with an activation barrier dependent on defect concentration. Once polarons interact to form bipolarons in PA the charged defects can then move essentially independently as charged solitons. (However, the dopant counterions would tend to pin the charged defects near the ionization site.) In PPP the

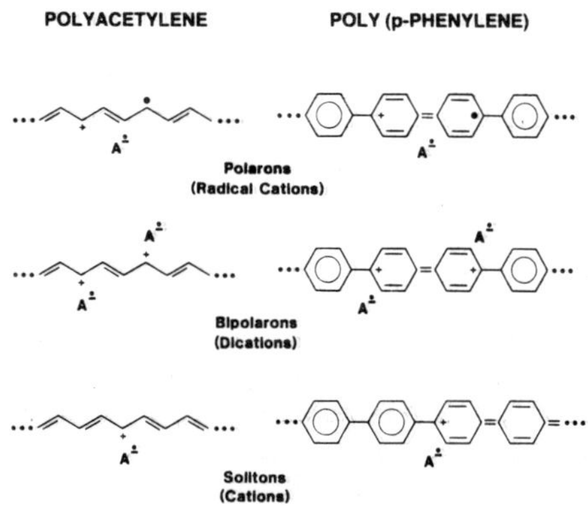

Figure 9. Charged defect structures on trans polyacetylene and poly(p-phenylene).

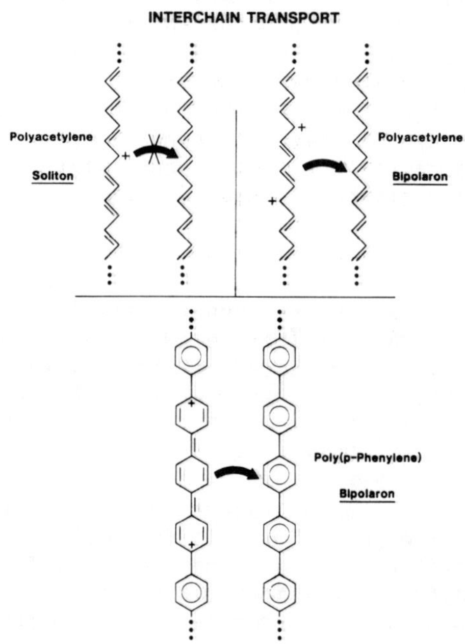

Figure 10. Schematic illustration of interchain transport(26) in polyacetylene and poly(p-phenylene).

bipolaron would remain as a well-defined entity since the two charged defects are bound (or in the language of field theory confined) due to the nondegenerate ground state.

How can one get spinless transport in these systems? In the PPP case the only likely possiblity is the bipolaron species depicted in Figure 9. Spinless transport in PA has been attributed to charged solitons(8); this assignment has immediate limitations when interchain transport is considered. It is now well-established that disorder in the PA system requires that interchain transport be facile for efficient carrier migration. Consider an infinite PA chain containing one charged soliton with a neighboring, defect-free chain, as depicted in Figure 10. This soliton cannot jump to the next chain since an infinitely large activation barrier is required in the reorganization of bond lengths that would ensue from such a jump. The soliton mechanism is also unattractive for finite chains since the soliton cannot jump isoenergetically in that case. Kivelson(78) has discussed a model for interchain transport that gets around the above problems by requiring a neutral soliton on the next chain. This model would be applicable only at very low doping levels, since any neutral solitons that are present in PA (trapped between crosslinks) would be immediately ionized on further doping. Thus Kivelson's model has limited applicability and cannot address the problem of spinless transport observed at doping levels as high as 2%(49). In any case, Kivelson's theory ignores the fact that a neutral and a charge soliton form a bound state, a polaron, with a binding energy of about 0.3 eV (24,25).

The above problems are eliminated by a model in which there are two solitons (a bipolaron) on an infinite chain. The soliton pair can jump to the next chain isoenergetically and with a reasonable activation barrier (Figure 10). The jump probability for two solitons separated by x carbon atoms, P(x), will contain the following elements:

$$P(x) = F(x)\, Q_1(x)\, Q_2(x) \qquad (1)$$

where F(x) is the Franck Condon factor (or square of the vibrational overlap) for the jump; $Q_1(x)$ is the probability of finding a second charged soliton x carbon atoms from the first on a single chain; and $Q_2(x)$ is the probability of finding a suitable (unoccupied) site on chain 2 to receive the x-length bipolaron. F(x) would be expected to decrease rapidly as x increases. $Q_1(x)$ is expected to be peaked at roughly 1/C, where C is the concentration of solitons, due to the coulombic repulsion between solitons of like charge. $Q_2(x)$ is derivable from $Q_1(x)$, and would lead to saturation of the spinless (bipolaron) contribution to the conductivity at high C. The spinless bipolaron contribution to the conductivity, σ_{++}, can be computed by integrating Eq. 1

over x and multiplying by C:

$$\sigma_{++} = C \int P(x)\, dx \qquad (2)$$

We have modeled this process and computed conductivity versus C curves which are very similar to those obtained experimentally (Figure 3). For the purposes of this discussion we will only consider qualitative aspects of the results of this model. At low C, there will be few soliton pairs of sufficiently small x for transport to take place due to the F(x) term; i.e., F(x) at the peak in the $Q_1(x)$ distribution will be small. As C increases, σ_{++} will increase superlinearly since the peak in $Q_1(x)$ will shift to lower x values where F(x) is larger. At still larger C values, σ_{++} will level off and eventually decrease due to the small value of the $Q_2(x)$ term, as there will now be few sites available for accepting the bipolaron. Thus, the net result from this model, qualitatively, is an S-shaped curve for the spinless bipolaron contribution to the conductivity. The detailed calculations we have carried out with this model(26) bear out this qualitative discussion and yield results quite similar to the experimental data shown in Figure 3.

For PPP, the situation is quite different. In that case, the two charged defects making up the bipolaron are bound. Our calculations(24,25) indicate that the bipolaron will extend over about five rings in PPP (though we ignore any coulombic effects). Thus, in our treatment the bipolaron is a well-defined entity of limited spatial extent in PPP, so that the F(x) term becomes a constant as does the $Q_1(x)$ term. Therefore, the spinless contribution to the conductivity will rise in proportion to C, eventually saturating and decreasing when the Q_2 term becomes small. No experimental data for conductivity versus doping level exists for PPP--the problem being the inhomogeneity of the doping process. There is some experimental indication of the existence of bipolarons in AsF_5-doped PPP(79).

Bipolaron transport offers a reasonable explanation for spinless transport in PA and PPP. The model is not specific to PA, in that it does not require a degenerate ground state. Our model for charge transport via bipolarons in PPP is applicable to the broad range of conducting polymer systems. It should be emphasized, however, that we have made no explicit consideration of the effect of the dopant ion array on the transport process.

Application of this model obviously assumes the reliability of the observation of spinless transport, for which the evidence is strong(49,80) but not universally accepted. In particular, Epstein et al.(81) suggest that in the I_2 doping of PA the magnetic and electrical data can be adequately explained by a model based on variable range hopping. Their model requires a finite density of states at the Fermi level and is based on the theoretial work of Mele and Rice(82) who argue that a finite density of states will be induced at the Fermi level by disorder. The most recent data from Chung et al.(80) argues quite strongly that in Na doped PA the density of states is much too small to account for observed conductivities. Thus, though the question is not entirely resolved, the experimental evidence still favors spinless transport.

To summarize, we offer the following qualitative picture of the doping and transport process of general applicability to conducting polymer systems. First, the dopant, which we will take to be an acceptor, ionizes the chain to produce a polaron (radical cation). This polaron is pinned to the ionization site and does not contribute significantly to conductivity. As more dopant is added, the chain is ionized further and a higher concentration of polarons is formed. Also, the polarons can be further ionized to produce a bipolaron (dication), the polaron being more easily ionized than the polyacetylene chain. As the polaron concentration increases, polarons interact to produce bipolarons, which are uncorrelated charged solitons in the PA case and correlated dications in PPP and other cases. The bipolarons transport charge via interchain hoping, and are responsible for the observed spinless conductivity. At still higher doping levels, there is little left of the electronic structure of the original polymer; geometric distortions due to the charged defects overlap spatially along the polymer chain so that the energy gap between occupied and unoccupied states is eliminated(83). At this point conventional conductivity involving carriers with spin sets in. Note again that this overall picture applies equally well to all conducting polymer systems.

REFERENCES

1. H.Shirakawa, E.J.Louis, A.G.MacDiarmid, C.K.Chiang, and A.J.Heeger, J. Chem. Soc. Chem. Commun. 578 (1977).
2. K.Bechgaard, Mol. Cryst. Liq. Cryst. 79, 357 (1982).
3. S.C.Gau, J. Milliken, A. Pron, A.G.MacDiarmid, and A.J.Heeger, J. Chem. Soc. Chem. Commun. 1979, 662 (1979).
4. J.E.Fischer, Mat. Sci. Eng. 31, 211 (1977).
5. P.J.Nigrey, A.G.MacDiarmid, and A.J. Heeger, J. Chem. Soc. Chem. Commun. 1979, 593 (1979).
6. D.MacInnes, Jr., M.A.Druy, P.J.Nigrey, D.P.Nairns, A.G. MacDiarmid, and A.J.Heeger, J. Chem. Soc. Chem. Commun. 316 (1981).
7. L.W.Shacklette, R.L.Elsenbaumer, R.R.Chance, J.M.Sowa, D.M. Ivory, G.G.Miller, R.H.Baughman, J. Chem. Soc. Chem. Commun. 361 (1982).
8. W.P.Su, J.R.Schrieffer, and A.J.Heeger, Phys. Rev. Lett. 42, 1698 (1979); Phys. Rev. B 22, 2209 (1980).
9. C.B.Duke and A.Paton, in Conductive Polymers, edited by R.B. Seymour Plenum, New York, 1981), p.155; W.K.Ford, C.B.Duke, and A.Paton, J. Chem. Phys. 77, 4564 (1982).
10. A.Karpfen and J.Petkov, Theor. Chim. Acta 53, 65 (1979).
11. J.A.Pople and S.H.Walmsley, Mol. Phys. 5, 15 (1962).
12. D.M.Ivory, G.G.Miller, J.M.Sowa, L.W.Shacklette, R.R.Chance, and R.H.Baughman, J. Chem. Phys. 71, 1506 (1979).
13. R.R.Chance, L.W.Shacklette, G.G.Miller, D.M.Ivory, J.M.Sowa, R.L.Elsenbaumer, and R.H.Baughman, J. Chem. Soc. Chem. Commun. 348 (1980).
14. J.F.Rabolt, T.C.Clarke, K.K.Kanazawa, J.R.Reynolds, and G.B. Street, J. Chem. Soc. Chem. Commun. 347 (1980).
15. K.K.Kanazawa, A.F.Diaz, R.H.Geiss, W.D.Gill, J.F.Kwak, J.A. Logan, J.F.Rabolt, and G.B.Street, J. Chem. Soc. Chem. Commun. 854 (1979).
16. A.F.Diaz, J.Crowley, J.Bargon, G.P.Gardini, and J.B. Torrance, J. Electroanal. Chem. 121, 355 (1981).
17. Y.S.Papir, V.P.Kurkov, and S.P.Current, in Extended Abstracts (Electrochemical Society Meeting, San Francisco, 1983) p.820.
18. L.W.Shacklette, R.R.Chance, D.M.Ivory, G.G.Miller, and R.H. Baughman, Synth. Met. 1, 307 (1979).
19. M.J. Rice, Phys. Lett. A 71, 152 (1979); M.J.Rice and J. Timonen, Phys. Lett. A 73, 368 (1979).
20. J.Tinka Gammel and J.A.Krumhansl, Phys. Rev. B 24, 1035 (1981).
21. A.R.Bishop, D.K.Campbell, and K.Fesser, Mol. Cryst. Liq. Cryst. 77, 253 (1981).
22. Y.P.Lin-Liu and K.Maki, Phys. Rev. B 22, 5754 (1980).
23. J.P.Albert and C.Jouanin, Phys. Rev. B 26, 955 (1982).
24. J.L.Brédas, R.R.Chance, and R.Silbey, Mol. Cryst. Liq. Cryst. 77, 319 (1982).

25. J.L.Brédas, R.R.Chance, and R.Silbey, Phys. Rev. B 26, 5843 (1982).
26. R.R.Chance, J.L. Brédas and R.Silbey, to be published.
27. M.H.Whangbo, R.Hoffmann, and R.B.Woodward, Proc. Roy. Soc., London, A 366, 23 (1979).
28. J.L.Brédas, B.Themans, and J.M.Andre, Phys. Rev. B 26, 6000 (1982).
29. J.L.Brédas, R.R.Chance, and R.Silbey, J. Phys. Chem. 85, 756 (1981).
30. S.Suhai and J.Ladik, Mol. Cryst. Liq. Cryst. 83, 199 (1982).
31. G.Nicolas and Ph.Durand, J. Chem. Phys. 70, 2020 (1979); 72, 463 (1980).
32. J.M.André, L.Burke, J.Delhalle, G.Nicolas, and Ph.Durand, Int. J. Quantum Chem. Symp. 13, 283 (1979).
33. J.L.Brédas, R.R.Chance, R.Silbey, G.Nicolas, and Ph.Durand, J. Chem. Phys. 75, 255 (1981).
34. J.L.Brédas, R.R.Chance, R.H.Baughman, and R.Silbey, J. Chem. Phys. 76, 3673 (1982).
35. J.L.Brédas, R.R.Chance, R.Silbey, G.Nicolas, and Ph.Durand, J. Chem. Phys. 77, 371 (1982).
36. J.L.Brédas, B.Themans, and J.M.Andre, J. Chem. Phys. 78, 6137(1983).
37. J.L.Brédas, R.L.Elsenbaumer, R.R.Chance, and R.Silbey, J. Chem. Phys. 78, 5656 (1983).
38. R.R.Chance, D.S.Boudreaux, J.L.Brédas, and R.Silbey, Org. Coat. Appl. Polym. Sci. 48, 536 (1983).
39. J.L.Brédas, R.Silbey, D.S.Boudreaux, and R.R.Chance, J. Amer. Chem. Soc. in press.
40. T.Ito, H.Shirakawa, and S.Ikeda, J. Polym. Sci. Polym. Chem. Ed. 12, 11(1974).
41. B.Francois, M.Bernard, and J.J.André, J. Chem. Phys. 75, 4142(1981).
42. F.B.Shuegerl and H.Kuzmany, J. Chem. Phys. 74, 953(1981); S.Lefrant, Proceed. Int. Conf. Cond. Polym., Les Arcs (1982) J. Physique, in press.
43. H.Eckhardt, J. Chem. Phys. in press.
44. R. Elsenbaumer, unpublished results. See, also, ref. 56.
45. C.R.Fincher, C.E.Chen, A.J. Heeger, A.G. MacDiarmid, and J. B.Hastings, Phys. Rev. Lett. 48, 100(1982).
46. T.C.Clarke, R.D.Kendrick, and C.S.Yannoni, Proceed. Int. Conf. Cond. Polym., Les Arcs (1982); J. Physique, in press.
47. R.R.Chance, D.S.Boudreaux, J.L.Brédas, and R.Silbey, Phys. Rev. B 27, 1440(1982).
48. Y.W.Park, A.J.Heeger, M.A.Druy,and A.G.MacDiarmid, J. Chem. Phys. 73, 946(1980).
49. D.Moses, A.Denenstein, J.Chen, A.J.Heeger, P.McAndrew, T. Woerner, A.G.MacDairmid, and Y.W.Park, Phys. Rev. B 25, 7652(1982).
50. P.Kovacic and A.Kyriakis, J. Am. Chem. Soc. 85, 454(1963); P.Kovacic and J.Oziomek, J. Org. Chem. 29, 100(1964).

51. L.W.Shacklette, H.Eckhardt, R.R.Chance, G.G.Miller, D.M. Ivory, and R.H.Baughman, J. Chem. Phys. 73, 4098(1980).
52. J.C.W. Chen, J.D. Capistran, F.E. Karasz, L.C. Dickerson, and M.A. Schen, J. Polym. Sci. Polym. Phys. Ed. 21,93(1983).
53. J.T.Edmonds,Jr. and H.W.Hill,Jr., U.S.Patent #3354129, 1967.
54. L.W.Shacklette, R.L.Elsenbaumer, R.R.Chance, H.Eckhardt, J.E.Frommer, and R.H.Baughman, J. Chem. Phys. 75, 1919 (1981).
55. G.B.Street, T.C.Clarke, M.Kroumbi, K.Kanazawa, V. Lee,P. Fluger, J.C.Scott, and G.Weiser, Mol. Cryst. Liq. Cryst. 83, 253(1982).
56. A.G.MacDiarmid and A.J.Heeger, Synth. Met. 1, 101(1980).
57. L.W.Shacklette, unpublished results.
58. H.Eckhardt, P.Delannoy, G.G.Miller, N.S.Murthy, and R.H. Baughman, Proceed. 12th NATAS/38th Calorimetry Conf. Williamsburg VA, Sept. 1983.
59. J.E.Frommer, R.L.Elsenbaumer, and R.R.Chance, Org. Coat. Appl. Polym. Sci. 48, 552(1983).
60. For measurement technique, see H.H.Wieder in Laboratory Notes on Electrical and Galvanomagnetic Measurements, Materials Science Monographs (Elsevier, Amsterdam,1979).
61. M.Peo, S.Roth, K.Dransfeld, B.Tieke, J.Hocker, H.Gross, A.Grupp, and H.Sixl, Solid State Comm. 35, 119(1980).
62. J.E.Frommer, R.L.Elsenbaumer, H.Eckhardt, and R.R.Chance, J. Polym. Sci. Polym. Lett. Ed. 21, 39(1983).
63. J.E.Frommer, R.L.Elsenbaumer, to be published.
64. S.Etemad, A.J.Heeger, and A.G.MacDiarmid, Ann. Rev. Chem. 33, 443(1982).
65. J.J.Ritsko, Phys. Rev. B 26, 2192(1982).
66. M.Tanaka, A.Watanabe, and J.Tanaka, Bull. Chem. Soc. Jpn. 53, 3430(1980).
67. S.Etemad, A.Feldblum, A.G.MacDiarmid, T.C.Chung, and A.J. Heeger, Proceed. Int. Conf. Cond. Polym., Les Arcs (1982); J. Physique, in press.
68. M.J.S.Dewar and W.Thiel, J. Am. Chem. Soc. 99, 4899(1977).
69. G.Herzberg, Electronic Spectra of Polyatomic Molecules (Van Nostrand, NY, 1976) pp.629,654,656.
70. B.S.Hudson, B.E.Kohler, and K.Schulten in Excited States Vol. 5 (ed. by E.C.Lim, Academic Press, 1982) pp. 1-95.
71. J.Maier and D.W.Turner, Farad. Disc. 54, 149(1972).
72. A.F.Diaz, J.Crowley, J.Baryon, G.P.Gardini, and J.B.Torrance J.Electroanal. Chem. 121, 355(1981).
73. W.Ford, C.B.Duke, and W.R.Salaneck, J. Chem. Phys. 77, 5020 (1982).
74. B.S.Hudson, J.Ridyard, and J.Diamond, J. Am. Chem. Soc. 98, 1126(1976).
75. L.L.Miller, G.D.Nordblom, and E.A.Mayeda, J. Org. Chem. 37, 916(1972).
76. P.J.Nigrey, A.G.MacDiarmid, and A.J.Heeger, Mol. Cryst. Liq. Cryst. 83, 309(1982).

77. G.B.Street, personal communication. The reduction potential of polypyrrole is known to be more negative than 2.9V versus SCE, since sodium naphthalide solution does not act as a doping agent for this polymer.
78. S.Kivelson, Phys. Rev. Lett. 46, 1344(1981).
79. G. Crecelius, M. Stamm, J. Fink, and J.J. Ritsko, Phys. Rev. Lett. 50, 1498(1983).
80. T.C.Chung, F.Moraes, J.D.Flood, and A.J.Heeger, preprint.
81. A.J.Epstein, H.Rommelmann, R.Bigelow, H.W.Gibson, D.M. Hoffmann, and D.B.Tanner, Phys. Rev. Lett. 50, 1866(1981).
82. E.J.Mele and M.J.Rice, Phys. Rev. B 15, 5397(1981).
83. J.L.Brédas, B.Thermano, J.M.André, R.R.Chance, D.S. Boudreaux, and R.Silbey, Proceed Int. Conf. Conducting Polymers, Les Arcs (1982); J. Physique, in press.

RESONANT INTERACTION BETWEEN LASER PULSES AND SURFACE LAYERS

Joachim Heidberg

Institut für Physikalische Chemie und Elektrochemie der Universität Hannover, D-3000 Hannover 1, F.R.Germany

Localized chemical processes, such as desorption and ablation, stimulated by resonant laser pulse-surface layer interaction have been discovered recently. In this lecture the essential theoretical features of the desorption induced by resonant excitation of adsorbate vibrations with laser infrared and their influence on yield, rate, and quantum efficiency are presented. Results on selective damage to pigmented biological structures by short resonant optical and ultraviolett laser pulses are briefly reported.

INTRODUCTION

In the interaction between laser radiation and matter, localisation and substance specificity can be achieved by appropriate selection of wavelength, intensity and duration of interaction. Also the yield of the photoreaction is determined by these quantities. Especially vibrational molecular excitation exhibits a rich structure rendering high specificity in radiation-molecular coupling, as is well established in the chemical analysis of matter. This holds particularly for surface layers. The possibility of using this high specificity in the selective excitation of chemical reactions has been enhanced by the discovery that monochromatic intense and coherent infrared radiation can deposit tens of photons in isolated molecules and induce isotope-selective processes. To elucidate whether a specific process may be promoted over others by exciting a particular molecular vibration which is related to the reaction coordinate

J. Ladik et al. (eds.), Quantum Chemistry of Polymers - Solid State Aspects, 249–258.

is of significance indeed.

Very fast energy dissipation after vibrational excitation in the electronic ground state, however, has been inferred from picosecond relaxation measurements on large molecules in the gas phase. Moreover the view has been adopted that in solids even at low temperature rapid energy dissipation prevents multiphoton excitation and vibrational ladder climbing in matrix isolated molecules. On the other hand, in recent experiments surface reactions, such as desorption, evaporation and molecular decomposition, stimulated by vibrational multi-quantum excitation with resonant laser infrared have been observed at moderate threshold laser intensity and with high frequency selectivity.

Selectivity in laser light-matter interaction can be accomplished also in the visible and ultraviolet spectral region. In any case, an absolute requirement is that the targets have greater optical absorption at the employed wavelength than the surrounding medium. In photochemistry it can be advantageous that for electronic excitation the absorption cross section is in general higher and the randomization of energy proceeds slower than for vibrational excitation. Moreover, focusing of light and local macroscopic control of reactions is the sharper the shorter the wavelength. Substance specificity, however, is less pronounced.

The laser intensity determines the rate of the excitation process. Certain threshold intensities are required in order to promote chemical surface processes at observable rates and yields. For evaporation and desorption of neutrals with CO_2 laser pulses threshold intensities of about 10^5 Wcm^{-2} were measured. At 10^7 to 10^9 Wcm^{-2} very fast desorption and evaporation takes place at rates of about 10^7 monolayers per second and more, followed by ionization probably just above the surface due to chemical reactions such as proton transfer and attachment of alkali metal ions. Saturation is expected at a certain intensity, above which the process considered cannot be promoted further.

No matter how judiciously one has selected the laser wavelength, specificity will suffer, if the duration of laser light-matter interaction is too long. During long exposures, energy relaxation occurs and the entire system is heated relatively uniformly, causing non-specific thermal processes, even though energy absorption occurs initially at certain localized sites.

It is the intricate competition between excitation, tunnelling from the excited state into the reaction coordinate and relaxation which determines selectivity and yield of the photon induced process. In this lecture we will confine our attention to describing the essential features of photodesorption induced by resonant laser-adsorbate vibrational coupling as well as presenting some illustrative experimental examples for the desorption from ionic and metal surfaces. Finally a brief perspective of applications of resonant interactions between laser pulses and biopolymers will be presented.

LASER INFRARED INDUCED DESORPTION

A laser infrared pulse striking a surface covered with an adsorbate can pump energy into

1) the adsorbate,
2) the adsorbate - adsorbent bond,
3) the adsorbent, inducing a reaction.

Process (1) was first observed in the evaporation and desorption of SF_6 molecules from NaCl surfaces induced by resonant CO_2 laser pulses of 200 nanoseconds duration in 1978.(1) To understand process (2) we bear in mind that desorption takes place, when the instantaneous vibrational energy in the bond, which ties the molecule (adsorbate) to the surface (adsorbent), exceeds the binding energy. In thermal desorption phonon energy is transferred from the adsorbent to the vibration of the adsorptive bond. (2,3) In principle, also a laser infrared pulse should be able to deposit sufficient energy into the vibration of the adsorptive bond so that the molecule desorbs. However, the vibration of the adsorptive bond will be very anharmonic, particularly at high excitation, so that desorption induced by monochromatic infrared will be very unlikely, except at enormous laser intensities, calculated to be in the order of $> 10^{13}$ Wcm^{-2}.(3) In addition there must be considerable broadening of the vibrational energy levels of the adsorptive bond. Then process (2) is of little resonant character. In general process (3), the direct interaction between the adsorbent and laser infrared, will heat up the system leading to thermal desorption. This is particularly true for metals, where extremely fast relaxation of electronic excitation into the phonon bath within 10^{-13}s takes place. Laser induced thermal desorption is possible at all laser frequencies at which appreciable absorption of light occurs in the adsorbent.(4)

DESORPTION BY RESONANT LASER-ADSORBATE VIBRATIONAL COUPLING

Let us consider process (1) in which laser infrared is resonantly coupled into an internal vibration of the adsorbed molecule followed by desorption. This process has been investigated in the systems CH_3F on NaCl, SF_6 on NaCl by Heidberg et al. (5,1) and for pyridine on KCl and silver by Chuang et al.(6) It has been theoretically treated in an illuminating way by Kreuzer and coworkers (7) and Ewing and coworker.(8) We shall partly adopt Kreuzer's description of the CH_3F-NaCl experiment. According to the experiment the frequency of the internal vibration of the adsorbed molecule is much larger than the frequency of the vibration of the adsorptive bond, i.e. the energy $\hbar\omega$ of the internal vibrational fundamental is much larger than the spacing between the energies $E_{i+1} - E_i$ of bound states in the adsorption potential. Describing the internal mode as a harmonic oscillator for the lowest excited states in concert with experiments on CH_3F trapped in solid rare gas matrices, the energy of the adsorbed molecule is

$$E_i^v = E_i + (v + \frac{1}{2})\hbar\omega \ , \tag{1}$$

where v counts the number of quanta in the internal vibration of the adsorbed molecule. A laser tuned to the oscillator's frequency (see fig.1) can induce transitions $v \rightarrow v \pm 1$ through absorption or emission of a photon. At low temperature the adsorbed molecule will be in the ground state of energy

$$E_o^o = E_o + \frac{1}{2}\hbar\omega \ . \tag{2}$$

Upon absorption of a laser photon the molecule will be in a state E_o^1 from which it can either absorb more photons to go up to higher

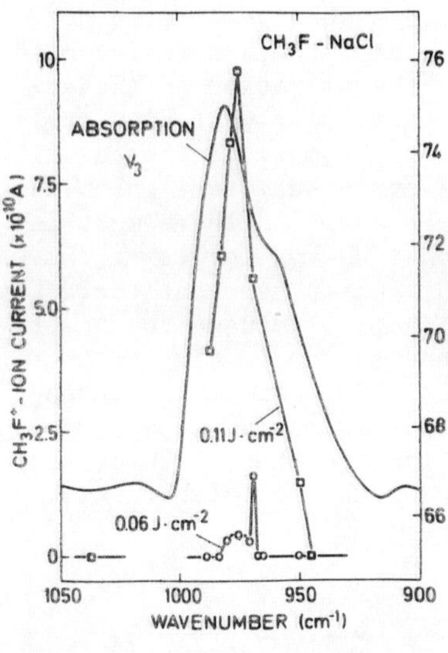

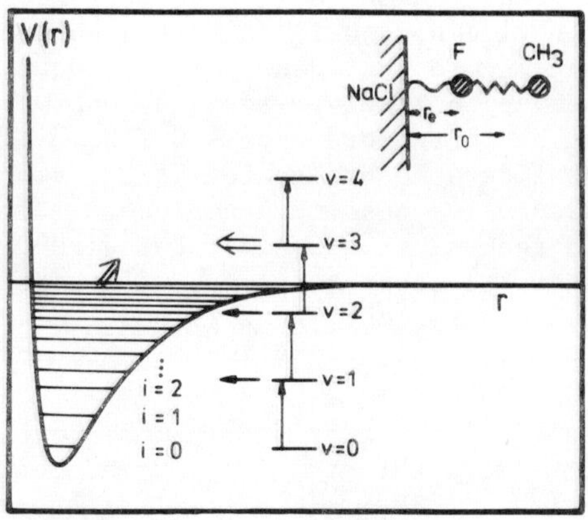

Fig. 1. (left) A laser tuned to the adsorbed oscillator's frequency induces desorption: Linear absorption band ν_3 of CH_3F on NaCl and yield of desorption (CH_3F^+ ion current) of CH_3F from NaCl versus laser frequency (in cm^{-1}). NaCl temperature 77K. Basic pressure 1.6 x 10^{-9}mbar (5c).

Fig. 2. (right) Schematic energy diagram of a molecule (CH_3F) adsorbed on a surface (NaCl). (5e) Energy level systems of two vibrational modes between which transitions can occur. The tunnelling ($\Rightarrow$) represents desorption.

v or emit a phonon of energy $\hbar\Omega$ such that

$$E_i^o + \hbar\Omega = E_o^1 \quad . \qquad (3)$$

It is then in an excited state E_i^o of the adsorption potential from which it can cascade down to the state i = 0, emitting more phonons and thus heating up the solid. This process is called resonant heating.

If the adsorbed molecule can be excited into a higher internal vibrational state such that its total energy

$$E_o^v = E_o + (v + \frac{1}{2})\hbar\omega = \frac{p^2}{2m} + (v' + \frac{1}{2})\hbar\omega \qquad (4)$$

is degenerate with some continuum state of momentum p with v' possibly zero, then elastic transfer into the latter state represents the desorption of the molecule. (A possible change of the rotational state is neglected here.) Since the vibrational wavefunctions of the two states (i=0,v) and (p,v') overlap substantially, a way to express this state of affairs is to say that the transition occurs on account of tunnelling. We call this tunnelling process predesorption by vibration. See fig. 2.

Tunnelling into continuum states can be an elastic process, which has been examined by Lucas and Ewing. (8) It can also be inelastic, i.e. aided by the emission or absorption of phonons, investigated by Kreuzer and Lowy. (8)

The occupation functions n_i^v of the states with energy E_i^v change in time at random. There are two forces driving the random walk of the molecule up and down the energy states. One is due to the thermal vibrations of the lattice atoms, which the adsorbed molecule perceives as a time-dependent force. If the molecule is in a state (i,v), this force induces transitions to other states (i',v') with a transition probability $P_{i'i}^{v'v}$ per unit time. For $(E_i^v - E_{i'}^{v'}) < 0$ the transition occurs via absorption of a phonon, otherwise via stimulated or spontaneous emission of a phonon. In Kreuzer's theory one-phonon processes are considered only. If the spacing of the energy levels $E_i^v - E_o^v$ is larger than the Debye energy, multiphonon processes must be considered, if relaxation into the ground state is to be included.(7) The other driving force is the field of the laser which induces transitions $v \to v \pm 1$ with a transition probability $L_{i'i}^{v'v}$ per unit time.

We assume that the evolution in time of the occupation functions $n_i^v(t)$ is given by a master equation with the rate coefficients $R_{i'i}^{v'v}$ being the sum of the laser and the phonon effect

$$R_{i'i}^{v'v} = L_{i'i}^{v'v} + P_{i'i}^{v'v} \quad , \qquad (5)$$

and $P_{ci}^{v'v}$ and $Q_{ci}^{v'v}$ denoting the transition probabilities for inelastic phonon-mediated and elastic tunnelling into continuum states, respectively:

$$\frac{dn_i^v(t)}{dt} = \sum_{v'=0}^{v_{max}} \sum_{i'=0}^{i_{max}} \left[R_{ii'}^{vv'} \, n_{i'}^{v'}(t) - R_{i'i}^{v'v} \, n_i^v(t) \right] - \sum_{v'=0}^{v_{max}} (P_{ci}^{v'v} + Q_{ci}^{v'v}) \, n_i^v(t), \tag{6}$$

$(i_{max} + 1)$ is the number of bound states in the adsorption potential, chosen to be a Morse potential,

$$V_o(x) = U_o \left[e^{-2\gamma(x-x_o)} - 2e^{-\gamma(x-x_o)} \right], \tag{7}$$

for which the wavefunctions are known in analytic form; $(x-x_o)$ is the displacement of the atom from the equilibrium position x_o. U_o is the potential well depth. One could express the adsorption potential as a sum of Lennard-Jones potentials (Heidberg et al.) (9), which, however, would imply extensive numerical work; the simpler analytical approach is preferred here. The harmonic model for the internal oscillator is useful only up to a certain v_{max}, which in general will be larger than v_{th}, given by $v_{th} \cdot \hbar\omega \approx |E_o|$ and being the threshold value. The rate coefficients R, P, and Q can be calculated according to Fermi's golden rule.

The application of the master equation invokes a number of assumptions, the main one concerns the neglection of effects due to the excitation of the adsorbate with coherent light. Eq. (6) may be used for systems with a finite heterogeneous width of the energy levels owing to slightly inequivalent occupied adsorption sites as is expected for real systems.

The adsorbate is formed by molecules trapped in bound states of the adsorption potential; molecules in the continuum states build up the gas. Then the adsorbate develops in time according to

$$\frac{\Theta(t)}{\Theta(0)} = \frac{\sum_{v=o}^{v_{max}} \sum_{i=o}^{i_{max}} n_i^v(t)}{\sum_{i',v'} n_{i'}^{v'}(0)} = \sum_k A_k \, e^{+\lambda_k t} \tag{8}$$

starting with a thermal occupation

$$n_i^v(0) = B \exp\left[-E_i^v / k\,T \right] . \tag{9}$$

The $\lambda_k < 0$ are the eigenvalues of the matrix (L + P + Q). In general the largest eigenvalue λ_1 is the only close to zero. Then after sufficiently long time the contributions from all other terms $\exp[\lambda_k t]$ may be neglected, which defines a "steady state".

Thus we have in steady state

$$- \frac{d \ln[\Theta(t)/\Theta(0)]}{dt} = - \lambda_1, \tag{10}$$

which may be written for constant intensity I during irradiation

$$- \frac{d \ln[\Theta(t)/\Theta(0)]}{I\, dt} = - \frac{\lambda_1}{I} \tag{11}$$

or

$$- \frac{d \ln[\Theta(t)/\Theta(0)]}{d\phi} = - \frac{\lambda_1}{I}, \tag{12}$$

where ϕ is the fluence of the laser pulse defined by

$$\phi(t) = \int_0^t I(t')dt'. \tag{13}$$

Numerical evaluation of eq. (12) shows that λ_1/I is only slightly dependent upon intensity within a range of intermediate intensities. Then the desorption yield is dependent only upon the fluence ϕ and is independent of the shape I(t) of the laser pulse. Under these conditions λ_1 can be evaluated from the limiting slope of the plot yield versus ϕ. Fig. 3

RESONANT HEATING AND QUANTUM EFFICIENCY

When the adsorbed molecule makes a transition $(v,i) \rightarrow (v',i')$ such that

$$(E_i^v - E_{i'}^{v'}) > 0 \tag{14}$$

a phonon is emitted into the adsorbent, when

$$(E_i^v - E_{i'}^{v'}) < 0 \tag{15}$$

a phonon is absorbed. As a result of these processes the phonon occupation numbers change during the laser induced desorption. Since phonon relaxation processes are very fast even on the time scale of nanosecond laser pulses, a local equilibrium is established implying that a local temperature at time t, T(r,t), can be defined, being determined by Fourier's law of heat conduction

$$\frac{\partial T}{\partial t} - \chi \nabla^2 T = \dot{E}/k, \tag{16}$$

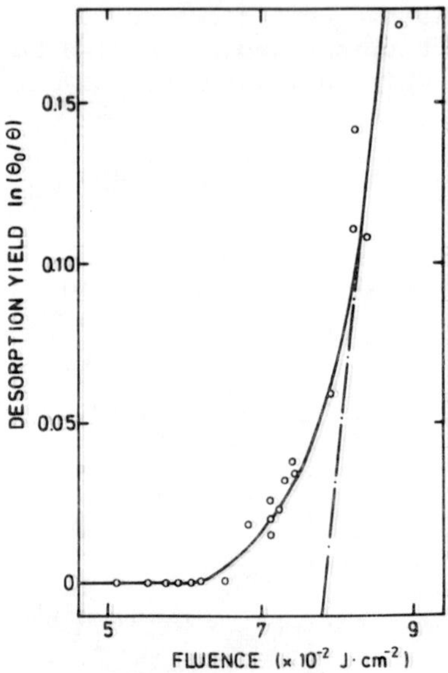

Fig. 3. Yield of resonant desorption of CH_3F molecules from NaCl surfaces versus laser fluence. NaCl temperature 77 K. Basic pressure 1.6×10^{-9} mbar (5c).

where χ is the thermal diffusivity and $\dot{E}$ is the rate at which energy is transferred to or from the solid, being proportional to the total number of adsorbed molecules and to terms of the form

$$n_i^v(t) \cdot (E_i^v - E_{i'}^{v'}) \, P_{i'i}^{v'v} \, . \tag{17}$$

Thus the rate coefficients P, which are independent of the laser intensity, determine resonant heating. But it is not only the coupling to the phonon heat bath, which determines resonant heating, it is also determined via the $n_i^v(t)$ by the laser coupling L and the tunnelling. It occurs practically only during the desorption process. Simularly the quantum efficiency is determined by the rates of pumping, of transitions induced by the phonon heat bath and of tunnelling, $L_{i'i}^{v'v}$, $P_{i'i}^{v'v}$ and $P_{ci}^{v'v}$, $Q_{ci}^{v'v}$, respectively. Up to a saturation level, where tunnelling $(i,v) \rightarrow (cv')$ becomes rate determining, the quantum efficiency is very small below threshold intensity and increased by increasing the laser intensity.

According to Fourier's law a thermal relaxation time t_r can be introduced, which is inversely proportional to χ and depends on a geometrical factor b of the system absorbing the laser radiation:

$$t_r = b/\chi \, , \tag{18}$$

t_r being a measure of the cooling rate after a heat pulse, striking the target in its surrounding medium, i.e. the time required for the target temperature to decrease by 50 %. If the laser pulse action can be described as a pure thermal effect, and if the initial laser light-matter interaction is a localized process, the transition from specific to non-specific thermal effect occurs as the laser exposure duration exceeds the thermal relaxation time. If the laser pulse length t_L is short, thermal diffusion may be neglected during exposure and the temperature rise ΔT can roughly be estimated according to

$$\Delta T = \frac{\alpha \cdot \phi}{L \cdot c_v / V_m} \tag{19}$$

from the total fluence ϕ of the pulse, the spectral absorption α of the target, its molar heat c_v, molar volume V_m and thermal penetration depth $L = \sqrt{\chi \cdot t_L}$, if the light absorption occurs in the surface. For resonant heating equation (18) can give wrong results. Generally, in resonant surface reactions the total reaction rate may not be separable into the sum of the thermal reaction rate in the absence of laser and the laser induced reaction rate in the absence of thermal effects.

Laser pulses of suitable duration, which are selectively absorbed, can cause selective damage to pigmented biological structures, cells and organelles in vivo. Selective damage to cutaneous microvessels and to melanosomes within melanocytes has been reported (10) to occur during irradiation with XeF excimer laser pulses (351 nm) of 20 nanoseconds duration and dye laser pulses (577 nm) of 300 nanoseconds duration, the fluence ranging from 0.5 to 2.0 J cm^{-2}, the pulse length appreciably below the thermal relaxation time. Instead of applying the approximate eq. (18) it is proposed to use a master equation of the type (6) for optimizing laser intensity and pulse length in resonant micro surgery.

ACKNOWLEDGEMENTS

I should like to thank my coworkers Dr. H. Stein, E. Riehl and Dr. I. Hussla for their scrutiny in the experimental work, Prof. H.J. Kreuzer for fruitful discussions and for sending papers prior to publication. The support by Land Niedersachsen, Deutsche Forschungsgemeinschaft and Fonds der Chemischen Industrie is gratefully acknowledged.

REFERENCES

(1) Heidberg, J., Stein, H., Nestmann, A., Hoefs, E. and Hussla, I. 1978, "Symposium Laser-Solid Interactions and Laser Processing", Materials Research Society, Boston. "Laser-Solid Interactions and Laser Processing" 1979, Eds. Ferris, S.D., Leamy, H.J. and Poate, J.M., American Institute of Physics, New York, pp. 49 - 54.

(2) Gortel,Z.W., Kreuzer, H.J. and Spaner, D. 1980, J.Chem.Phys. 72, pp. 234.
Kreuzer, H.J. 1981, Intern. J. Quantum Chem. 15, pp. 683-694.

(3) Jedrzejek, C., Freed, K.F., Efrima, S. and Metiu, H. 1981, Surface Sci. 109, pp. 191 - 206.
Murphy, W.C. and George, T.F. 1981, Surface Sci, 102,pp.L46.

(4) Wedler, G. and Ruhmann, H. 1982, Surface Sci. 121, pp. 464-486

(5) a) Heidberg, J., Stein, H., Riehl, E. and Nestmann, A. 1980 Z. Physik. Chem. N.F. 121, pp. 145 - 164.
b) Heidberg, J., Stein, H. and Riehl, E. 1982, "Vibrations at Surfaces". Eds. Caudano, R., Gilles, J.-M. and Lucas, A.A., Plenum Press, New York, pp. 17 - 38.
c) Heidberg, J., Stein, H. and Riehl, E. 1982, Phys.Rev. Lett. 49, pp. 666 - 669.
d) Heidberg, J., Stein, H. and Riehl, E. 1983, Surface Sci., 126, pp. 183 - 191.
e) Heidberg, J., Stein, H., Riehl, E. and Hussla, I. 1983, "Surface Studies with Lasers", Eds. Aussenegg, F.R., Leitner, A.,Lippitsch, M.E., Springer Verlag, Berlin pp. 226 - 229.

(6) a) Chuang, T.J. 1982, J. Chem. Phys. 76, pp. 3828.
b) Chuang, T.J. and Seki, H. 1982, Phys.Rev.Lett. 49, pp. 382.
c) Chuang, T.J. 1983, "Vibrations at Surfaces", Eds. Brundle, C.R. and Morawitz, H., Elsevier, pp. 125 - 138.

(7) Gortel, Z.W., Kreuzer, H.J., Piercy, P. and Teshima, R. 1983, Phys. Rev. B, 27, pp. 5066 - 5083.
Kreuzer, H.J. and Lowy, D.N. 1981, Chem. Phys.Lett. 78, pp.50 - 53.

(8) Lucas, D. and Ewing, G.E. 1981, Chem. Phys. 58, pp. 385 - 393.

(9) Heidberg, J., Singh, R.D. and Chen, C.F. 1978, Z. Physik. Chem. N.F. 110, pp. 135 - 158.

(10) Anderson, R.R. and Parrish, J.A. 1983, Science 220, pp. 524 - 527.

ENERGY TRANSFER AND MOLECULAR WEIGHT EFFECTS ON POLYMER LUMINESCENCE

Charles E. Swenberg and Robert T. Devine

Armed Forces Radiobiology Research Institute, Bethesda, USA.

Energy transfer in, and luminescence from polymers in dilute rigid glasses and solid polymer films are subjects of extensive current research. This interest stems partly from the vast differences in optical and electronic properties which polymers exhibit and their usefulness. A microscopic description of how electronic energy is transported and trapped and the various routes of dissipation is essential for understanding inhibition of photodegradation processes in polymers. This chapter will have the limited objectives of (a) reviewing the experimental evidence for intramolecular energy transfer in polymers as demonstrated initially by Fox and Cozzens (1), (b) summarizing experimental evidence for dependence of delayed fluorescence and phosphorescence on the molecular weight of the polymer, (c) developing a simple model which accounts for the enhancement of the delayed fluorescence with increasing degree of polymerization and its saturation for high molecular weight polymer, and (d) discussing the time dependence of polymer intramolecular excimer and monomer luminescence using the spectral function formalism for a one-dimensional transport model. The chapter is not intended to be exhaustive in its treatment but rather selective, illustrating the general concepts involved.

I. EVIDENCE FOR ENERGY MIGRATION IN POLYMERS

Singlet and triplet energy transport in polymers with pendent chromophores has been actively studied since the pioneering work of Eisinger and Shulman (2) and Cozzens and Fox (1). Intermolecular transfers of triplet excitation energy in fluid solutions (3) and triplet and singlet exciton migration in crystalline solids such as anthracene (4) were well established prior to conclusive experimental proof of intramolecular energy migration among the chromophores of a polymer. Intramolecular

J. Ladik et al. (eds.), Quantum Chemistry of Polymers - Solid State Aspects, 259–277.

energy transfer in linear organic polymers, such as vinyl polymers, is expected since their adjacent chromophores have separations on the order of 4 Å and favorable relative orientations such that long sequences of closely spaced pendent groups exist. Before presenting experimental and theoretical results of delayed emission dependence on polymer molecular weight we review the classical work of Cozzens and Fox (1) on delayed emission from poly(1-vinylnaphthalene) (P1VN) in tetrahydrofuran-diethyl ether at 77°K. For dilute rigid glasses, where the concentration of the pendent chromophore(s) is generally less than 10^{-3}M, each polymer is spatially well separated from all others thereby precluding intermolecular energy transfer. A property of any dilute polymer glass is that the emission spectrum (at 77°K) exhibits a fluorescence and phosphorescence action spectrum nearly identical to its chromophore analog. For P1VN this implies that 1-ethylnaphthalene and P1VN spectrum are similar although delayed fluorescence is not observed for 1-ethylnaphthalene at equal molar concentrations. Dilute polymer glasses at low temperatures generally do not exhibit excimer fluorescence. Following Stevens (5) we use the term excimer to denote an excited dimer which is dissociative in its ground state. Intramolecular excimer formation occurs when the interaction of an excited chromophore with another chromophore on the same molecule have the proper configurational arrangement; hence excimer fluorescence in solid matrices is sensitive to the temperature at which the matrix is formed as demonstated by the Frank and Harrah (6) study of poly(2-vinylnaphthalene). A typical delayed emission spectrum of P1VN is shown in Fig. 1.

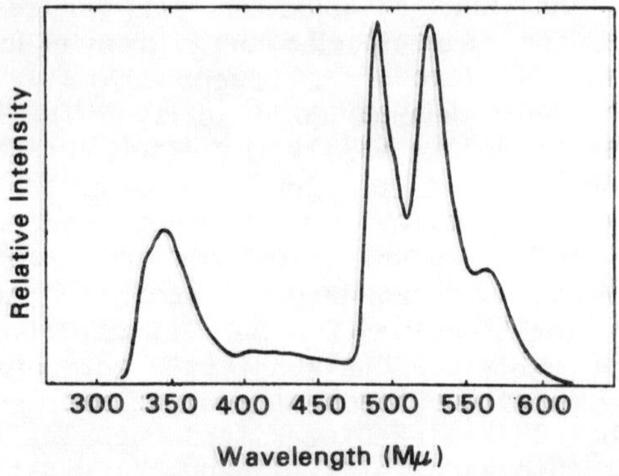

Fig. 1. Delayed emission spectrum of poly(1-vinylnaphthalene) in tetrahydrofuran-diethyl ether at 77°K; excitation wavelength, 290 nm; approximately 10^{-3}M. From Cozzens and Fox (1).

Characteristic of the delayed emission spectrum is the delayed fluorescence band centered at 346 nm and the structured phosphorescence emission between 490 nm and 570 nm. That mobile triplet states and their mutual annihilation are involved in the origin of the delayed band centered at 346 nm can be shown as follows. Suppose we add to the P1VN-tetrahydrofuran-diethyl ether solution molecules M having the following properties: (1) its lowest singlet absorption band is below that of 1-ethylnaphthalene, (2) its triplet state lies energetically higher than 1-ethylnaphthalene and (3) it has a high intersystem crossing rate so that a large fraction of excited singlet molecules result in triplet excited molecules, then convincing evidence for triplet state involvement is shown if upon excitation of M an identical delayed fluorescence spectra to that resulting from direct excitation of P1VN is observed. Using benzophenone and an exciting at 366 nm, a wavelength at which P1VN does not absorb, Cozzens and Fox (1) observed delayed fluorescence identical to that observed using 290 nm light. Fig. 2 illustrates schematically the energy pathways involved.

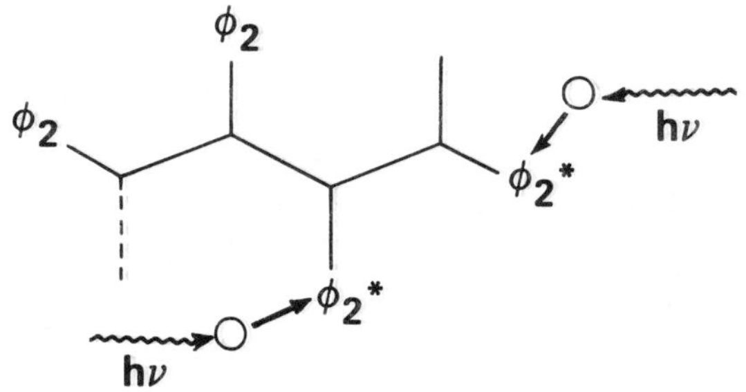

Fig. 2. Schematic model of extended poly(1-vinylnaphthalene) with triplet donor molecules M (O). ϕ_2 denotes a naphthalene moiety and hν the incident photon absorbed by M. M* intersystem crosses before transferring its excitation energy to pendent groups.

Observation that I_{DF} varied approximately as the square of the incident intensity suggests the delayed fluorescence arises from triplet (T) - triplet (T) annihilation; symbolized as:

$$T + T \rightarrow S^* + S_o \rightarrow S_o + S_o + h\nu \qquad (1)$$

Additional experimental evidence for triplet state involvement is provided by the quenching of both phosphorescence and delayed fluorescence by piperylene, a well known triplet quencher. In fact a study of its quenching effects on the delayed emission demonstrates that at least one of the triplet excitons involved in the fusion process is mobile. This is evident from the phosphoresence quenching curves (Stern-Volmer graphs) of P1VN and 1-ethylnaphthalene illustrated in Fig. 3.

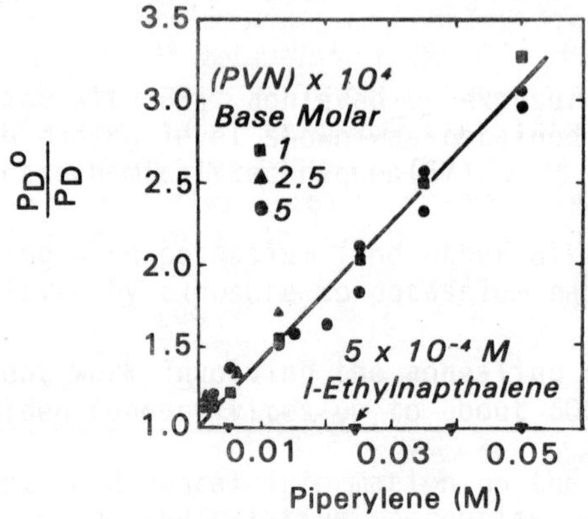

Fig. 3. Quenching poly(1-vinylnaphthalene) and 1-ethylnaphthalene phosphorescence at 77°K as a function of piperylene molarity. From Cozzens and Fox (1).

Since the probability of quenching is proportional to the concentration of quenchers and the effective quenching volume, the enhanced decrease in P1VN phosphorescence as compared to 1-ethylnaphthalene implies a larger cross-sectional volume; this can only be obtained if some of the pendent triplet excitons are mobile. An alternative experimental technique for proving intramolecular energy transport is offered by studying the luminescence from copolymer glasses. One of the first experiments was Fox and Cozzens (7) study of sensitized phosphorescence from copolymers of styrene and 1-vinylnaphthalene, in which the groups derive from the latter act as energy sinks with respect to chain segment derived from styrene. The delayed emission spectra for a mixture of polystyrene and poly(1-vinylnaphthalene) and the corresponding copolymer is illustrated in Fig. 4. The single broad band centered at 400 nm is attributed to polystyrene whereas the delayed emission from poly-(1-vinylnaphthalene) consists of a group of phosphoresence bands centered about 500 nm and a delayed fluorescence band at 346 nm. Note that in the copolymer, curve A in Fig. 4, no delayed fluorescence is observed, a consequence of the lack of neighboring cluster formation of 1-vinylnaphthalene moieties within the polystyrene chain. Comparison of the spectrum of the copolymer (curve A) relative to an equivalent homopolymer glass mixture (curve B) provides conclusive proof of energy transfer within the host polymer chain since the intensity of the styrene-derived phosphorescence has decreased whereas the 1-vinylnaphthalene-derived phosphorescence band has been correspondingly enhanced.

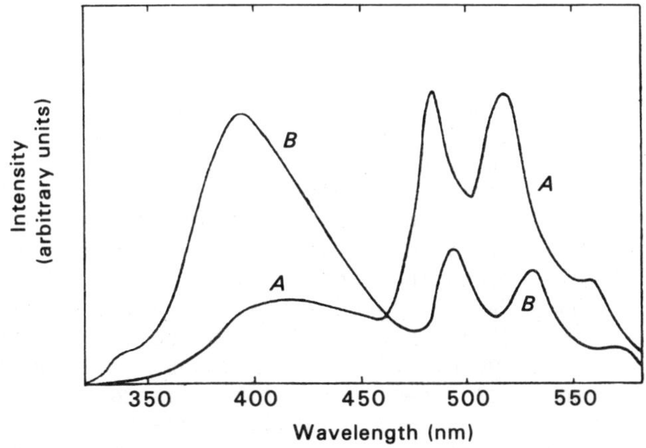

Fig. 4. Delayed emission spectra in 1:1 diethylethertetrahydrofuran glasses. Curve A: Styrene-1-vinylnaphthalene copolymer containing 1 mol percent 1-vinylnaphthalene derived groups. Curve B: Mixture of polystyrene and poly(1-vinylnaphthalene) polymers; 1 mol percent poly(1-vinylnaphthalene). Excitation wavelength 260 nm, 77°K. Modified from Fox and Cozzens (7).

II. MOLECULAR WEIGHT EFFECTS ON DELAYED EMISSION

Low temperature (77°K) emission from isolated polymers in glass matrices, e.g. in 2-methyltetrahydrofuran (MTHF) and tetrahydrofuran: diethyl ether (THF:Et_2O), is characterized by both phosphorescence and delayed fluorescence bands. The intensity of these two emission bands have been shown by a number of investigators (8, 9) to be dependent on the molecular weight of the polymers; the delayed fluorescence intensity (I_{DF}) increasing and the phosphorescence intensity (I_P) decreasing with increasing degree of polymerization. Fig. 5 illustrates this tremendous influence of the average molecular weight of the polymer on the delayed fluorescence and phosphorescence intensity for glasses of poly(N-vinylcarbazole) in MTHF. Similar effects have been reported for poly(2-vinylnaphthalene) in THF:Et_2O (8) and poly(2-naphthylmethacrylate) in matrices of MTHF (10) for luminescence measured at liquid nitrogen temperatures. Although the ratio of I_{DF}/I_P increases with the (average) degree of polymerization ($\bar{P}$) at low molecular weights a saturation in this ratio is observed when $\bar{P} > 100$, where the effective chain length at which saturation occurs depends on the particular polymer. Since this molecular effect on the delayed spectrum is observed in dilute rigid glasses (polymer concentrations less than 10^{-3}M) the fusion process occurs between triplet excited chromophores within a single polymer, ie.

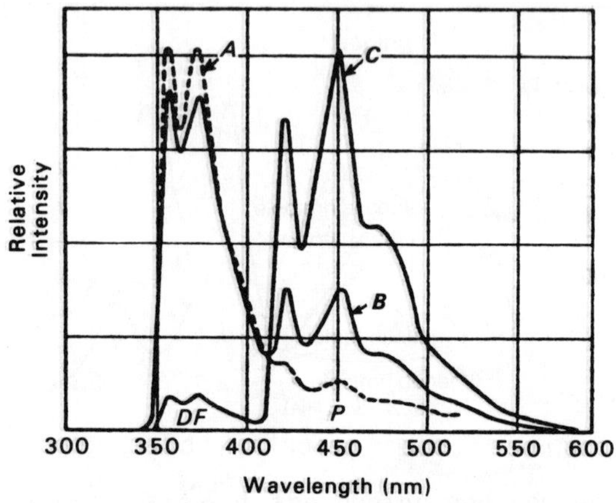

Fig. 5. Delayed emission spectra of three different poly(N-vinylcarbazole) fractions, 6 x 10^{-4} mol basic unit/liter in 2-methyltetrahydrofuran at 77°K. Molecular weights: A) 2 X 10^6; B) 3 X 10^5; C) 9 X 10^4. Excitation wavelength 330 nm. Modified from Klopffer et al (9).

intramolecular exciton fusion as opposed to intermolecular annihilation. A qualitative understanding of enhancement and eventually saturation of I_{DF} with increasing polymer chain length was initially given by Pasch and Webber (8). Their reasoning was as follows. We note first that for a fixed excitation intensity larger polymers have a larger average number of triplet excitons per molecule. If the length (L) of the polymer is comparable to or less than the exciton diffusion length (L_d) and if some of the triplet excitons are mobile, then an increase in L produces a corresponding increase in the delayed fluorescence since a mobile triplet exciton has a nonzero probability of annihilating with either a trapped triplet or another mobile exciton. This increase in I_{DF} with L occurs at the expense of the polymer's phosphorescence. When $L > 2L_d$ no further effects on I_{DF} are expected since excitons separated at distances greater than $2L_d$ do not annihilate. Although this model possesses the essential qualitative feature observed for low temperature polymer glasses a quantitative theory is exceedingly difficult since triplet-triplet annihilation is known to be affected by chromophore separations and intrachain, non-nearest neighbor contacts.

A semiquantitative model of the effects of L on I_{DF} and I_P was given by Webber and Swenberg (11). These authors analyzed the dependence of exciton annihilation on parameters such as lattice size, lattice dimensionality and the exciton fusion rate using a Pauli-Master equation. Their treatment is quite similar to the formalism of Paillotin et al (12) used in modeling the fluorescence quantum yield decreases observed in photosynthetic membranes (13) with increasing excitation pulse intensity. Let $P_n(t)$ denote the probability that a lattice contains n excitons at time t. The (Ansatz) master equation obeyed by the components $P_n(t)$ for a lattice of size L is

$$\frac{dP_n}{dt} = S(t) + D(t) + A(t) \tag{2}$$

where

$$S(t) = \beta\left(\frac{L}{L_o}\right)P_{n+1}(t) - \beta\left(\frac{L}{L_o}\right)P_n(t) \tag{3}$$

is the excitation rate,

$$D(t) = (n+1)\beta P_{n+1}(t) - n\beta P_n(t) \tag{4}$$

denotes the net monomolecuar decay and

$$A(t) = G_{n+1}\frac{L_c\beta}{L^2}P_{n+1}(t) - G_n\frac{L_c\beta}{L^2}P_n(t) \tag{5}$$

is the decrease in the nth state caused by exciton annihilation. L_o is the parameter describing the excitation intensity, L_o is inversely proportional to the excitation intensity, β is the unimolecular exciton decay rate, L_c is the second order dimensionless rate constant for exciton fusion and G_n describes the dependence of the annihilation on the occupation number, n, of the lattice. Previous formulations of exciton annihilation (12) assumed that $G_n = n(n-1)$ which corresponds to twice the number of exciton pairs on the lattice. In Webber and Swenberg (11) formulation, however, the general form, namely

$$G_n = [n(n-1)]^{\alpha} \tag{6}$$

was assumed. The constant α was taken to differ from unity and a value of 1.5 was adapted on the basis of computer simulations. This is discussed later in the text. The addition of the double annihilation events, events where both excitons disappear simultaneously upon annihilation, are already included in equation (2) since Paillotin et al (12) have shown that inclusion of level transitions, $n \rightarrow n-2$ simply alter the scaling factors of the final equations; their explicit inclusion is therefore unimportant in semiquantitative modeling of I_{DF} dependence on chain length, L, discussed herein.

In the absence of annihilation, the occupation distribution, P_n, obeys Poisson statistics,

$$P_n = \left(L/L_o\right)^n \frac{1}{n!} \exp\left(-L/L_o\right) \tag{7}$$

For $L_c \neq 0$

$$P_n = \frac{\left(L/L_o\right)^n \prod_{m=1}^{n}\left(m + G_m L_c/L^2\right)^{-1}}{\sum_{K=1}^{\infty}\left(L/L_o\right)^k \prod_{m=1}^{k}\left(m+G_m L_c/L^2\right)^{-1}} \tag{8}$$

and the value of n for which P_n is maximal decreases as L_C increases; the density of excitons on a polymer diminishes relative to the Poisson limit ($L_C = 0$). The effects of bimolecular annihilation on the steady-state distribution is illustrated in Fig. 6. Experiments displaying the increase in I_{DF} and decrease in I_P with increasing molecular weight reported by Webber and coworkers (8, 10) and Klöpffer et al (9) were performed under steady-state conditions, hence the experimental measureables in terms of the distribution functions P_n are

$$I_P \propto \frac{\langle n \rangle}{L} = \frac{1}{L} \sum n P_n \tag{9}$$

and

$$I_{DF} \propto \frac{\langle n(n-1) \rangle}{L^3} = \frac{1}{L^3} \sum n(n-1) P_n \tag{10}$$

The additional L^{-1} dependence in equation (10) arises because under the experimental conditions polymers in low temperature glasses had constant chromophore densities. For simplicity we have neglected in equations (9) and (10) the weighted sum over the polydispersiveness of the sample and assume that L in the above equations corresponds to an average degree of polymerization, $\bar{P}$. How I_{DF} behaves for L large can

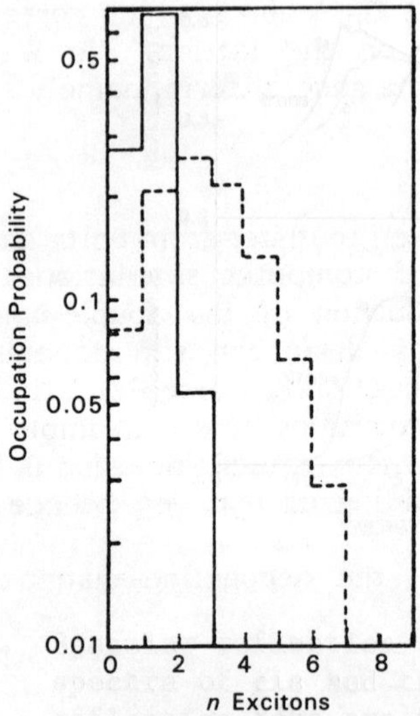

Fig. 6. Comparison of normal Poisson distribution (---) and modified Poisson distribution (—) for $L_C = 10^5$, $\alpha = 1$. Both are for $L = 250$, $L_O = 100$. From Webber and Swenberg (11).

be seen by noting that for large L_c and large L the second order term, G_n, dominates. In this limit P_n exhibits a maximum at

$$\bar{n} = \left(L^3 / L_c L_o \right)^{1/2\alpha} \tag{11}$$

Hence the asymptotic limit of Ip depends on the numerical value of the parameter α . For

$$\alpha > 3/2, \; I_p \rightarrow 0 \quad \text{as } L \rightarrow \infty \tag{12}$$

$$\alpha = 3/2, \; I_p \rightarrow (L_c L_o)^{-1/3} \quad \text{as } L \rightarrow \infty \tag{13}$$

$$\alpha < 3/2, \; I_p \rightarrow \infty \quad \text{as } L \rightarrow \infty \tag{14}$$

In contrast to Ip, the delayed fluorescence is relatively insensitive to α for large L_c and α values near unity. Fig. 7(A) illustrates the theoretical dependence of I_{DF} and Ip on chain length L as calculated by Webber and Swenberg (11). It is evident that the predictions of the model are qualitatively similar to the delayed emission data from poly(2-vinylnaphthalene) reported in Fig. 7(B).

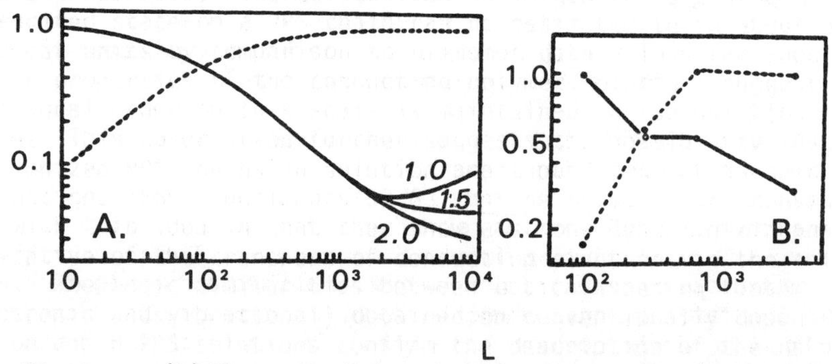

Fig. 7. (A) Theoretical plot of triplet density ($L_o I_p$), for α = 1.0, 1.5, and 2.0 (——), and annihilation density ($L_o I_{DF}$) (---), where Ip and I_{DF} are given by eqs. (9) and (10) respectively. The latter curve is essentially independent of α for L_c large. $L_c = 10^8$ for these plots. (B) Plots of experimental phosphorescence and delayed fluorescence intensity for poly(2-vinylnaphthalene) versus L. From Webber and Swenberg (11).

It is important to emphasize that the fusion model described by Eq. (2) explicitly neglects the spatial variables of the excitons, an assumption which is strictly valid only when the randomization time is rapid compared to the annihilation rate (12, 14). For singlet migration within the photosynthetic membrane where the hopping rate is greater than 10^{11}sec^{-1} this assumption is known to be valid (12), however triplet transport among the pendent groups of a polymer is quite slow, less than 10^{10}sec^{-1}, and suggest the inapplicability of the randomization approximation. Inclusion of non-randomized events unfortunately necessitates extensive computer simulations. A further difficulty with the formalism is that the effects of a finite diffusion length (L_d) of the exciton has not been properly incorporated into the theory. Although α not equal to unity decreases the importance of bimolecular annihilation when $L_d < L$ by shifting the peak in the distribution P_n to smaller n values there are nevertheless annihilating events contributing to I_{DF} for separations (on the average) greater than $2L_d$.

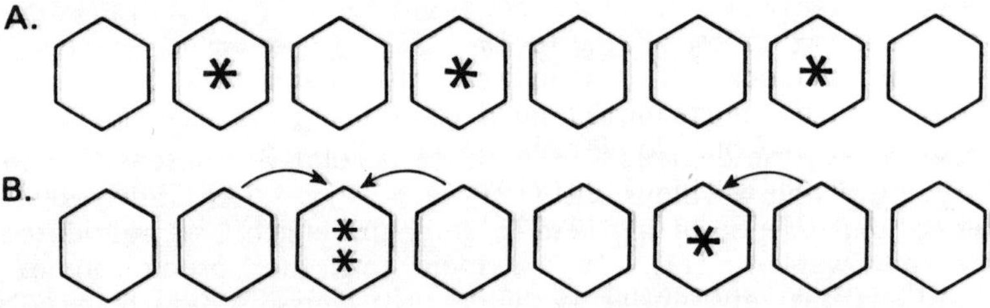

Fig. 8. Schematic of model used to derive rate constant k_n^x. A) At t = 0 there are 3 excitons. B) At t = τ (one hopping time unit) there is a collision which is indicated by double asterisk on single molecule.

A rationalization for α = 1.5 was based on an effective decay constant determined from numerical simulations of the probability distribution of <u>first collision time</u>, $\Psi_n(t)$ for lattices containing L sites and n excitons. A collision was defined as either two excitons arriving at the same site or neighboring sites at the same time. The function $\Psi_n(t)$ was constructed by repeated simulations for a fixed number of excitons. Fig. 8 illustrates the case where n = 3. In this case a collision occurred after only a single hopping time unit. In terms of the simulated distribution function and in the absence of unimolecular decay, the occupation distribution is given by

$$P_n(t) = 1 - \int_0^t \Psi_n(t)\,dt \qquad (15)$$

The effects of annihilation on $P_n(t)$ is apparent in Fig. 9 and illustrates the highly nonexponential behavior expected for the delayed fluorescence. The <u>effective annihilation rate</u> was defined by Webber and Swenberg (11) as

$$k_n^x = -\ln(x)/t_x^n \qquad (16)$$

where $P_n(t_x^n) = x$. The dependence of k_n^x for x = 0.5 illustrated in Fig. 10 is suggestive of a nonintegral value for α. An α somewhat smaller than the simulated value might be expected because of near intrachain crossings; the parameter α includes crudely the effects of nonintegral dimensionality effects. Disregarding the enhancement in excimer emission α is expected to be lower in solvents where chain contraction occurs.

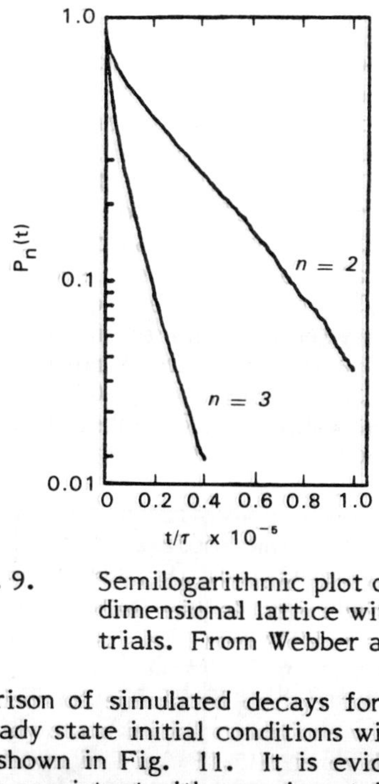

Fig. 9. Semilogarithmic plot of $P_n(t)$ for n = 2 and 3 for a one dimensional lattice with L = 400 constructed from 4000 trials. From Webber and Swenberg (11).

A comparison of simulated decays for lattices with 100 and 500 sites under steady state initial conditions with experimental decay profiles of P2VN is shown in Fig. 11. It is evident that the quality features of $I_{DF}(t)$ are consistent with experimental data in the sense that (1) there is a strong L dependence, (2) decays are nonexponential, and (3) first order processes do not dominate the kinetics. Only when first order kinetics dominate is $I_{DF}(t)$ exponential with a decay time (T_{DF}) one-half the phosphorescence lifetime (T_P). Unfortunately a direct comparison of lifetime data is difficult since polymer phosphorescence usually arises from shallow traps, e.g. in P1VN T_{DF} = 80 msec whereas T_P = 1.9 sec (1).

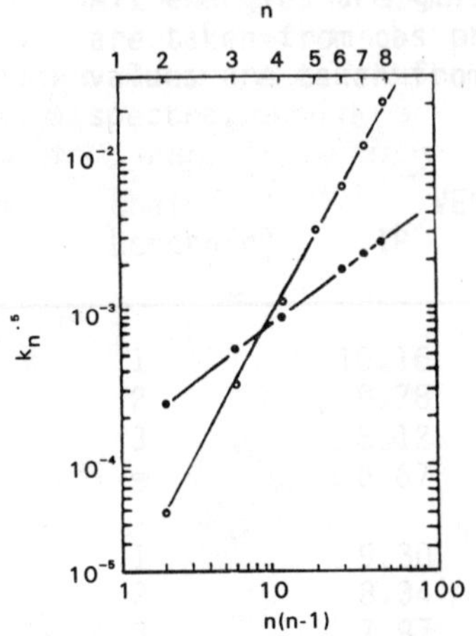

Fig. 10. Computer calculated decay constants for one dimensional (O) and two-dimensional lattices (●). See text equation (16). The slopes are 1.8 and 0.7, respectively. From Webber and Swenberg (11).

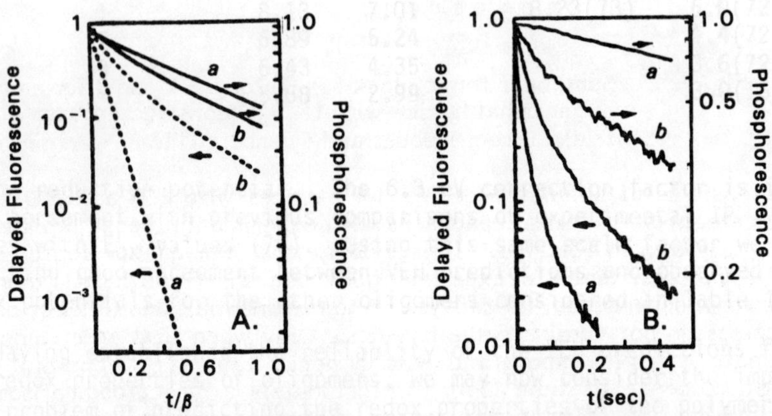

Fig. 11. (A) Semilogarithmic plot of I_P (—) and LI_{DF} (---) versus t/β. For L = 100 (a curves) and L = 500 (b curves) with $L_C = 5 \times 10^4$, $\alpha = 1.5$. (B) Experimental decay curves for poly(2-vinylnaphthalene) phosphorescence and delayed fluorescence. The a-curve corresponds to L = 100 and the b-curve corresponds to L = 3250. Date reported in Webber and Swenberg (11).

Kim and Webber (16) have shown that even films and powders of P2VN at 77°K exhibit molecular weight effects on the delayed emission spectrum similar to its manifestation in dilute polymer glasses. In solid matrices, however, both the delayed fluorescence and phosphorescence are broad, structureless, and red shifted (~700 cm^{-1} for I_P and ~1800 cm^{-1} for I_{DF} in P2VN) relative to their positions in glasses. Thus the delayed emission, as Fig. 12 illustrates, is chiefly excimer in nature (17). Presumably the steric arrangement of neighboring polymer chains inhibit interpolymer excitonic migration. This could arise either from unfavorable chromophore distances, thereby lessening the likelihood of interchain Forster transfer or points of contact are excimer-forming sites (18). The assignment of the phosphorescence to intra- or intermolecular excimers is not easy in solid vinyl polymers although the delayed excimer fluorescence necessitates both mobile and trapped triplet excitons.

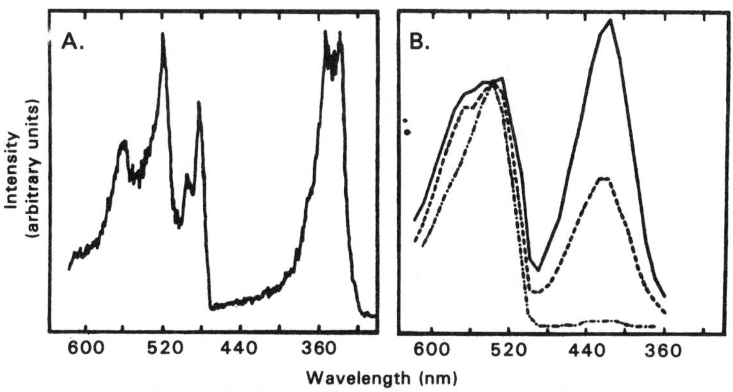

Fig. 12. (A) Delayed emission spectrum of poly(2-vinylnaphthalene) in 2-methyltetrahydrofuran at 77°K. (B) Delayed emission spectra for powdered poly(2-vinylnaphthalene) of different molecular weights: (——) 505,000; (----) 100,000; (-·-·-) 49,000. Spectra scaled to equal phosphorescence intensity maxima. From Kim and Webber (16).

III. Fluorescence Decay Profiles

A quantitative analysis and microscopic assignments of the time resolved fluorescence emission components from polymers in the solid or solution phase to specific physical entities is generally exceedingly difficult. This stems partly from (1) the intrinsic heterogeneity of the emission which in some cases arises from the preponderance of shallow traps, (2) the formation of intramolecular and in solid films or powders intramolecular excimers and (3) the effects of dimensionality and size.

The monomer fluorescence intensity decay for many polymers, such as poly(N-vinylcarbazole) (19), P1VN and P2VN (20), but not polystyrene in CH_2Cl_2 solution at room temperature (21), is highly nonexponential. Although it is tempting to associate components in any decay profile as originating from separate distinquishable molecular species it is important to know that any identification need not be unique. A point well illustrated by the fact that recent theoretical studies of excitation energy transport (among monomer units) and trapping have demonstrated that size and dimensionality of a physical system can result in nonexponentiality in the monomer emission (23, 24). This is apparent for one dimensional systems, an appropriate model for dilute polymers solutions in a good solvent where the polymer is primarily extended with few intramolecular neighbor contacts.

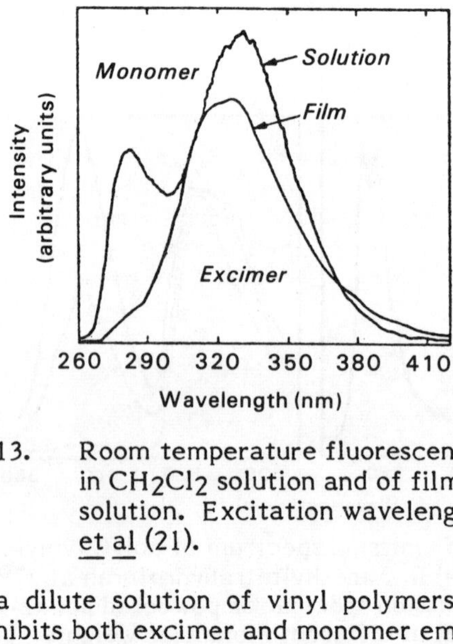

Fig. 13. Room temperature fluorescence spectra of polystyrene in CH_2Cl_2 solution and of film cast from CH_2Cl_2 solution. Excitation wavelength 250 nm. From Gupta et al (21).

For a dilute solution of vinyl polymers the prompt fluorescence spectra exhibits both excimer and monomer emission. For films emission is primarily excimer in character. Fig. 13 illustrates these general features for polystyrene. The temporal characteristics of these emission bands provide information on the formation process of the excimer band and the intramolecular singlet exciton migration rate. This can be seen as follows. Neglecting chain crossings, excimers results either (1) from rotation of a pendent group into a favorable position for excimer formation with its neighbor while in its excited singlet state or (2) from the trapping of monomer excitation energy at excimer forming sites (EFS), i.e.: properly performed chromophore pairs. The routes available for monomer decay and excimer formation are illustrated in Fig. 14. Two limiting cases can be distinquished: (1) rotational alignment in the excited state is fast compared to the rate of excited singlet energy migration, i.e.: $K_R > K_E$ and (2) where $K_E > K_R$. In the first case the excimer rise time provides an estimate of the rotation rate of the pendent groups. In case (2), EFS can be considered as partitioning the extended polymer into finite disjoint monomer segments bounded by absorbing excimer forming sites. We restrict our discussion to case (2) as this will illustrate how nonexponential monomer decays can occur in the absence of special moieties. Consider a chain segment N monomer units in length as illustrated in Fig. 15.

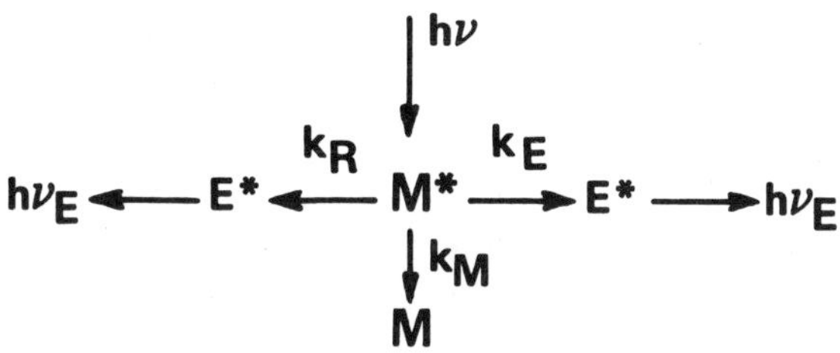

Fig. 14. Excited state kinetic scheme for a single monomer (M) and excimer (E) species. K_R denotes the rate constant for direct formation of excimers, k_E is the trapping rate function for monomer excitation energy at excimer-forming sites. See text for details.

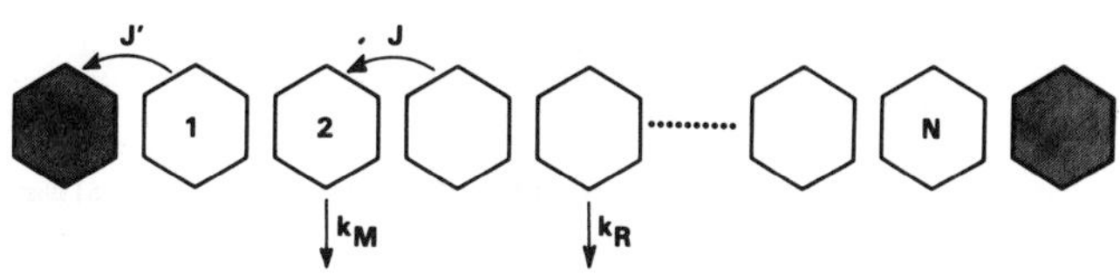

Fig. 15. One dimensional model employed in calculating effective exciton trapping function, k_E, at excimer forming sites. J denotes the monomer nearest neighbor intermolecular transfer energy, J′ is the transfer energy from a neighboring monomer to an excimer forming site, K_R is the rate constant for direct formation of excimers.

Let $Q_i(t)$ denote the probability excitation energy resides at site $i(1 \leq i \leq N)$ at time t, then in the incoherent limit, a valid approximation since both phase and spatial coherence time is considerably less than 10^{-11} sec for disordered systems (4), the equations describing the evolution of $Q_i(t)$ are:

$$\frac{dQ_i}{dt} = J(Q_{i+1} - 2Q_i + Q_{i-1}) - Q_i/\tau \tag{17}$$

$$\frac{dQ_1}{dt} = J(Q_2 - Q_1) - J'Q_1 - Q_1/\tau \tag{18}$$

$$\frac{dQ_N}{dt} = J(Q_{N-1} - Q_N) - J'Q_N - Q_N/\tau \tag{19}$$

in the absence of biomolecular annihilation. The definition of the parameters are evident from Figs. 14 and 15 and $\tau = (k_M + k_R)^{-1}$. In the limit where $J = J'$ and for uniform initial conditions the solution (25) to eqs. (17) - (19) gives the monomer fluorescence for an N site system as

$$FM(t) = C \exp(-t/\tau)\, G_N(t) \tag{20}$$

where C is an appropriate construct and $G_N(t)$ is the spectral response function for an N sites;

$$G_N(t) = \frac{2}{N(N+1)} \sum_{k=1}^{[(N+1)/2]} \cot^2\beta_k \exp\left[-4Jt \sin^2\beta_k\right] \tag{21}$$

and

$$2\beta_k = (2k-1)\pi/(N-1) \tag{22}$$

The monomer fluorescence measured is given by an appropriate average of the spectral function times $\exp(-t/\tau)$. It is apparent from equation (21) that decay profiles are non-exponential. Under thermodynamic equilibrium conditions, Fredrickson and Frank (26) have shown in the average t matrix approximation (27) and for small dyad trap concentrations, $q \ll 1$, and long times, $tJ > 1$,

$$FM(t) \propto \exp\left[4q^2Jt - t/\tau\right] \operatorname{erfc}\left[2q(Jt)^{1/2}\right] \tag{23}$$

Although the one-dimensional diffusion problems with randomly distributed deep traps solved exactly by Movaghar et al (28) predicts a long time decay proportional to $\exp(-\alpha t^{1/3})$ the above solution is sufficient in illustrating the nonexponentiality of the decay profiles. Fig. 16.(A) shows that the effects of increasing the intramolecular transfer rate J or decreasing the rotational rate, k_R, enhances the nonexponential character of the decay profiles. Furthermore it is evident from the kinetic scheme in Fig. 14 that the trapping rate function is given by

$$k_E = -\frac{d}{dt} \ln G(t) \tag{24}$$

where G(t) denotes the ensemble average spectral function. At low excimer concentrations it follows from eq. (23) that $k_E \propto t^{-1/2}$. This particular functional form has been utilized by Smith et al[g](29) to analyze the transient luminescence properties in isotopically doped 1,4-dibromonaphthalene. The excimer fluorescence, FE(t), can also be formulated (26) in terms of the spectral function, namely

$$FE(t) \propto \mathcal{L}^{-1}\{[1-(s+k_M)\tilde{G}(s+1/\tau)]/(s+k)\} \quad (25)$$

where $\mathcal{L}^{-1}$ denotes the inverse Laplace transform and $\tilde{G}$ is the Laplace transform of G(t). Model calculations by Fredrickson and Frank (26) are shown in Fig. 16. The important point to note about Fig. 16.(B) is that the excimer risetime decreases as the energy transfer rate is increased and an enhancement in the excimer intensity is expected. A similar increase in intensity is predicted for increases in k_R; Fig. 16. (C) illustrates this point.

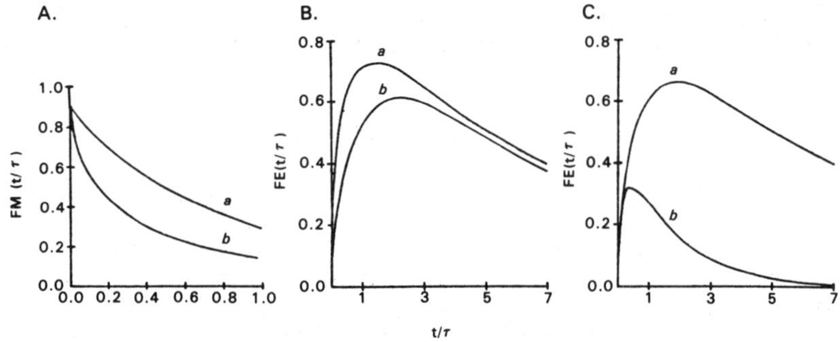

Fig. 16. Time dependence of monomer and excimer fluorescence using eqs. (23) and (25) at a fixed dyad trap fraction q = 0.05. A) Monomer fluorescence; a) $\tau J = 10$, b) $\tau J = 100$; B) Excimer fluorescence with monomer (τ_m) and excimer (τ_E) lifetimes 50 nsec and 80 nsec, respectively, $\tau = 10$ nsec; a) $J = 10^{10}s^{-1}$, b) $J = 10^7 s^{-1}$; C) Excimer fluorescence, $\tau_m J = 50$, $\tau_E J = 80$, a) $\tau J = 10$, b) $\tau J = 50$. Modified from Fredrickson and Frank (26).

The Aspler et al (30) study of time resolved fluorescence from copolymers poly(1-naphthylmethacrylate)-9-vinylanthracene provides an illustration of how an estimate of the transfer rate within the naphthyl chromophores can be inferred from the risetime of 9-vinylanthracene (9-VA) moieties. In this case intramolecular excimers are replaced by traps and $K_R = 0$. This can be seen as follows. The main observation is that 9-VA emission at 425 nm is within experimental error the same whether the copolymer is excited in the naphthyl absorption band, 280 nm, or at 380 nm where only 9-VA absorbs. To be consistent with the rapid risetime in trap emission, $k^{-1} < 10^{-9}$ sec, where $k^{-1} = k_E^{-1} + \tau_M$. Since $\tau_M = 1.5 \times 10^{-9}$ sec, it follows that $k_E > 3 \times 10^9 \text{sec}^{-1}$, a value which should be viewed as a lower limit.

For generalization of the time dependence luminescence properties for more complex donor and acceptor configurations and higher dimensionality effects, the papers by Fayer and coworkers (22 to 24) should be consulted.

REFERENCES

1. Cozzens, R.F., and Fox, R.B., J. Chem. Phys., 50, 1532 (1969).

2. Eisinger, J., and Shulman, R.G., Proc. Natl. Acad. Sci. USA, 55, 1387 (1969).

3. See e.g., Photophysics of Aromatic Molecules, Birk, J.B. (John Wiley and Sons, Inc., London, 1970).

4. See e.g., Electronic Processes in Organic Crystals, Pope, M., and Swenberg, C.E., (Clarendon Press, Oxford University Press, New York, 1982).

5. Stevens, B., Spectrochim. Acta, 18, 439 (1962).

6. Frank, C.W., and Harrah, L.A., J. Chem. Phys., 61, 1526 (1974).

7. Fox, R.B., and Cozzens, R.F., Macromolecules, 2, 181 (1969).

8. Pasch, N.F., and Webber, S.E., Chem. Phys., 16, 361 (1976).

9. Klöpffer, W., Fischer, D., and Naundorf, G., Macromolecules, 10, 450 (1977).

10. Pasch, N.F., and Webber, S.E., Macromolecules, 11, 727 (1978).

11. Webber, S.E., and Swenberg, C.E., Chem. Phys., 49, 231 (1980).

12. Paillotin, G., Swenberg, C.E., Breton, J., and Geacintov, N.E., Biophys. J., 25, 513 (1979).

13. Geacintov, N.E., Breton, J., Swenberg, C.E., and Paillotin, G., Photochem. Photobiol., 26, 629 (1979).

14. Paillotin, G., J. Theor. Biol. 58, 219 (1976).

15. Paillotin, G., J. Theor. Biol. 58, 237 (1976).

16. Kim, N., and Webber, S.E., Macromolecules, 13, 1233 (1980).

17. Klöpffer, W., in Organic Molecular Photophysics, Vol. 1 pp. 357 to 400, ed. J.B. Birks (John Wiley and Sons, Inc. London, 1973).

18. Klöpffer, W., J. Chem. Phys., 50, 2337 (1969).

19. Roberts, A.J., Cureton, C.G., and Phillips, D., Chem. Phys. Lett., 72, 554 (1980).

20. Phillips, D., Roberts, A.J., and Soutar, I., Polymer, 22, 293 (1981).

21. Gupta, M.C., Gupta, A., Horwitz, J., and Kliger, D., Macromolecules, 15, 1372 (1982).

22. Gochanour, C.R., Andersen, H.C., and Fayer, M.D., J. Chem. Phys., 70, 4254 (1979).

23. Ediger, M.D., and Fayer, M.D., J. Chem. Phys. 78, 2518 (1983).

24. Loring, R.F., Andersen, H.C., and Fayer, M.D., J. Chem. Phys. 76, 2015 (1982).

25. Pearlstein, R.M., J. Chem. Phys., 56, 2431 (1972).

26. Fredrickson, G.H., and Frank, C.W., Macromolecules, 16, 572 (1983).

27. Richards, P.M., Phys. Rev., B20, 2964 (1979).

28. Movaghar, B., Sauer, G.W., and Wurtz, D., J. Stat. Phys., 27, 473 (1982).

29. Smith, D.D., Powell, R.C., and Zewarl, A.H., Chem. Phys. Lett., 68, 309 (1979).

30. Aspler, J.S., Hoyle, C.E., and Guillet, J.E., Macromolecules, 11, 925 (1978).

CALCULATIONAL METHODS FOR DISORDERED QUASI-ONE-DIMENSIONAL SYSTEMS

Frank Martino

Laboratory of the International Foundation for Cancer Research at the Physics Department, The City College of the City University of New York, N.Y 10031.

I. INTRODUCTION

The transport of entities such as electrons, excitons, and phonons in disordered media is affected greatly by the nature of the randomness. If the degree of randomness is sufficiently large, the electronic wave function is localized in a small region in a sense described by Anderson[1] and transport of electrons takes place by hopping among these localized states. Even if we limit our consideration to quasi-one-dimensional systems the calculation of the density of states and the wave functions in long disordered chains is a formidable task. In what follows we shall describe a number of different approaches to the problem and present the results of recent calculations on various model systems. In all cases we are aiming at approximate methods for systems sufficiently complex that massive self consistent field calculations are at present completely out of the question.

In section II we review a method particularly well suited to take advantage of the quasi-one-dimensionality of our problem, the method of negative factor counting (NFC)[2-4]. The techniques for dealing with banded matrices introduced here prove to be of general value in other contexts developed below (sections III and IV).

In particular, in section III we apply these techniques to the calculation of Green function matrix, both for the purpose of studying the localization of an

J. Ladik et al. (eds.), Quantum Chemistry of Polymers - Solid State Aspects, 279–306.

eigenstate in a given energy region and for explicit calculation of an eigenstate. This calculation requires only an expansion in a tight-binding basis set that is confined to a local region spatially, and is of value whenever explicit calculation of the eigenvectors, even by inverse iteration, is inpracticable.

Another approximation technique which has a considerable history and literatuare going back over 30 years[5-8] is the coherent potential approximation (CPA). There are many ways to derive this approximation, especially in its single site form, and a number of its generalizations. We shall deal with the fundamental idea here in two forms. In section IV we consider the application of the cluster CPA (CCPA) theory[8-13] to quasi-one-dimensional chains. The CCPA seems a natural choice for studying the electronic spectrum of aperiodic polymers where one has complicated units as well as a Hamiltonian matrix with disordered elements in both its diagonal and off-diagonal matrix elements. In section V we consider developments from the original notion of Scher and Lax [14] to construct a tractable model system on a discrete lattice which simulates actual disordered systems and employs the continuous-time random walk (CTRW) formalism of Montroll and Weiss[15]. These lead to a new approximate procedure of Odagaki and Lax called the coherent-medium approximation (CMA)[16] which can be used to calculate the frequency depended conductivity due to hopping in random media. This is applied to a number of models in random chains with different types of random distributions, including a bond percolation model, and finally to hopping among randomly distributed localized centers.

II. Banded Matrices and Negative Factor Counting

We shall consider quasi-one-dimensional systems, such as polymer chains, whose Hamiltonians in a local orbital basis set are represented by a banded matrix. A banded matrix H has matrix elements H_{ij} that vanish as $|i-j|$ becomes large. Dealing with banded matrices does not mean that we restrict ourselves to polymers or linear chain systems. If one is willing to sacrifice dealing with a local basis set, then mappings[17-19] can be performed which take an arbitrary Hamiltonian H in a atomic basis, to a tridiagonal form (i.e., $H_{ij}=0$, for $|i-j|>1$) in another basis. We will, for the sake of generality, deal with banded matrices which can be partitioned into block-tridiagonal subblocks rather than tridiagonal matrices or strictly one-dimensional systems. This will give some flexibililty in the choice of a local basis set.

One cannot hope to do an actual SCF LCAO HF MO calculation for a large polymer chain involving more than a hundred or so atoms. It is possible though to do calculations of Fock (F) and overlap (S) matrices for molecular subunits of the polymer. Then, providing that the polymer consists of repetitions of these subunits, the Fock and overlap matrix for the full polymer can be extrapolated or built up from those of the subunits. The problem therefore is to approximate the F and S matrices for a large polymer chain from the F and S matrices for much smaller chains that are structurally and chemically identical to local segments of the chain. This approach assumes that the matrix elements of F are primarily a function of the local environment. That this is not always the case is owing to the SCF nature of the electronic exchange and Coulomb potentials. However it is relatively safe to extrapolate from a small to a large regular chain, or from a subunit with a single defect to a long chain with a low concentration of point defects. It is the local nature of the atomic orbital basis functions that enables one to construct F and S in a building block fashion.

The technical procedure used to contruct the polymer F from the subunit Fock matrices $F_1, F_2 \ldots$ is as follows. First one does SCF HF LCAO MO calculations for two or more subunits The specific case of a line of 50 molecules will be considered for the sake of illustration. In this case we might consider two 50 molecule subunits where all the interactions are identical except those involving molecules at the center of the chain. The first subunit could have molecule 25 displaced and the second subunit could have 24 and 25 displaced from the regular array. If the Fock matrices of the two defect units were compared to that of a regular subunit one would observe differences primarily in the middle of these three matrices. However as one went away from the defects the matrix elements would become equal, or nearly so, for the regular and defect subunits. The idea then is to pick blocks of elements in F_1, F_2, centered on the defects, such that the elements on the boundary of the blocks are either small or identical. By translating these blocks along the main diagonal in a random sequence we can construct a Fock matrix that describes a polymer H_N with two types of structural defects that occur at "random" sites along the chain.

Having described the construction of F and S above, let us now discuss how the density of electronic states, $\rho(\lambda)$, and the eigenvectors, $|\Psi_i\rangle$, can be extracted from the secular equation. Due to the nonorthogonality of the atomic basis functions the eigenvalue problem is written in the following general form

$$(F - \lambda_i S)|\Psi_i\rangle = 0. \tag{1}$$

The overlap matrix S and the eigenvector are defined in terms of the atomic orbitals

$$|\Psi_i\rangle = \sum_j C_{ij}\,|\phi_j\rangle, \tag{2}$$

$$S_{ij} = \langle\phi_i|\phi_j\rangle. \tag{3}$$

We first describe the generalized NFC technique which is used to obtain $\rho(\lambda)$, the distribution function for the eigenenergies λ_i . We then deal with the inverse iteration technique, which is used to obtain $C(\lambda_i)$, the vector whose components are the expansion coefficients C_{ij} for the j th egenvector.

The roots of the secular matrix, $M(\lambda)$, (4)

$$M(\lambda) = (F - \lambda S)$$

are located by the zeroes of the determinant of $M(\lambda)$. It is fairly straightforward matter to show that,

$$|M(\lambda)| = |S|\,|\tilde{F} - \lambda I|, \tag{5a}$$

where

$$\tilde{F} = S^{-1/2} F S^{-1/2}. \tag{5b}$$

The determinant of M can be written in terms of the roots S_i of S and λ_i of F

$$|M(\lambda)| = \prod_{i=1}^{N} s_i \prod_{j=1}^{N} (\lambda_j - \lambda). \tag{6}$$

In the above N is the rank of the secular matrix. It is obvious that any sign changes in the determinant of M as a function of λ ,are due to λ passing though a root of the secular equation. The presence of S in the secular equation only serves to multiply the characteristic poloynomial by a constant factor independent of λ . Thus the normal NFC technique used for secular equations with S=I is applicable in the present circumstances with S not equal to I. We briefly outline the essential features of the NFC method below.

The negative factor counting method is applicable to any real symmetric matrix. It is most efficient when the matrix has the additional property of being banded. The reader is refered to Dean[2] for additional details. One starts the computation of $\rho(\lambda)$ by subdividing M into quadrants

$$M(\lambda) = \begin{vmatrix} X_1 & Y_1 \\ Y_1' & Z_1 \end{vmatrix}. \tag{7}$$

The prime denotes transpose and X,Y,Z are respectively a scalar, rowvector and matrix

$$X_1 = M(1, 1), \tag{8a}$$

$$Y_1(j) = M(1, j+1), \tag{8b}$$

$$Z_1(j, k) = M(j+1, k+1). \tag{8c}$$

The letters in brackets imply a vector or matric component. The above quantities are initial values which are used in the following recursion relations

$$X_i = Z_{i-1}(1, 1) - Y_{i-1}^2(1)/X_{i-1}, \quad i = 2, \dots N, \tag{9a}$$

$$Y_i(j) = Z_{i-1}(1, j+1) - Y_{i-1}(1)\, Y_{i-1}'(j+1)/X_{i-1}, \quad j = 1, \dots N-i, \tag{9b}$$

$$Z_i(j, k) = Z_{i-1}(j+1, k+1) - Y_{i-1}'(j+1)\, Y_{i-1}(k+1)/X_{i-1}, \quad j, k = 1, \dots N-i. \tag{9c}$$

The scalars X_i are of particular interest since one can show that the number of negative X_i is equal to the number of λ_i less than λ in the scalar equation. Thus by counting the number of negative X_i as a function of λ, one can construct the cumulative density of states. If M was not banded, the generation of the above sequences would be impractical. However for a polymer $M(\lambda)$ is banded so that most of the elements of the Y_i are zero which greatly simplifies the above recursive calculations.

The inverse iteration technique requires an arbitrary trial vector $|b_1\rangle$ and a good estimate, λ_e, to an eigenvalue as imput. The quantity $\lambda_e = \lambda_k + \epsilon$ should be chosen such that ϵ is small compared to the spacing between eigenvalues. In practice, the NFC technique in conjunction with a bisection procedure can be used to obtain λ_e so that it is good approximation to some eigenvalue. Using $|b_i\rangle$ as a starting value we generate a sequence $|b_i\rangle$ from the recursion

$$M(\lambda_e)|b_n\rangle = S|b_{n-1}\rangle, \tag{10}$$

where it may be verified explicitly that

$$|b_n\rangle = \sum_j C_j|\Psi_j\rangle/(\lambda_j - \lambda_e)^{n-1} \tag{11}$$

if

$$|b_1\rangle = \sum_j C_j|\Psi_j\rangle. \tag{12}$$

Thus one sees from the above that $|b_n\rangle$ will be a reliable representation of $|\Psi_k|$ if λ_e is closer to λ_k than any other eigenvalue. The above procedure differs from the usual inverse iteration technique[4], only by the presence of S on the r.h.s. of Eq.(10). The solution of the equation can be done by making a factorization of M(λ) into a product of lower and upper triangular matrices. The trangular factorization is straightforward when M has a banded structure.

The techniques have been successfully applied to such systems as polyacetylene[2], treated as $(CH)_{350}$, both regular and with distorted C-H bands randomly distributed thoughout the polymer. The energy bands of the regular system compared well with crystal orbital calculations. Inverse iteration was used to calculate eigenstates, which were found to be localized owing to dissorder starting at the band edges.

III. Green matrix Studies of Eigenstates and Localization

If we wish to study the tendency of electrons to localize spatially owing to disorder in larger or more complex systems, the above kind of explicit calculation of the eigenstates by inverse iteration, or some similar technique, become prohibitively difficult. We have developed a technique based upon the ideas of Herbert and Jones[20] that allows us to deduce the qualitative nature of the eigenstates in some energy region (for a polymer system) without having to calculate the eigenstates explicitly. The only condition for the method to work is that our tight-binding basis functions should be confined to a local spatial region and should be ordered in a manner that ensures a banded matrix.

Let us consider a linear chain containing N units which is described by the blocks H_{ij} of a banded hamiltonian matrix:

$$H_{ij} = A_i\,\delta_{ij} + B_{i,i+1}\,\delta_{i+1,j} + B_{i,i-1}\,\delta_{i-1,j}\,. \tag{13}$$

where the A's and B's are square subblocks.
Due to the hermiticity of the real H, A_i must be symmetric and $B_{i,i-1} = B^{tr}_{i-1,i}$ for any i. The eigenvalues and eigenvectors of H can be defined by the poles of the Green matrix.

$$G = (z1 - H)^{-1} \tag{14}$$

and as the solution of the homogeneous matrix equation

$$G^{-1} c = 0 \tag{15}$$

respectively, where G is given by the following recursion relation[21]:

$$G_{ii} = (z1 - A_i - B_{i,i+1} P_{i+1}^{(+)} B_{i+1,i} - B_{i,i-1} P_{i-1}^{(-)} B_{i-1,i})^{-1}$$

$$G_{ij} = \left(\prod_{k=i}^{j \pm 1} P_k^{(\pm)} B_{k,k\mp 1} \right) G_{jj}, \qquad \text{upper sign: } i > j, \quad \text{lower sign: } i < j \tag{16}$$

and

$$P_i^{(\pm)} = (z1 - A_i - B_{i,i\pm 1} P_{i\pm 1}^{(\pm)} B_{i\pm 1,i})^{-1}, \quad P_1^{(-)} = (z1 - A_1)^{-1}, \quad P_N^{(+)} = (z1 - A_N)^{-1} . \tag{17}$$

The meaning of the product symbol $\Pi_{l=m}^{n} X_l$ is, for $n < m$, $X_m X_{m-1} \ldots X_{n+1} X_n$. The recursion relation (16) and(17) can only be applied when H is a banded matrix.

These formulae are ideally suited for evaluation on the computer. We may obtain the N,N block of G as a by-product of evaluating the corner block element. Since all the quantities in the above formulae are matrices, one must be careful to take the products in the correct order, but otherwise these formulae are similar to the scalar continued fractions encountered in the recursion algorithm[22] or in negative factor counting [2]. We have never found $G_{(1,N)}$ as computed by (16) to grow exponentially large. Thus this formula seems to be immune to the usual instabililty shown by one-dimensional recursive formulae with respect to exponentially growing solutions.

We can rearrange the system (15), starting with the nth row:

$$\begin{aligned} &-B_{n,n-1} c_{n-1} + (z1 - A_{nn}) c_n - B_{n,n+1} c_{n+1} = 0, \\ &-B_{n\pm 1,n} c_n + (z1 - A_{n\pm 1}) c_{n\pm 1} - B_{n\pm 1,n\pm 2} c_{n\pm 2} = 0, \\ &-B_{n\pm 2,n\pm 1} c_{n\pm 1} + (z1 - A_{n\pm 2}) c_{n\pm 2} - B_{n\pm 2,n\pm 3} c_{n\pm 3} = 0, \quad \ldots . \end{aligned} \tag{18}$$

In matrix form the last N-1 equations are

$$\begin{pmatrix} z1 - A_{n\pm 1} & -B_{n\pm 1,n\pm 2} & 0 & \cdots & 0 \\ -B_{n\pm 2,n\pm 1} & z1 - A_{n\pm 2} & -B_{n\pm 2,n\pm 3} & \ddots & \\ \vdots & & \ddots & & \ddots \\ 0 & & & & \end{pmatrix} \begin{pmatrix} c_{n\pm 1} \\ c_{n\pm 2} \\ \vdots \end{pmatrix} = \begin{pmatrix} B_{n\pm 1,n} c_n \\ 0 \\ \vdots \\ 0 \end{pmatrix}, \tag{19}$$

or

$$\mathsf{M}x = x^0 . \tag{20}$$

where x = $(c_{n\pm1}, c_{n\pm2}, \ldots, c_{n\pm i}, \ldots)^{tr}$, x^0 is a vector whose components are all zero except the first m ones which are $B_{n+1,n}c_n$, and M is the matrix with the explicit form given in (19). It has the banded form necessary to apply Butler's method to invert it. Then the solution for x is

$$x = \mathsf{M}^{-1} x^0 , \tag{21}$$

or

$$c_{n\pm i} = \mathsf{M}_{i1}^{-1(\pm)} \mathsf{B}_{n\pm1,n} c_n . \tag{22}$$

from relations (16) and (17) we obtain

$$\begin{aligned}
\mathsf{P}^{(+)}_{n+i} &= (z\mathbf{1} - \mathsf{A}_{n+i} - \mathsf{B}_{n+i,n+i+1}\mathsf{P}^{(+)}_{n+i+1}\mathsf{B}_{n+i+1,n+i})^{-1} ,\\
\mathsf{P}^{(-)}_{n-i} &= (z\mathbf{1} - \mathsf{A}_{n-i} - \mathsf{B}_{n-i,n-i-1}\mathsf{P}^{(-)}_{n-i-1}\mathsf{B}_{n-i-1,n-i})^{-1} ,\\
\mathsf{M}^{-1(+)}_{11} &= (z\mathbf{1} - \mathsf{A}_{n+1} - \mathsf{B}_{n+1,n+2}\mathsf{P}^{(+)}_{n+2}\mathsf{B}_{n+2,n+1})^{-1} = \mathsf{P}^{(+)}_{n+1} ,\\
\mathsf{M}^{-1(-)}_{11} &= (z\mathbf{1} - \mathsf{A}_{n-1} - \mathsf{B}_{n-1,n-2}\mathsf{P}^{(-)}_{n-2}\mathsf{B}_{n-2,n-1})^{-1} = \mathsf{P}^{(-)}_{n-1} ,\\
\mathsf{M}^{-1(\pm)}_{i1} &= \left(\prod_{k=n\pm i}^{n\pm2} \mathsf{P}^{(\pm)}_{k}\mathsf{B}_{k,k\mp1}\right)\mathsf{P}^{(\pm)}_{n\pm1} ,
\end{aligned} \tag{23}$$

and therefore

$$c_{n\pm i} = \left(\prod_{k=n\pm i}^{n\pm2} \mathsf{P}^{(\pm)}_{k}\mathsf{B}_{k,k\mp1}\right)\mathsf{P}^{(\pm)}_{n\pm1}\mathsf{B}_{n\pm1,n}c_n = \prod_{k=n\pm i}^{n\pm1} \mathsf{P}^{(\pm)}_{k}\mathsf{B}_{k,k\mp1}c_n . \tag{24}$$

since

$$\mathsf{G}_{n\pm i,n} = \left(\prod_{k=n\pm i}^{n\pm1} \mathsf{P}^{(\pm)}_{k}\mathsf{B}_{k,k\mp1}\right)\mathsf{G}_{nn} ,$$

we obtain finally

$$c_{n\pm i} = \mathsf{G}_{n\pm i,n}\mathsf{G}^{-1}_{n,n}\, c_n . \tag{25}$$

Equation (25) is of interest for two reasons. It is an iterative method of calculating explicit eigenvectors (the Nth component is determined by normalization, $c^2 = 1$. The second interesting aspect of equation (25) is that it indicates that the ratio of the Green function matrix elements, as a function of energy, should be a good measure of which regions in the energy bands contain localized eigenstates. This assumes, of course, that localised and extended states do not coexist at the same energies, and that in an energy region in which there are localized states these will be spatially distributed thoughout the system. With these assumptions it is clear that if site n+i is sufficiently far from site n that the ratio of the matrix elements will peak or drop exponentially in energy regions in which there are localized states. As a practical matter it is usually sufficient to consider only the two ends of the chain.

Let us consider a simple illustrative example, with orthogonal basis functions, and apply the method to both ordered and disordered systems. The Hamiltonian for these systems has zero matrix elements, H_{ij}, when $|i-j| > 4$. It is trivial to modify the computer programs to consider a greater or lesser number of diagonals. The matrix half-bandwidth of 5 was chosen arbitrarily to demonstrate the technique for something other than the usual trivial case of the tridiagonal one-dimensional Hamiltonian.

The Hamiltonian of the ordered system has matrix elements given by the following prescription:

$$H_{ij} = \begin{cases} 0 & i = j \\ -2/(|i-j|+1) & i \neq j. \end{cases}$$

the matrix elements of the disordered system are

$$H_{ij} = \begin{cases} (0.5 - R_j)\Gamma & i = j \\ -2/(|i-j|+1) & i \neq j \end{cases}$$

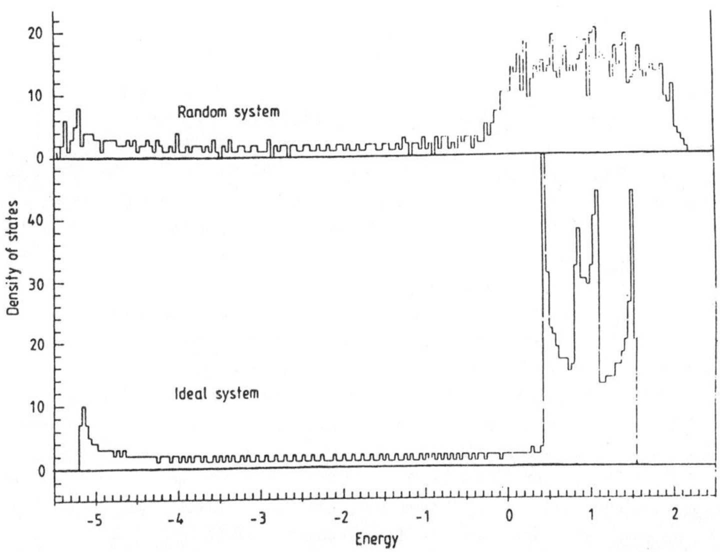

Figure 1. Density of states for the random and ideal periodic Hamiltonians defined in text with N=1000.

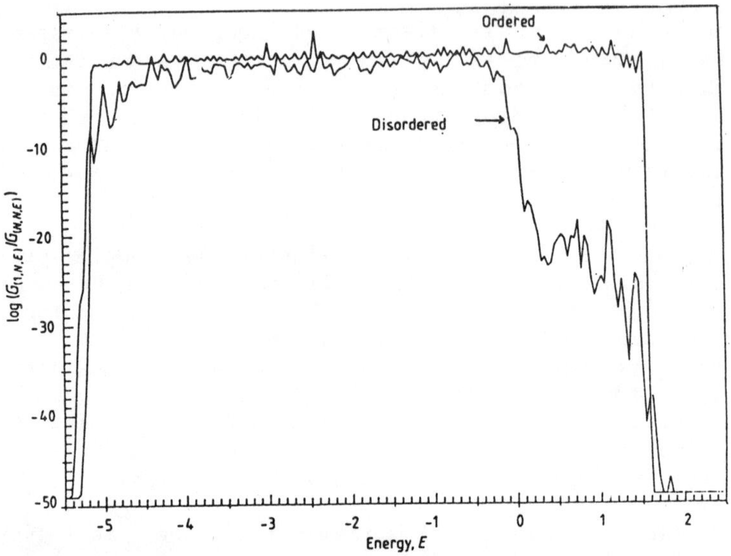

Figure 2. Comparison of Log $G(1,N,E)/G(N,N,E,)$ for the same ordered and disordered systems as in figure 1. Strong localisation at the band edges of the disordered system and complete delocalised behavior in the ordered system is predicted.

where R_j is a random number distributed uniformly on the interval (0-1), and Γ is a constant that can be changed to yield various degrees of disorder. In figure 1 we present the density of states for a regular and disordered (Γ =2) system of 1000 sites, obtained by negative factor counting[2]. Figure 2 compares the ratio of the N,N, and 1,N Green function elements in the two systems. Note that in the ordered system the ratio is orders of magnitude larger on the average than in the disordered system. One expects the ratio to be of the order of unity in the ordered system when we are within the 'allowed' band of states. In the disordered system there is considerable variation in the ratio corresponding to different degrees of localisation of the eigenstates. Indeed one can see from the eigenstates obtained by inverse iteration in figures 3(a)-(d) that there is good correlation between the ratios in figure 2 and the localisation apparent in the figures. Though we never looked at all the eigenstates in any particular system due to the excessive amount of computer time required, we did sample a variety of systems, and did not see any contradictions between the explicit eigenstates and the localisation measure involving the Green function.

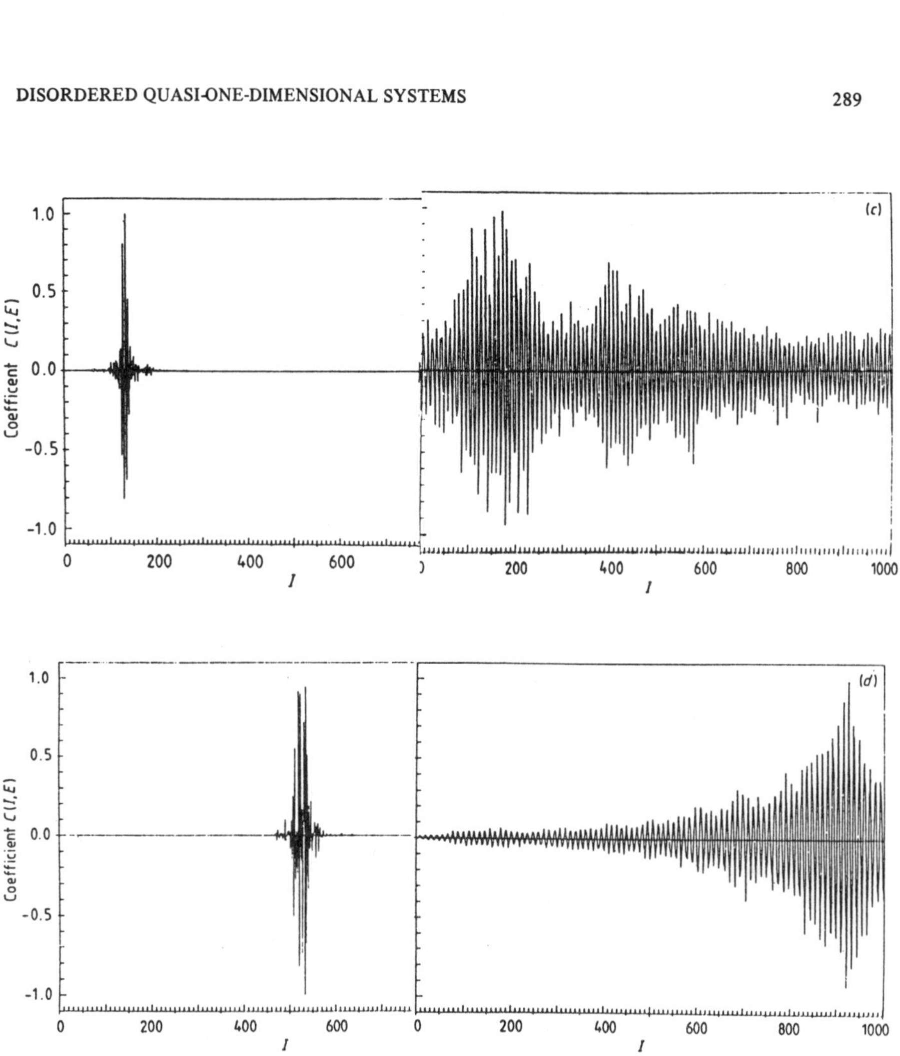

Figure 3. Some eigenstates obtained by inverse iteration for the disordered system having a uniform distribution of the diagonal elements of the Hamiltonian. The expansion coefficients versus orbital or site index, I, for the local basis orbitals are plotted. The eigenenergy is: (a)1.18015;(b)0.18448;(c)-0.81422;(d)-1.81044. The ratio of the 1,N to N, N elements of the Green function in these energy regions was (a)10^{-26},(b)10^{-18},(c)10^{-2},(d)10^{-2}.

IV Cluster Coherent - Potential Approximation

The purpose of the CPA method is to develop a periodic, self-consistent, energy-dependent effective-medium Hamiltonian $\overline{H}$ that is in some sense equivalent to the configuration average of a exact Hamiltonian H. In particular, $\overline{H}$ is determined self-consistently so that the configuration average of the exact Green function G is approximately equal to $\overline{G}$. The Hamiltonians, $\overline{H}$ and H, are taken to be banded matrices that are defined in a Wannier or tight binding basis. The exact Hamiltonian H, written in block-tridiagonal form, and its Green's function G are, repectively

$$H=\sum_{\alpha}|\alpha\rangle\underline{\Sigma}_{\alpha}\langle\alpha|+\sum_{\alpha}(|\alpha\rangle\underline{W}^{t}_{\alpha+1}\langle\alpha+1|+|\alpha+1\rangle\underline{W}_{\alpha+1}\langle\alpha|) \quad (26)$$

and

$$G(z)=(zI-H)^{-1}. \quad (27)$$

In (26) superscript t denotes transpose, and $\underline{\Sigma}_{\alpha}$ and $\underline{W}_{\alpha+1}$ are, respectively, a diagonal and off-diagonal subblock of H whose matrix elements may be "random." The operator H has been expressed in a matrix notation where the ket vectors $|\alpha\rangle$ indicate which portion of the basis space the subblocks $\underline{\Sigma}_{\alpha}$ and $\underline{W}_{\alpha+1}$ act upon. The basis orbitals are numbered from $-\infty$ to ∞, and a sequence of a p number of orbitals is arbitrarily defined as a "cell." Hence, the ket vector $|\alpha\rangle$ for the αth cell is associated with p basis orbitals.

$$|\alpha\rangle\equiv|\alpha,1,2,\ldots,p\rangle, \quad (28)$$

and the $\underline{\Sigma}_{\alpha}$ and $\underline{W}_{\alpha+1}$ are p x p matrices,

$$\underline{\Sigma}_{\alpha}=\begin{bmatrix}\epsilon_{11} & \cdots & \epsilon_{1p}\\ \vdots & & \vdots\\ \epsilon_{p1} & & \epsilon_{pp}\end{bmatrix}_{\alpha} \quad (29)$$

We now introduce the periodic effective medium $\overline{H}$ and its corresponding Green Function $\overline{G}$:

$$\overline{H}=\sum_{\alpha}|\alpha\rangle\underline{\Sigma}\langle\alpha|+(|\alpha\rangle\underline{\Gamma}^{t}\langle\alpha+1|+|\alpha+1\rangle\underline{\Gamma}\langle\alpha|), \quad (30)$$

$$\overline{G}(z)=(zI=\overline{H})^{-1}. \quad (31)$$

The Hamiltonian H describes a perfect periodic system, and we will see later how $\underline{\Sigma}$ and $\underline{\Gamma}$ are to be determined self-consistently.

The exact Green's function G(z) can be expanded in an infinite perturbation series in terms of $\bar{G}$ with the aid of a T matrix,

$$G(z)=\bar{G}(z)+\bar{G}(z)T(z)\bar{G}(z)\ , \tag{32}$$

$$T(z)=[I-(H-\bar{H})\bar{G}(z)]^{-1}(H-\bar{H})\ . \tag{33}$$

The T matrix in (33) is the T matrix for the whole system, and thus a very unwieldy object. It is relatively easy to show that T can be written as an infinite summation of smaller cluster T matrices that act upon appropriate subspaces of our basis-function space. Breaking the perturbation H-$\bar{H}$ into a sum over appropriate subspaces ,

$$H-\bar{H}\equiv\sum_{\alpha}V_{\alpha}\ , \tag{34}$$

and one may show that (33) becomes

$$T(z)=\sum_{\alpha}Q_{\alpha}\ , \tag{35}$$

where

$$Q_{\alpha}=T_{\alpha}(z)\left[I+\bar{G}(z)\sum_{\beta\neq\alpha}Q_{\beta}\right]\ , \tag{36}$$

and the cluster T matrix T is

$$T=[I-V_{\alpha}\bar{G}(z)]^{-1}V_{\alpha}\ . \tag{37}$$

Up until now no approximations have been made. The coherent-potential approximation consists of setting the ensemble-averaged cluster T matrix to zero,

$$\langle\langle T_{\alpha}\rangle\rangle=0\ . \tag{38}$$

With this condition the average Green's function becomes approximately equal to $\bar{G}$(z),

$$\langle\langle G\rangle\rangle\simeq\bar{G}(z)\ . \tag{39}$$

The error inherent in (39) is of the fourth order in T_a as has been pointed out by Soven[6].

There are many ways of peforming the partitioning implicit in Eq.(34). The usual single-site CPS approximation implies H and $\overline{H}$ are identical except for their diagonal matrix elements, and thus ignores nonperiodicity in all off-diagonal matrix elements of H. This is too simplistic for our purposes so we must introduce better approximations. The next-simplest approximation is to set the off-diagonal blocks of H and $\overline{H}$ identical($\underline{W}_\alpha = \underline{\Gamma}$ for all α) so that V_a is the pxp matrix given by

$$\langle \alpha' \mid V_\alpha \mid \alpha \rangle = (\epsilon_\alpha - \underline{\Sigma})\delta_{\alpha\alpha'} . \tag{40}$$

The approximation leading to (40) ignores non-periodicity in the off-diagonal blocks of H, but does include diagonal cluster disorder. The disorder or nonperiodicity in the off-diagonal subblocks of the block-tridiagonal H can be included in a variety of ways. The most straightforward way is to let the rank of $\underline{\Sigma}(\underline{\Gamma})$ be an integer multiple of the rank of the $\underline{\Sigma}_\alpha(\underline{W}_\alpha)$. This points out the fact that the effective-medium $\overline{H}$ can have subblocks which are of the same size as the subblocks in H.

Another very interesting means of including the off-diagonal cluster disorder is the so-called homomorphic CPA (HCPA) of Odagaki and Yonezawa[11]. The HCPA, like the simpler molecular CPA (MCPA) of Eq.(40) ensures an analytic Green's function, G(z). The simplest example of the HCPS, in our notation, is shown in Eq. (41),

$$\langle \alpha,\alpha+1 \mid V_\alpha \mid \alpha,\alpha+1 \rangle \equiv \begin{bmatrix} (\underline{\epsilon}_\alpha - \underline{\Sigma})/2 & \underline{W}_{\alpha+1} - \underline{\Gamma} \\ \underline{W}^t_{\alpha+1} - \underline{\Gamma}^t & (\underline{\epsilon}_{\alpha+1} - \underline{\Sigma}')/2 \end{bmatrix} . \tag{41}$$

In (41), V_a is now a 2px2p matrix and the factor 0 or 1 is to ensure that there is no double counting of interactions in (34). The factor x should, in practice, be chosen as 0 or 1 so that the V_a's in (34) do not operate on overlappimg subspaces, which will cause problems in the configuration averaging in (38). Note that (41) implies that $\underline{\Sigma}_\alpha$ and $\underline{\Sigma}_{\alpha+1}$ are independent or uncorrelated, though it is still possible that the matrix elements within $\underline{\Sigma}_\alpha$ or $\underline{\Sigma}_{\alpha+1}$ may be correlated.

The following technique was first suggested by Butler[23] for strictly one-dimensional systems. We would like to present a generalization to infinite quasi-one-dimensional systems, described by the previously mentioned banded matrices. The evaluation of (38) and (37) requires the computation of a diagonal cluster matrix element of $\overline{G}(z)$. The block-tridiagonal matrix

$$\bar{H}=\begin{bmatrix} \cdots & & & & \\ & \underline{\Gamma}^t & \underline{\Sigma} & \underline{\Gamma} & \\ & & \underline{\Gamma}^t & \underline{\Sigma} & \underline{\Gamma} \\ & & & & \end{bmatrix} \tag{42}$$

owing to its symmetry, can be inverted by k-space techniques, but because of its simple form it is also possible to do a real-space evaluation of G(z). First, one partions H into quandrants

$$(z-\bar{H})=\begin{bmatrix} \underline{x} & \underline{y} \\ \underline{y}^t & \underline{z} \end{bmatrix}, \tag{43}$$

and then one may verify directly that

$$\bar{G}(z)=\begin{bmatrix} (x-yz^{-1}y^t)^{-1} & -x^{-1}y(z-y^tx^{-1}y)^{-1} \\ -z^{-1}y^t(x-yz^{-1}y^t)^{-1} & (z-y^tx^{-1}y)^{-1} \end{bmatrix}. \tag{44}$$

By suitable repeated application of (44) one can show that a diagonal subblock of G(z) is given by the following expressions:

$$\bar{G}_{\alpha,\alpha}(z)=(zI-\Sigma-\Gamma^tP\Gamma-\Gamma Q\Gamma^t)^{-1},$$

$$P=(zI-\Sigma-\Gamma^tP\Gamma)^{-1}, \tag{45}$$

$$Q=(zI-\Sigma-\Gamma Q\Gamma^t)^{-1}.$$

The above forumlas also can be applied to finite disordered systems if one reworks their derivation with appropriate subscripts attached to all quantities. The matrix-continued fractions in (45) enable one efficiently to obtain diagonal blocks of the Green's function of banded matrices of large rank. It is also interesting to note the sequence P, for a finite-tridiagonal system, becomes the negative factors produced by the negative factor counting method of Dean[2].

Since we are dealing with an infinite system the recursions for P and Q represent matrix-continued fractions. As pointed out by Butler, these equations are highly nonlinear and there does not appear to be a closed form solution except for strictly one dimension. We have found that it is possible to generate self-consistent solutions for the matrices P and Q, by simple iteration, with a computer. That is to say, one starts with trial matrices and P and Q and uses Eqs.(45b) and (45c) to generate new trial matrices. The process is repeated (with averaging) until one converges to a desired accuracy on the matrix elements. That P and Q are complex symmetric and that we must perform a matrix inversion at each step is only a minor inconvenience if the rank is not too great. One can prove that the recursions preserve certain symmetries and definitness properties, but stability and uniqueness of the self-consistent solutions is an unresolved question. However, in calculations performed so far the results have been correct and in agreement with more standard k-space techniques and one has no difficuty obtaining 5 -10 decimal place accuracy.

Some off-diagonal blocks of $\overline{G}$ may be needed in HCPA calculations These are obtained trivially by noting the Eq.(44) yields, for an off-diagonal block of a periodic system.

$$\underline{\overline{G}}_{\alpha,\beta}(z)=(\underline{\mathscr{L}})^{\beta-\alpha}\underline{\overline{G}}_{\beta,\beta}(z),\quad \beta>\alpha$$

where

$$\mathscr{L}=P\Gamma. \tag{46}$$

This is a particularly useful result since it relates the off-diagonal clusters of G to the diagonal clusters and tells us that the asymptotic behavior is determined by the product PT. Again, we note in passing that (46) may be generalized quite readily to nonperiodice matrices.

We now can consider the details of the cluster averaging and self-consistency procedures for pure-cluster-diagonal disorder. Equations (37) and (38) are not in a very convenient form for iteration. However, by simple matrix algebra one obtains

$$\begin{aligned} \underline{\Sigma}&=\underline{\eta}-\langle\langle(\underline{\eta}-\underline{\mathscr{E}}_{\alpha})^{-1}\rangle\rangle^{-1}, \\ \underline{\eta}&=\underline{\overline{G}}_{\alpha,\alpha}^{-1}(z)+\underline{\Sigma}, \end{aligned} \tag{47}$$

which were the actual relations used in iterating for Σ. The generalizations to include off-diagonal cluster disorder ($\underline{W}_{\alpha}$ and Γ)are straightforward.

The cluster averages appearing in (38) or (47) are performed with the aid of a joint probablity function $P(l_1,\dots,l_p)$

$$\langle\langle h \rangle\rangle = \sum_{l_1 \cdots l_p} P(l_1, \ldots, l_p) h \,. \tag{48}$$

There is one index $l_1, \ldots, l_p$ in $P(l_1, \ldots, l_p)$ for each of the p-diagonal matrix elements in the cluster. Thus, the label l_n specifies the value of the diagonal matrix element at the nth position in the cluster. Whether the l_n are discrete or continuous labels depends on whether the matrix elements of H have discrete values or a continuous distribution. Nothing has been said about labels for the off-diagonal elements within a cluster, since it is assumed that specifying the diagonal elements also determines the off-diagonal elements. It would not be realistic to assume that the interaction between two "sites" is independent of the identity of the occupants of these sites. Note the word site is used loosely to mean a position in the basis-function space and not necessarily a physical location.

In a single-site CPA calculation the single matrix element of the 1x1 matrix Σ_α can take on several discrete values, or perhaps a continuous range of values. However, if Σ_α encompasses more than one site, then something new is added to the averaging procedure. The l_n's are now not necessarily independent random variables. Instead, it is more likely that they are correlated so that l_n is a function of $l_{n-1}, l_{n+1}, l_{n-2}$, etc. We demonstrate explicitly below just how significant the above considerations can be in determining the cluster-CPA density of states for a random alloy. This will be done within the context of a system with diagonal cluster disorder only.

The system is a polymer chain constructed from atoms of type A and B, with relative concentrations P_A=X and P_B=1-X. Suppose atom A is associated with a single basis function and atom B with two basis functions. A and B could thus correspond to hydrogen and lithium in some minimal basis. A crude approcimation is to assume that the scalar-diagonal terms of H associated with A have the value a; independent of neighbors. Similarly, we approximate the scalar diagonal terms of H associated with B by b_1 or b_2, independent of neighbors The off-diagonal terms will be a function of the diagonal occupants. The cluster-cluster interactions Γ have been approximated by the virtual crystal average of the $\underline{W}_\alpha$ in our calculations. The matrix elements H_{ij} of our model Hamiltonian are zero unles $|i-j| \leq 2$.

Atoms A and B can occur in any random sequence so a is uncorrelated with respect to the pair (b_1, b_2). However, since b_1 and b_2 are associated with the same atom they will occur in adjacent sites along the diagonal of H in the order $b_1 b_2$ and are therefore correlated. (B does not have to be an atom It could also be envisioned as a two-atom cluster or molecule which we associate with some Huckel parameters.) Consequently, in the case of two-site clusters, instead of there being nine clusters to average over, there are only five clusters with nonzero P. With three-site clusters the comparable numbers are 27 and 8. Without the existence of this correlation we would be dealing with a random three-component alloy. In what follows we will compare calculations for three-component alloys, with calculations including the above type of simple correlation.

The single-site probabilities for a_1, b_1, and b_2 are easily shown to be given by:

$$P(a_1) = \frac{x}{2-x}, \quad P(b_1) = P(b_2) = \frac{1-x}{2-x} \,. \tag{49}$$

The joint probability function for two-, three-, etc., site clusters needed in the T- matrix averaging procedure can easily be computed. The matrix $\underline{\Sigma}$ has less symmetry than in the case of uncorrelated sites, since the clusters averaged over do not occur in symmetric pairs, as in the case when the joint probability P is given by a produce of independent random variables. Specifically, Σ_{11} does not have to equal Σ_{22} (for a two-site cluster), now that the sites are correlated.

Upon performing calculations for the random alloy, we found that the correlations made drastic changes in the resulting density of states, as one might expect. Figure 4 shows a series of calculations done for various alloy concentrations with a two-site cluster CPA. The solid curves in Fig. 4 show the density of states for a three-component random ally which resulted when the cluster averaging was done with independent random variables. The site-diagonal energies could take the values -1.5, -0.5, and -1.0 and the site-off-diagonal energies were dependent on the site-diagonal occupants. The dashed curves in Fig. 4 are the density of states for a system identical in all aspects to that of the solid curves except that the cluster averaging included correlations of the type described above. With the presence of the correlation betwen the elements of H, the dashed curves in Fig. 4 represent a binary alloy. The concentration of two of the species for the three- component alloy (-1.5and -0.5) in Fig. 4 was kept equal to afford a better comparison between the two calculations.

The differences in the calculations will disappear in the limit of a nearly pure system with one orbital per atom (i.e., if we increase the concentration of "H" type components). At higher alloy concentrations the correlations are very important. These results have important consequences for certain types of random systems One is the random molecular alloy where molecular groups can be substituted instead of atoms. In this case the relationship of the basis orbitals within a substitutional group is nonrandom, and is correlated in precisely the manner we have considered. The other type of system is a simple random alloy with more than one orbital per atom. The above calculations lead us to believe that other, more subtle long-range correlations could also be very important in determining the density of states in a random alloy.

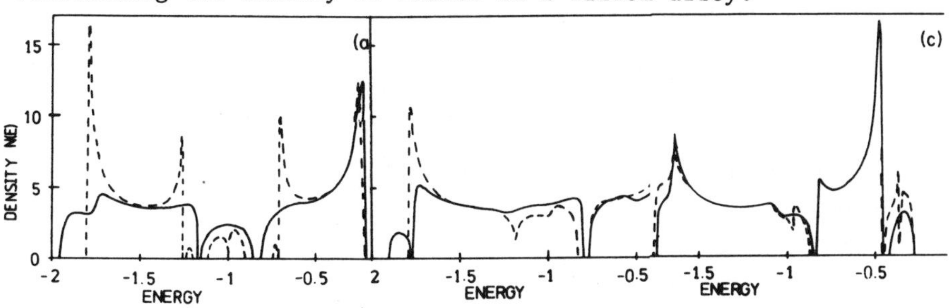

Fig. 4 Density of states vs energy as computed by a two-site cluster-CPA calculation. The solid curves represent a random three-component alloy. The relative concentrations of the three components, having site-diagonal energies (-1.5,-0.5,-1.0) are (a) (45,45,10), (b) (25,25,50), and (c) (5,5,90). The dashed curves represent a binary alloy whose components A and B have diagonal energies A= -1.0, B=(-1.5,-0.5). The relative concentrations of A and B were (a) (B=90,A=10), (b) (B=50,A=50), and (c) (B=10, A=90).

V. Coherent-Medium Approximation and Frequency Dependent Hopping

Hopping conduction in disordered media has attracted a wide range of theoretical interest. So far, the approach based on the continuous time random walk[14] and the coherent medium approximation[16] are the most successful approximate methods to deal with the stochastic transport problem. Both methods have succeeded in reproducing the frequency dependence of the ac conductivity observed in doped semiconductors. Some exact results have been obtained in one-dimensional chains. In particular, the exact frequency dependence of the ac conductivity has been determined for the one-dimensional percolation model, where an infinite chain is decomposed into a set of finite chunks. However, little is known exactly for general cases.

As is widely done in the literature, Odagaki and Lax[16], employ a master equation to describe the stochastic motion of carriers in random media[14]. Let $P(\vec{s},t|\vec{s}_0,0)$ be the probability that a carrier arrives at site $\vec{s}$ at a time t when it was at site $\vec{s}_0$ at time t = 0. We assume that $P(\vec{s},t|\vec{s}_0,0)$ obeys

$$\frac{\partial P(\vec{s},t\,|\,\vec{s}_0,0)}{\partial t} = -\Gamma_{\vec{s}} P(\vec{s},t\,|\,\vec{s}_0,0) + \sum_{\vec{s}\,'\neq\vec{s}} w_{\vec{s},\vec{s}\,'} P(\vec{s}\,',t\,|\,\vec{s}_0,0)\,, \tag{50}$$

where

$$\Gamma_{\vec{s}} = -\sum_{\vec{s}\,'\neq\vec{s}} w_{\vec{s}\,',\vec{s}}\,, \tag{51}$$

which insures conservation of probability. The elementary jump rate $w_{\vec{s},\vec{s}}$ is assumed to have the property (detailed balance)

$$w_{\vec{s},\vec{s}\,'} f(E_{\vec{s}\,'}) = w_{\vec{s}\,',\vec{s}} f(E_{\vec{s}})\,, \tag{52}$$

where E_s is the energy of the carrier at site $\vec{s}$ and $f(E_s)$ denotes the equilibrium distribution function for the localized carrier. The ac conductivity $\sigma(\omega)$ is given by a generalized Einstein relation

$$\sigma(\omega) = \frac{ne^2}{kT} D(\omega)\,, \tag{53}$$

where n is the number density of carriers, e is the charge of a carrier, k is the Boltzmann constant and T is the absolute temperature. The diffusion constant $D(\omega)$ for isotropic systems is written in terms of the Laplace transform of $P(\vec{s},t|\vec{s}_0,0)$ as

$$D(\omega)=-\frac{\omega^2}{2d}\left\langle \sum_{\vec{s},\vec{s}_0} (\vec{s}-\vec{s}_0)^2\tilde{P}(\vec{s},i\omega \mid \vec{s}_0) f(E_{\vec{s}_0})\right\rangle, \tag{54}$$

where

$$\tilde{P}(\vec{s},u \mid \vec{s}_0)=\int_0^\infty e^{-ut}P(\vec{s},t \mid \vec{s}_0,0)dt \tag{55}$$

and $\langle \quad \rangle$ expresses an ensemble average and d is the dimensionality of the system.

The Laplace transform $P(\vec{s},u \mid \vec{s}_0)$ obeys

$$(u+\Gamma_{\vec{s}})\tilde{P}(\vec{s},u \mid \vec{s}_0)+\sum_{\vec{s}\,'\neq\vec{s}} \tilde{P}(\vec{s}\,',u \mid \vec{s}_0)=\delta_{\vec{s},\vec{s}_0}. \tag{56}$$

therfore, a formal solution for $P(s,u\ s_0)$ is readily written as

$$\tilde{P}(\vec{s},u \mid \vec{s}_0)=[(u\overleftrightarrow{1}-\overleftrightarrow{H})^{-1}]_{\vec{s},\vec{s}_0}, \tag{57}$$

where $\overleftrightarrow{H}$ is a matrix of an infinite dimension whose elements are

$$\overleftrightarrow{H}_{\vec{s},\vec{s}}=-\Gamma_{\vec{s}} \tag{58}$$

and

$$\overleftrightarrow{H}_{\vec{s},\vec{s}\,'}=w_{\vec{s},\vec{s}\,'},\quad \vec{s}\neq\vec{s}\,' \tag{59}$$

and $\overleftrightarrow{1}$ is a unit matrix. We call $\overleftrightarrow{H}$ a random-walk matrix and $(u\overleftrightarrow{1}-\overleftrightarrow{H})^{-1}$ a random-walk propagator.

A number of exact properties of hopping conduction can be derived from this formulation[24], but although the expressions for the ac conductivity (53) and (54) have a simple form, it is still in general difficult to evaluate $\langle P(\vec{s},i\omega \mid \vec{s}_0)\rangle$ on the basis of first principles.

Instead one can develope a coherent-medium approximation for equation (50). Fisrt we introduce the operators[16]

$$\hat{P}(u) = \sum_{\vec{s},\vec{s}_0} |\vec{s}\rangle P(\vec{s},u|\vec{s}_0)\langle\vec{s}_0| \ , \tag{60}$$

$$\hat{H} = -\sum_{\vec{s}} |\vec{s}\rangle \Gamma_{\vec{s}} \langle\vec{s}| + \sum_{\vec{s}\neq\vec{s}'} |\vec{s}\rangle w_{\vec{s},\vec{s}'} \langle\vec{s}'| \ . \tag{61}$$

Then Eq (56) can be written in a form

$$(u\hat{1} - \hat{H})\hat{P}(u) = \hat{1} \tag{62}$$

which make $\hat{P}(u)$ behave as a propagator or resolvent, and $\hat{H}$ plays the role of the corresponding Hamiltonian. Of course, $\hat{H}$ bears no simple relationship to the Hamiltonian H of the underlying problem. One may call $\hat{P}(u)$ a random-walk propagator.

However, one may exploit the analogy to the usual Hamiltonian formulation to construct a coherent-medium approximation for equation (50). In particular, the formal solution for $\hat{P}$ is $(u\hat{1}-\hat{H})^{-1}$ so that $P(\vec{s},u|\vec{s}_0)$ is given by the $\vec{s}\ \vec{s}_0$ matrix element of the propagator:

$$P(\vec{s},u|\vec{s}_0) = \langle\vec{s}|(u\hat{1} - \hat{H})^{-1}|\vec{s}_0\rangle \ . \tag{63}$$

To evaluate the ac conductivity by Eqs. (53) and (54), we must calculate the ensemble average of $P(\vec{s},u|\vec{s}_0)$ which is written formally as

$$\langle P(\vec{s},u|\vec{s}_0)\rangle = \langle\vec{s}|(u\hat{1} - \hat{\Sigma})^{-1}|\vec{s}_0\rangle \ , \tag{64}$$

where the coherent Hamiltonian operator $\hat{\Sigma}$ is defined by

$$\langle(u\hat{1} - \hat{H})^{-1}\rangle \equiv (u\hat{1} - \hat{\Sigma})^{-1} \ . \tag{65}$$

So far, the structure of the coherent medium described by $\hat{\Sigma}$ has not been specified. The structure of $\hat{\Sigma}$ should be determined by the nature of the actual system which we treat. We restrict our discussion to a case where the set of sites $\{\vec{s}\}$ forms a regular lattice and the transition rate $w_{s,s}$, is zero unless two sites $\vec{s}$ and $\vec{s}'$ are nearest neighbor to each other Though this model system is a very simplified one, it will work well if the hopping probabilities $w_{s,s}$ are given a suitable distribution.

Now, let us consider an approximation to obtain the coherent medium. Suppose that in a coherent (average) medium all the transition probabilities associated with a given pair of nearest-neighbor sites (say 1 and 2) are given their specific rather than average values. The effective Hamiltonian $\hat{H}_A$ for this system is

$$\hat{H}_A = \hat{\Sigma} + \hat{V} \ , \tag{66}$$

where $\hat{\Sigma}$ is the coherent part, defined in Eq. (65) and

$$\hat{V} = |1\rangle(w_C - w_{21})\langle 1| + |1\rangle(w_{12} - w_C)\langle 2| + |2\rangle(w_{21} - w_C)\langle 1| + |2\rangle(w_C - w_{12})\langle 2| \ , \tag{67}$$

is the localized pertubation produced by using real rather than average values on the 1-2 bond. The as yet unknown coherent transition rate $w_C \equiv w_C(u)$ is introduced to describe the effective Hamiltonian as

$$\hat{\Sigma} = -zw_C \sum_{\vec{s}} |\vec{s}\rangle\langle\vec{s}| + w_C \sum_{\substack{\vec{s}' \neq \vec{s} \\ (\text{n.n.})}} |\vec{s}\rangle\langle\vec{s}'| \ , \tag{68}$$

where z is the coordination number of the lattice We determine the unknown parameter w_c in a self-consistent manner by the condition

$$\langle (u\hat{1} - \hat{H}_A)^{-1} \rangle = (u\hat{1} - \hat{\Sigma})^{-1} \ , \tag{69}$$

where$\langle \ldots \rangle$ denotes the average over the distribution of w_{12}. The situation is schematically shown in Fig 5. In this figure, circles and bonds represent, respectively, the diagonal and off-diagonal matrix elements of $\hat{H}_A$ and $\hat{\Sigma}$. The diagonal element of a site is a sum of off-diagonal element of bonds connected to the site, which is designated by a partition of the diagonal element.

The expicit form of condition Eq. (69) is rewritten as a matrix equation:

$$\langle \bar{V}(\bar{1} - \bar{P}\bar{V})^{-1} \rangle = 0 \ , \tag{70}$$

where

$$\bar{V} = \begin{pmatrix} w_C - w_{21} & w_{12} - w_C \\ w_{21} - w_C & w_C - w_{12} \end{pmatrix} \ , \tag{71}$$

$$\bar{P} = \begin{pmatrix} \bar{P}_{11} & \bar{P}_{12} \\ \bar{P}_{21} & \bar{P}_{22} \end{pmatrix} \tag{72}$$

with $\bar{P}_{ij} \equiv \bar{P}(\vec{i},u|\vec{j}) = \langle \vec{i}|(u\hat{1}-\hat{\Sigma})^{-1}|\vec{j}\rangle$ and $\bar{1}$ denotes the 2x2 unit matrix. Since $\bar{P}_{ij}$ is a function of w_c, Eq. (70) must be understood as a self-consistent condition for w_c. A straightforward calculation shows that the matrix equation (70) ends up as a single condition

$$\left\langle \frac{w_C - w_{12}}{1-2(\bar{P}_{11}-\bar{P}_{12})(w_C-w_{12})} \right\rangle = 0 \ . \tag{73}$$

Here the symmetry $\bar{P}_{12}=\bar{P}_{21}, \bar{P}_{11}=\bar{P}_{22}, w_{12}=w_{21}$ is used. From the definition of $\bar{P}_{ij}$ we have an identity

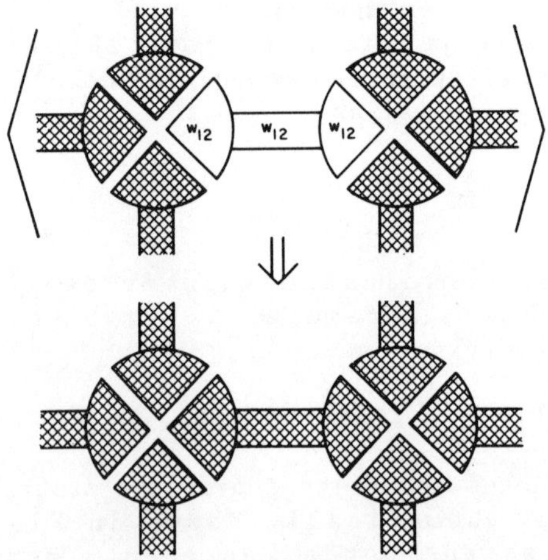

Fig. 5 Schematic illustration of the coherent-medium approximation. The hatched region denotes the coherent medium. The partition of the diagonal part (circle) denotes that the diagonal part is minus the sum of the off-diagonal parts (bonds) connected to it.

$$\bar{P}_{11} - \bar{P}_{12} = (1 - u\bar{P}_{11})/zw_C \ . \tag{74}$$

It is easy to see that $\bar{P}(\vec{s},u|\vec{s}_0)$ is given by

$$\bar{P}(\vec{s},u|\vec{s}_0) = \frac{1}{N}\sum_{\vec{k}} \frac{e^{-i\vec{k}\cdot(\vec{s}-\vec{s}_0)}}{u+zw_C(1-f_{\vec{k}})} \tag{75}$$

with

$$f_{\vec{k}} = \frac{1}{z}\sum_{\vec{s}} e^{i\vec{k}\cdot(\vec{s}-\vec{s}_0)} \ , \tag{76}$$

where $\sum_{\vec{s}}$ sums over nearest neighbors of site $\vec{S}_0$ and N is the total number of sites in the system.

Specifically, it is apparent that the generalized diffusion constant $D(\omega)$ in the coherent medium approximation becomes

$$D(\omega)=a^2 w_C(i\omega)\ , \tag{77}$$

where a is the lattice constant.

The present formalism for the admittance is quite general and applicable to various quantities. As a representative example, we shall consider the ac conductivity, due to hopping which is governed by the master equation (50) with several kinds of distributions of $w_{s,s'}$. We are interested in the frequency dependence of the conductivity $\sigma(\omega)$. Since $\sigma(\omega)$ always bears a factor $ne^2a^2w_0/kT$, one is concerned with the dimensionless ac conductivity $\tilde{\sigma}(\omega)=\sigma(\omega)/(ne^2a^2w_0/kT)$ or equivalently the dimensionless diffusion constant $\tilde{D}(\omega)=D(\omega)/a^2w_0$, where w_0 is a certain scaling factor of $w_{s,s'}$.

Let us consider the bond-percolation model in a one-dimensional chain in which $w_{s,s'}$ is distributed according to the following function:

$$P(w_{\vec{s},\vec{s}'})=p\delta(w_{\vec{s},\vec{s}'}-w_0)+(1-p)\delta(w_{\vec{s},\vec{s}'})\ . \tag{78}$$

In other words, a bond is broken randomly with probability (1-p) and loses its ability to transfer electrons through it.

Since the diagonal element $P(\vec{s},u|\vec{s})$ of the propagator for the coherent medium in one dimension is given by

$$\bar{P}(\vec{s},u|\vec{s})=[u(u+4w_C)]^{-1/2}\ , \tag{79}$$

the self consistency condition (73) with distribution function (78) yields

$$\begin{aligned} w_C/w_0=\frac{1}{\tilde{u}}\{&\tilde{u}+2(1-p)^2-(1-p)\\ &\times[(\tilde{u}+2)^2+4p(p-2)]^{1/2}\}\ , \end{aligned} \tag{80}$$

where $\tilde{u}=u/w_0$. Therefore, the dimensionless diffusion constant is given by

$$\tilde{D}(\omega) = \frac{1}{\tilde{\omega}}\{\tilde{\omega} - 2(1-p)^2 i + (1-p) \times [(2+i\tilde{\omega})^2 + 4p(p-2)]^{1/2} i\} \tag{81}$$

with $\tilde{\omega} = \omega/w_0$. The real and imaginary parts of $\tilde{D}(\omega)$ are plotted in Fig. 6 for various values of the probability p. As is expected, the static limit of the diffusion constant is always zero except for p=1. It is straightforward to observe the limiting behavior of $\tilde{D}(\omega)$, namely,

$$\tilde{D}(\omega) \rightarrow \frac{p(2-p)}{8(1-p)^4}\tilde{\omega}^2 + \frac{p(2-p)}{4(1-p)^2}\tilde{\omega} i \quad (\tilde{\omega} \rightarrow 0) \tag{82}$$

and

$$\tilde{D}(\omega) \rightarrow p - \frac{2p(1-p)(2-p)}{\tilde{\omega}^2} + \frac{2p(1-p)}{\tilde{\omega}} i \quad (\tilde{\omega} \rightarrow \infty) . \tag{83}$$

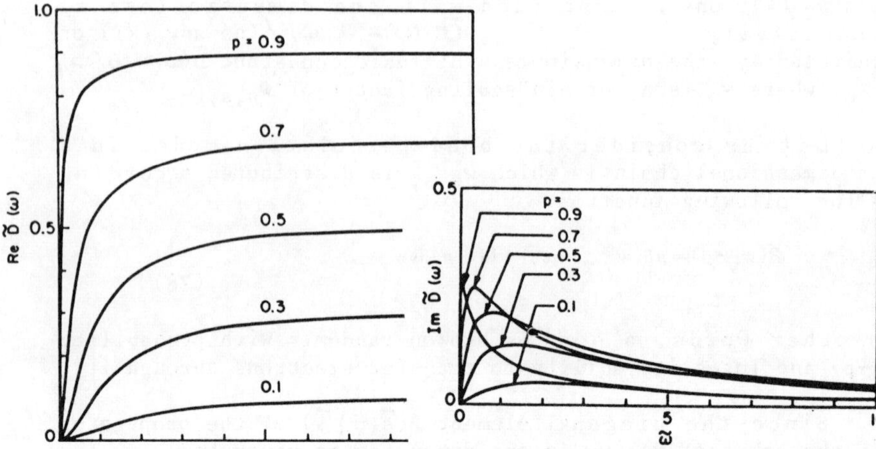

Fig.6 Frequency and probability dependence of the real and imaginary parts of the dimensionless ac conductivity for the one-dimensional bond-percolation model.

The dimensionless diffusion constant, Eq. (81), is in good agrement with exact results for both the frequence dependence and the critical behavior[25].

The present approach essentially requires a cluster treatment, since the random distribution of $w_{s,s'}$ in the Hamiltonian (61) cannot be expressed by single site disorder. It is well known that a certain type of the cluster CPA (coherent-potential approximation) is not analytic. The cluster treatment employed here, however, is identical with the homomorphic cluster CPA[11], the analyticity of which has been rigorously proved. It is easy to extend the present formulation to the multibond cluster. A number of applications of the method to multiple dimensions have also been made[16,24,25].

REFERENCES

1. P.W. Anderson, Phys. Rev. 109, 1492 (1958).
2. P. Dean, Rev. Mod, Phys. 44 (1972) 127.
3. M. Seel, Chem. Phys. 43 (1979) 103;
4. R.S. Day and F. Martino, Chem. Phys. Letters 84 (1981) 86.
5. M. Lax, Rev. Mod. Phys. 23, 237 (1951).
6. P. Soven, Phys. Rev. 156, 809 (1967).
7. P. Soven, Phys. Rev. 178, 1136 (1969).
8. M. Tsukada, J. Phys. Soc. Jpn. 26, 684 (1969).
9. M. Tsukada, J. Phys. Soc. Jpn. 32. 1475 (1971).
10. F. Duscastelle, J. Phys. C 7, 1795 (1974).
11. T. Odagaki and F. Yonezawa, Solid State Commun. 27, 1203 (1978), J. Phys, Soc. Jpn. 47, 379 (1979).
12. R.S. Day and F. Martino, Phys. Rev. 1325, 3482 (1982).
13. R.S. Day and J. Ladik, Int. J. Quantum Chem, 21, 917 (1982)
14. H. Scher and M. Lax, Phys. Rev. B 7, 4491 (1973);7, 4502 (1973).
15. E.W. Montroll and G.H. Weiss, J. Math. Phys. 6, 167 (1965).
16. T. Odagaki and M. Lax., Phys. Rev. B24, 5284 (1981).
17. R. Haydock, V. Heiner, and M.J. Kelly, J. Phys. C 8,2591 (1975).
18. R. Haydock, V. Heine, and M.J. Kelly, J. Phys. C 5, 2845 (1972).
19. Solid State Physics, edited by H. Ehrenreich, F. Seitz and D. Turnbull (Academic, New York, 1980), Vol. 35.
20. DC Herbert and R, Jones J. Phys. C:Solid State Phys. 4, 1145, 1971
21. R.J. Day and F. Martino J. Phys. C 14, 4247 (1981)
22. R. Haddock V. Heine and MJ Kelly J. Phys. C:Solid State Phys. 5 2845, 1972.
23. W.N. Butler, Phys. Rev. B8, 4499 (1973).
24. T. Odagaki and M. Lax, Phys. Rev. B26, 6480 (1982).
25. T. Odagaki and M. Lax, Phys. Rev. Lett 45, 847 (1980).

EFFECT OF LOCAL PERTURBATIONS ON THE ELECTRONIC STRUCTURE OF ORGANIC POLYMERS.

G. DEL RE

Cattedra di Chimica Teorica, Università di Napoli
via Mezzocannone 4, I-80134 Napoli (Italy)

ABSTRACT. These lectures notes treat the problem of local perturbations on a polymer with special reference to the actual range of formally strictly localized perturbation. They are divided into six sections:

1. The essential terms are defined.
2. A physically significant if crude analysis of the effects of a local perturbation is given in terms of a hierarchy of models: a simple semiempirical scheme for localized bonds, a pure charge--transfer scheme, and the case of a fully conjugated polymer
3. The last model is more extensively discussed with reference to Friedel oscillations and SCF procedures.
4. The general formalism for treating local perturbations in terms of Green's matrices is briefly presented.
5. A comment on dynamical aspects is given, to emphasize the difference between the instantaneous range of a perturbation and the distance to which it can travel.
6. Finally, a short bibliographical note is given.

In sections 2 and 3 it is shown that already in fully saturated chains a perturbation acting a single site induces secondary perturbations on the neighboring sites so that the range of the effect is much larger than the range of the cause due to polarization. It is shown how a perturbation can arise by simple juxtoposition of an appropriate donor or acceptor and electronegativity equalization.

The formal introduction of Green functions is important because they provide a formalism especially useful in the case of quasi continuous energy spectra quasi such as those of large polymer chains, and allow to set on clear formal grounds the question of the perturbation range.

The question of time-dependent perturbations is mentioned as the most important coming issue for p.t. applied to polymers.

J. Ladik et al. (eds.), Quantum Chemistry of Polymers - Solid State Aspects, 307–324.

1. DEFINITION OF TERMS

The problem we intend to treat in these lectures is best defined by reference to the terms used in the title.

In order to follow a rational hierarchic order we shall proceed from the general term 'polymer' down to the notion of local perturbation. Then we shall state the properties to be studied and their cause-effect relationship to a local perturbation.

a. Organic polymer.

We mean here by a polymer P an ordered quasi-one-dimensional array of identical monomers M - i.e. groups of atoms arranged to form a molecular unit with a number of free valencies saturated by the corresponding free valencies of neighboring monomers.

This definition is not as comprehensive as the current meaning of the word 'polymer': in particular. it does not apply to polymers whose monomer units are held together by intermolecular forces, say pi-pi coupling.

In other words, we shall assume that P is a periodic and has a back-bone consisting of sigma bonds, unless otherwise stated. This restriction is mainly dictated by reasons of simplicity. Also for reasons of simplicity, we shall normally assume that the number of monomers is finite, albeit very large, and that the Born--von Karmann boundary conditions apply.

The additional requirement that P should be organic amounts to assuming, in addition to the existence of a sigma frame, that M consists essentially of first row atoms (C,N,O,H), so that we are dealing neither with d orbitals nor with high co-ordination numbers.

b. Electronic structure.

The word structure, although currently used also in connection with electrons, corresponds in this case to a compact form of a rather elaborate notion. In general structure denotes the set of parts into which a system can be decomposed together with their relative positions and mutual interactions.

When we speak of 'electronic structure' we mean the structure of a molecule in the above sense (atoms, bonds, energy levels) in so far as it determined by or associated to the electronic part of the molecule or polymer.

It is important to recognize that this implies nothing else but the fixed-nuclei approximation: P (or, for that matter, any molecule) is described as a rigid nuclear framework whose geometry conformation, number and quality of bonds, charge and dipole moment distribution, etc. is entirely determined by the field of the fixed nuclei and the motion of the electrons in it.

If the nuclei are assumed to occupy their equilibrium positions as is usually done unless the potential energy surface is being constructed, this approximation corresponds to a total wavefunction where an electronic factor independent of the actual nuclear coordinates is multiplied by a vibrational factor determined according to some unspecified criterion. Its enormous success is its main justification.

c. Local perturbation.

The word perturbation is normally taken to signify a change in the potential-energy part of the Hamiltonian associated with the 'perturbed' system (here P). We shall give to it a more general meaning: a perturbation will correspond to a change in the Hamiltonian <u>either</u> because the forces acting on P change and/or because a new system A is added to P: in which case the Hamiltonian $\hat{H}(P)$ becomes

$$\hat{H}(P) \equiv \hat{T}(P)+\hat{V}(P) \rightarrow \hat{H}(A,P)= \hat{T}(P)+\hat{T}(A)+\hat{V}(P)+V'(A) \qquad (1)$$

$\hat{T}$ being the kinetic energy operator and $\hat{V}$ the potential energy one; the latter is affected by a prime if it is associated to P in the presence of A or viceversa.

It may be noted that, under the fixed-nuclei approximation, a perturbation consisting in the juxtaposition of a system A to the system P only involves additional terms associated to the kinetic energy of the additional electrons and an additional potential. But if an effective one-electron Hamiltonian is constructed, this reduces to a change in the potential under which the effective electron moves, and the difference in the number of electrons is taken into account in the Hamiltonian if and only if the effective potential is dependent on the density matrix - i.e. if the effective one-electron Hamiltonian is an SCF one. Otherwise, the only consequence of the change in number of electrons will be felt in the occupation scheme, and the Hamiltonian will not be affected by it.

This remark must be further elaborated when the one-electron scheme used rests on a matrix formalism corresponding to an AO basis . In this case, the matrix representing the effective one-electron Hamiltonian will have an order increased by the number of basis functions carried by A. Mathematically speaking, this is not necessary, but clearly, if the basis were kept the same as when P is treated alone, a poorer approximation should be expected. A perturbation is <u>local</u> in general if the range of the potential associated to it is small compared with the size of the perturbed system P.

If taken in this sense, the fact a perturbation is local does not imply that its effect on the properties of P is also confined to a small region of space. Suffice it to recall the mesomeric effect of substituents on aromatic hydrocarbons. Indeed, one of the most intriguing aspects of the problem under study is the possibility that a local perturbation may cause alterations in the electron distribution (and hence in the properties) of P far from its site.

In practice, we shall mean by a local perturbation

i) the juxtaposition of an atom or a small molecule A to the polymer P;

ii) the abstraction from the polymer of a few atoms and/or of a few electrons belonging to localized orbitals.

In the context of a treatment based on a single-particle basis set of the LCAO type the fact that the perturbation is local is reflected also in the addition to/or subtraction from the basis set of a number of basis functions localized in a particular region of space. This means that the two-electron integral set will also be modified in a limited region of space, so that for instance the effective one-electron potential to be used in the SCF version of the LCAO scheme has a limited range both as a consequence of the physical localization of A and a result of the nature of the basis set. This introduces a degree of arbitrariness in the range of the effective perturbing potential because, as is well known, no reliable physical criteria for the choice of an optimum basis set have been found so far, especially as concerns ab-initio calculations.

d. Effect of a local perturbation.

The rather vague word "effect" is shorthand for the permanent or transient changes induced by a local perturbation in the characteristics of a polymer P. These characteristics have already been limited in the present context to those associated with the equilibrium configuration of the nuclei inasmuch as they can be correlated to changes in the electronic distribution and energy spectrum. Our intention is to give here a conceptual analysis of the question, without much recourse to computational espects. This means that we shall largely rely on simple models which describe the features we want to emphasize. Of course, rigorous and reliable results depend on the possibility of performing a complete treatment: but, as is well known, such a treatment is not in itself a source of understanding, and is needed only provided that the interpretational scheme is available.

In other words, computations as such cannot show the effect of a local perturbation except by (often quite small) changes in a huge number of numerical data - the output of the computer program. It is the interpretation of those changes and their connection with specific characteristics of the unperturbad polymer and of the perturbation that we shall study by means of simple models.

A list of properties that can be affected includes the following:

i) charge distribution

ii) local and long-range geometry

iii) position of Fermi level.

2. NATURE OF EFFECTS AND A HIERARCHY OF MODELS

The simplest and most effective way to treat our subject along the lines indicated above is to see what the various models of the electronic structure of molecules can tell us. Of course, we can start from the simplest ones, and slowly climb up to the more complicated ones.

a. Inductive effect models.

Let us consider a fully localized bond polymer, like polyethylene, nylon, or a regular polypeptide, and the approach of an ion, say Na^+ (or its hydrated form). To a good approximation we can expect that electrostatic effects will be dominant. However, these effects do not consist only in an attraction or a repulsion between the ion and the polymer; because, as was suggested already in the old studies on the Hückel method, a change in the field around a given atom wil change its electronegativity, and thus modify the inductive effects transmitted along the bonds by the perturbed atom.

The easiest way to discuss this case is to refer to the old semi-empirical method proposed by the present author in 1958 for treating sigma systems. The idea of the method is that juxtaposed atoms adjust their atomic parameters (="effective electronegativities") δ to one another according to a sort of linear response

$$\delta_A = \delta_A^o + \sum_B \gamma_{AB} \delta_B \qquad (2)$$

regardless of the bond strength, but under the condition that B be a 'first neighbor' of A

The δ^o-values are related to Mulliken's electronegativities χ_A, according to Vandorffy by the equation

$$\delta_A^o = k(\chi_A - \chi_o)/(\chi_A + \chi_o) \qquad k=1.2797 \quad \chi_o=6.9461 \text{ eV} \qquad (2')$$

The charge separation caused by the electronegativity difference obtained from (2) is given, according to the standard Hund scheme, by

$$Q_{AB} = \sin \operatorname{arctg} \frac{\delta_B - \delta_A}{2\eta_{AB}} \qquad (3)$$

where - as is characteristic of Hückel-type approaches - η is at the same time a measure of the bond strength and of the resistance of the bond AB to become polarized.

Given this double rôle, the simple scheme under discussion only holds for standard bonds. In cases like Na^+ simply juxtaposed to the saturated system the effect can be evaluated by assuming that the potential created at the various sites by the juxtaposed charge is equivalent to an increase in Mulliken's electronegativity. Using expression (2'), we obtain from eqn (3) the modified set of charges for the region of the polymer affected by the ion.

The results obtained are illustrated by the example of polyethylene with a univalent ion located at 4 A perpendicular to the center of a CC bond and to the plane of a "trans" CCCC unit. If the bond close to the ion is the 67 one, the charge distribution is the following (in me):

	1	2	3	4	5	6	7	8	9	10	11	12
C	-66	-65	-47	+42	-72	-330	-330	-72	+42	-47	-65	-66
H_1	+33	+34	+36	+49	+40	+ 81	+ 81	+40	+49	+36	+34	+33
H_2	+33	+34	+36	+49	0	+119	+119	0	+49	+36	+34	+33
tot	0	+ 3	+19	+140	-31	-131	-131	-31	+140	+19	+ 3	0

At first sight, these results appear somewhat strange, for the effect of the neighboring ion seems to be not just an increase in the local charge, but an increase in the polarization of the CH bonds, the hydrogen atoms taking up a larger positive charge when the carbon atoms become more negative. On a more accurate inspection it is seen that this is not so surprising: indeed, it is a consequence of the fact that in the present model the electrons have no mobility, so that only polarization of the bonds is possible. In fact, the results given above represent a reasonable estimate of the polarization effect produced on an "insulator" polymer by a juxtaposed positive charge, and - as more reliable methods for the treatment of long distance interactions are not available - they can be used as a basis for a discussion of this type of local perturbation.

Two points of that discussion are worth mentioning in this brief summary. First, the fact that the hydrogen atoms appear to carry a positive charge is related to the use of Mulliken gross atomic populations for evaluating net charges. Therefore, especially in the case in point (hydrogen atoms) it cannot be interpreted as an indication that there will be repulsion between those atoms and the juxtaposed ion. As is well known, the protons of a methylene group are deeply embedded in the orbitals of the carbon atom, so that the only compact quantities that can be used for electrostatic considerations are the charge and possibly the dipole moment of the methylene group.

Secondly, the polarization effect just illustrated implies an effect on the conformation of the polymer chain. In the particular case of the polyethylene chain, the conformation where all the carbon atoms lie in the same plane will be twisted so as to bring the methylene groups 5 and 8 closer to the perturbing ion, while the methylene groups 2,3,4,9,10,11 will move away from it. This in turn will change the fields at the various atoms, and hence their polarizations. A more complete calculation would then have to follow the lines of the MCF method of Ladik and Otto.

b. Charge transfer.

The inductive effect just briefly described does not involve any kind of binding, except electrostatic forces. Another effect which is not connected with the presence of a strong coupling of the states of the perturbing and of the perturbed system is the charge transfer effect.

To be more precise, we assume that there is no coupling in the sense that the matrix element of the perturbation between states of the two systems is very small compared with current bond energies. Then, in addition to the polarization effect just illustrated, and provided that element is not exactly zero, an effect associated with the possibility of electron tunneling can take place: charge transfer.

There are two aspects of charge transfer that are currently considered: the dynamical and the static one. The former is associated to collision theory, and will not be considered here. The latter consists in the formation of a stable complex between the perturbing system and the polymer.

The basic idea is that, if the coupling between states of P and A is weak enough to be negligible with respect to other contributions to the energy but large enough to permit electron exchange within times below the measurable threshold (say, 1 ps), then electron redistribution is possible even if no real binding takes place. In other words, resonance between states of the types (A,P), (A^-,P^+), (A^+,P^-), and the like is possible. This phenomenon has been introduced in several studies both in solid state physics and in chemistry. However, a quantitative analysis of the problem has been only recently carried out. The questions which require an answer are at least the following: 1) What are the characteristic of the two partners responsible for charge transfer? 2) How does the mutual arrangement of the two partners come into play? 3) What is the formal frame in which the charge transfer problem can be treated?

The last question might seem preliminary to the other two. We prefer to consider it last because physical considerations must actually suggest the mathematical formalism, and not vice-versa. The general simple scheme adopted by many authors is a comparison of the so-called LUMO and HOMO energies. It is easy to see that this corresponds to an oversimplification. Suppose electrons are a continuous fluid. When no electron transfer takes place between A and P, this is because the overall energy E of the system (A,P) would increase upon transfer of a quantity of electrons from A to P, or viceversa. As is done in thermodynamics, we can say that the condition is thus

$$\frac{dE}{dn_1} dn_1 + \frac{dE}{dn_2} dn_2 = 0$$

and, as dn_1 must be equal to $-dn_2$, this means

$$\frac{dE}{dn_1} = \frac{dE}{dn_2} \qquad (4)$$

which says that the 'chemical potential' of the electrons in A must be equal to that of the electrons in P.

The question is now to show that a chemical potential can actually be defined, and that electrons can be treated as a fluid.

As there are several electronic levels both in a and P, let us first agree that we are indeed speaking of electrons that occupy the outer orbitals of either molecule.

Next, we consider the 'impurity' A. As its electronic states can be described by linear combinations of Slater determinants, it is clear that each orbital can have any integral or non-integral population, depending on the coefficients of the CI expansion. This being so, the assumptions that n_1 can be treated as a continuous variable is justified.

The case of P is somewhat different. If it is considered as a finite molecule, the same considerations hold as for A. On the other hand, if P is considered as a solid, the well-known difficulties associated with a many-electron description appear, and a more qualitative argument must be introduced. This argument is based on the consideration that the number of electrons being infinite, a derivative of the energy with respect to the number of electrons can always be defined - and, as is well known, it will yield the Fermi level as the chemical potential of the electrons in the solid.

We can assume without a deeper analysis that the case of a solid only differs from that of a molecule because the chemical potential of its electrons is practically insensitive to the occupation number of its highest occupied orbital - indeed, as mentioned above, coincides with its Fermi level. It is outside the scope of these notes to prove this statement, but we give as a hint the remark that the orbitals of a solid are extremely diffuse and extremely close in energy. (Of course, this applies to solids having a significant band structure: the question whether such considerations apply to an insulator like polyethylene is open, because such a system is physically best treated as a set of juxtaposed localized orbitals interacting only by polarization effects).

In short, the charge transfer problem in the case of a local perturbation on a polymer can be treated either in terms of transfer from and to a subunit of the polymer which behaves as an isolated molecule or as charge transfer from the system A to a system P having a fixed electron chemical potential. We illustrate briefly the latter case, taking as a polymer the well-known polyacetylene system, where charge transfer does seem to play an important rôle.

In such a simple system, the naïve Hückel scheme with overlap scheme suggests that the pi band has the form

$$\epsilon_k = \frac{e_o + \sum_n \beta_n \cos(kn\pi/(N+1))}{1 + \sum_n S_n \cos(kn\pi/(N+1))} \qquad (5)$$

so that Fermi level is located at e_o, the effective one-electron energy of an atom (pi orbital). The general form of the electron chemical potential for a finite molecule A in the presence of a second molecule P is approximately

$$\alpha_A = \alpha_E^o - q(U_A - U_{AP})/2 \qquad (6)$$

(q = net charge of A)

where U_A is the two-electron single-orbital integral, and U_{AP} is twice the mean two-electron Coulomb integral between the donating molecular orbital of the donor and the accepting molecular orbital of the acceptor. (Note that only one orbital per partner is involved in the actual electron transfer).

When the molecular orbital involved is extremely diffuse or delocalized, as is the case with very long polymers, U_P and U_{AP} are extremely small, unless some polarization takes place, so that the assertion that the electron chemical potential of an infinite system coincides with the orbital energy of its HOMO or LUMO is fully justified. Thus, in the case of a small perturbing system A and a very large perturbed system P, with a negligible polarization of the system, charge transfer can be treated as resulting from the process bringing the electron chemical potential of A to the same value as the Fermi energy of P.

The mathematical difficulties involved and the numerical aspects of the results obtained by such an adjustment can be illustrated on the very simple example of a two-electron diatomic molecule in the presence of polyacetylene.

The example in question may appear oversimplified: however, it must be kept in mind that it contains all the features of a more "realistic" one; moreover, a complete calculation is necessarily an MC-SCF one, whose characteristics are still to be explored.

As an example we shall choose a general molecule XY, and proceed using an SCF scheme corresponding to a Hubbard Hamiltonian. It has been demonstrated by the Author that the electron chemical potential of a molecule coincides with the electronegativity α of the highest orbital involved in electron exchange. Denoting by $|+\rangle$ (resp. $|-\rangle$) the orbitals that are HOMO (and LUMO) in the isolated molecule, we have *in situ*

$$\alpha_+ = \varepsilon_+ - q_+ U_{mol(+)} \qquad q_\pm = n_\pm^o - n_\pm$$
$$\alpha_- = \varepsilon_- - q_- U_{mol(-)} \qquad (7)$$

The quantities $U_{mol(\pm)}$ are effective two-electron integrals over the molecular orbitals concerned; a very crude estimate (based on Mulliken's approximation) gives

$$U_{mol(\pm)} = |c_{\pm X}*(c_{\pm X}+ c_{\pm Y}S_{XY}|^2 * U_{XX} + \\ + |c_{\pm Y}*(c_{\pm Y}+ c_{\pm X}S_{XY}|^2 * U_{YY} \tag{8}$$

where $c_{\mp X}$ and $c_{\mp Y}$ are the coefficients of the AO's of X and Y in the MO's under consideration.

Note that U_{AP} of eqn.(6) has not been introduced in eqns.(7). The reason is that a negligible two-electron effect is associated to a pair of orbitals one of which is localized and the other is extremely diffuse, as is the case with the orbitals of a large polymer. Denoting by (AA|PP) such a two electron effect, we have

$$(AA|PP) = \sum_{\tau\rho} c^*_{\tau P} c_{\rho P} (AA|\tau\rho)$$

where and denote centers of P. Now

$$c_{\tau P} = \frac{1}{\sqrt{2N+1}} \sin \frac{k\tau\pi}{2N+1}$$

whence

$$(AA|PP) = \frac{1}{2N+1} \sum_{\tau\rho} \sin \frac{k\tau\pi}{2N+1} \sin \frac{k\rho\pi}{2N+1} (AA|\tau\rho)$$

as $(AA|\tau\rho) \simeq 0$ when $\tau \neq \rho$, it follows that for N sufficiently large $(AA|PP) \simeq 0$.

In eqn.(7) $n^{\circ}_{\pm}$ is the occupation number of the HOMO (LUMO) in the isolated XY molecule, $n_{\pm}$ its value in situ.

The symbol ε denotes an orbital energy.

Just to make a realistic choice of parameters, let us refer to a $p\pi$-$p\pi$ bond in the iodine molecule. The ionization potential and the electron affinity of the $p\pi$ orbital of iodine are

$$I_p = 12.67 \text{ eV}, \quad A_p = 3.52 \text{ eV}$$

whence, according to the usual expressions:

$$\alpha_I = -8.095 \text{ eV}, \quad U_{II} = 9.15 \text{ eV}$$

Thus, with the overlap integrals taken equal to .186 from Clementi and Roetti Tables for an interatomic distance of 2.66 A, the following values for the elements of the Hamiltonian matrix are obtained:

$$\alpha_X = \alpha_Y = -8.095 - 4.575q_X \text{ eV}$$

$$\beta_{XY} = 1.75\ \alpha_X * S * (2-S) \qquad \text{(Cusachs formula)}$$

where $q_X = q_Y = (n^{\circ}_- - n_-)/2$ if $|+\rangle$ is full $(n_+ = 2)$

and $q_X = q_Y = (n^{\circ}_+ - n_+)/2$ if $|-\rangle$ is empty $(n_- = 0)$.

If we assume that the Fermi level of polyacetylene is the p electronegativity of carbon (4.625 eV), we find that the situation where the XY molecule has the same electronegativity is found when $n_+ = 1.925$ electrons, i.e. when the bond under consideration has lost .075 electrons to the polyacetylene system.

What is the validity of this conclusion? What does it teach us?

These questions can be discussed at two levels: one which only sees the crudeness and possible disagreement with facts of the model, and just dismisses it; the other which takes the results obtained as serious ones, and tries to analize the real situation by reference to the model used. We believe that the latter procedure is more consistent with the spirit and the program of theoretical physics, and will apply it here.

That the iodine molecule should act as a donor in the pair polyacetylene-iodine is by no means obvious even within the frame of the simple computational scheme adopted above. First, there is a question whether we should not have considered the species I_3^- and I^+ as the actual impurities to be analized. Second, it is doubtful whether the HOMO of the iodine molecule is really a sigma bonding orbital formed by two pure p atomic orbitals. Thus, in spite of the fact that we do find some charge transfer with our model even the problem of deciding what happens with the actual charge transfer lies remains open. It would not be so (apart from possibility of more sophisticated calculations) if we had used a more realistic description of the binding scheme of I_2, introducing parameters for tha various AO's and for the corresponding hybrids.

Another difficulty is that we have not allowed for the possibility that the presence of a charged molecule would polarize the pi system of the polymer. We have actually seen that even a saturated polymer would be affected by an electrostatic field. Here, we can consider both effects, but assuming that the pi electrons shield the sigma ones, the important point is to study the polarization of the delocalized system.

This takes us to considering a perturbation problem in the more familiar sense, for we want to discuss the modification on the bands brought about by the field of the positive impurity.

This problem has been extensively studied by Ladik and Seel, Del Re and Ladik, Seel Del Re and Ladik. We shall briefly review its conceptual status in the next section.

3. MESOMERIC EFFECT, FRIEDEL OSCILLATIONS, AND SCF PROCEDURES

In agreement with the general philosophy of these notes, let us refer again to the simple example provided by the π system of an ideal polyacetilene chain having a single bond distance a. Regardless of the degree of approximation used to treat the electrons of such a system, the law of alternating polarity is respected.

In an exhaustive work by Canuto, Calais and Goscinski, this law is seen as a manifestation of the same phenomenon which is responsible for the well-known Friedel oscillations. We have already seen an example of it in our evaluation of polarization in a polyethylene chain.

In the case of very mobile electrons, the oscillations propagate very far, indeed the charge density variation $\delta\rho(r)$ - r being the distance from the perturbed site - oscillates for large r's according to the expression

$$\delta\rho(r) = A\cos\left(\frac{2\pi N}{MA}r + \phi\right)/r, \qquad (9)$$

where

$$A = 2VN/Ma\cos\phi \quad , \qquad \phi = \arctan\frac{2VM}{N}a \qquad (10)$$

with V is the intensity of the perturbation localized (as a Dirac delta) at a given site, M the number of sites, a their spacing, N the number of electrons.

If M=N, as is the case for the neutral π system of a regular polyacetylene, then the oscillations of the charge density are congruent with the positions of the nuclei.

The above considerations provide the grounds for discussing the question of the 'range' of a localized perturbation.

There have been attempts to define the range outside which a local perturbation on an infinite system would have no effect. That such a finite range should exist is intuitively clear. We have seen that it is of the order of four lattice spacing in a saturated methylene chain. In the present case we are dealing with mesomeric effects which extend much further than sigma-polarization effects: in a 40-atom polyene chain, the variation in charge density induced in a perturbation at site 13 drops to 5% only roughly ten sites away. Now suppose that the atomic parameters α of the various atoms are given by the Hubbard expression

$$\alpha_\mu = \alpha_\mu^\circ - q_\mu U_\mu \qquad (11)$$

where q_μ is the net charge of the $2p\pi$ AO located at site μ and U_μ its effective Coulomb integral. Then α will change in the course of the iterations. At the beginning, of course, V is ignored, and

$$q_\mu = 0 \qquad \beta_{\mu\nu} = \beta \qquad (12)$$

$$\alpha_\mu = \alpha_\mu^\circ = \alpha \quad ; \qquad (13)$$

but the first iteration would give q_μ's (and hence α_μ's) significantly different from α at least on the first and second neighbor of the site $\mu_\circ$ on which the perturbation is localized. The next iteration will further expand the region where the q-values are significantly different from zero. Thus any argument on the range of a local perturbation must take into account the effect of an iterative process; which, in principle, may push it farther and farther away from the original center.

A formal analysis of this effect was proposed in 1980 and is briefly summarized in the next section. Here we add some more qualitative remarks.

As has been mentioned, the consideration that an SCF iteration based on eqn (11) will involve a change in the α-values of the centers close to the perturbed one may suggest that the range of the perturbation should be far larger at convergency than it appears to be at the first iteration.

This, however, is only a matter of principle. In practice, the situation may be much more favorable. The reason of this lies in the fact that the SCF procedure is equivalent to a negative-feedback scheme. The first iteration will generate, say, charges represented by the sequence

site	-3	-2	-1	0	1	2	3
charge	0.10	-0.13	0.19	-0.27	0.19	-0.13	0.10

for a perturbation localized at site 0. This means that the next iteration will see the α of atom 0 decrease, those of atoms -1 and 1 increase, those of atom -2 and 2 decrease, etc. Thus, the change in the net charge of, say, atom 3 at the second iteration will be opposite to that due to the first one. This alternating behavior will continue during the subsequent iterations, and therefore we can reasonably hope that the distance from site 0 to the site at which the initial perturbation is felt will be larger, but not far larger than estimated from the initial iteration.

We must leave the question as it stands, because numerical studies are still under way.

4. FORMAL ASPECTS: THE GREEN MATRIX AND LOCALIZED PERTURBATION.

Although many of the most interesting results on the effects and role of local perturbations can be described using very simple models and crude estimates there are cases when a more sophisticated formalism and possibly less crude calculations are advisable. This aspect of our subject has been treated in particular by Ladik and Seel (cf. bibliographical note). We shall briefly show here the essentials of that formal analysis, with some modifications whereby an adsorbed system is treated on a footing different from the corresponding local perturbation of the substrate.

We shall adopt the Green-matrix formalism in the LCAO one-electron model. The LCAO Hamiltonian matrix $\underline{\underline{H}}$ and the Green matrix $\underline{\underline{G}}$ of a polymer S plus and adsystem A can be written

$$\underline{\underline{H}} = \begin{vmatrix} \underline{\underline{H}}^{AA} & \underline{\underline{H}}^{AS} \\ \underline{\underline{H}}^{SA} & \underline{\underline{H}}^{SS} \end{vmatrix} \qquad \underline{\underline{G}} = (z\underline{\underline{S}}-\underline{\underline{H}})^{-1} = \begin{vmatrix} \bar{\underline{\underline{G}}}^{AA} & \underline{\underline{G}}^{AS} \\ \underline{\underline{G}}^{SA} & \bar{\underline{\underline{G}}}^{SS} \end{vmatrix} \qquad (14)$$

where z is a complex variable, $\underline{\underline{S}}$ is the overlap matrix; the bars on $\bar{G}^{AA}$ and $\bar{G}^{SS}$ remind that these are not Green matrices of the adsystem and substrate, which are

$$\underline{\underline{G}}^{AA} = (z\underline{\underline{S}}^{AA} - \underline{\underline{H}}^{AA})^{-1} \quad ; \quad \underline{\underline{G}}^{SS} = (z\underline{\underline{S}}^{SS} - \underline{\underline{H}}^{SS})^{-1} \qquad (15)$$

respectively; these matrices are not in general those of the isolated partners, because a perturbing field will be created by each of the partners on the other.

The matrix $\underline{\underline{H}}^{AS}$ will have, of course, non-zero elements connecting A to that or those monomers or unit cells of S where A is actually adsorbed.

By simple algebra it can be shown that

$$\bar{\underline{\underline{G}}}^{AA} = \underline{\underline{G}}^{AA} + \underline{\underline{G}}^{AA}\bar{\underline{\underline{H}}}^{AS}\underline{\underline{G}}^{SS}\bar{\underline{\underline{H}}}^{SA}\bar{\underline{\underline{G}}}^{AA} \qquad \text{where } \bar{\underline{\underline{H}}}^{AS} = (\underline{\underline{H}}^{AS} - z\underline{\underline{S}}^{AS}) = \bar{H}^{SA\dagger}$$

The matrix $\underline{\underline{G}}^{SS}$ can be related to the Green matrix $\underline{\underline{G}}^{\circ}$ of the free substrate by the usual Dyson equation

$$\underline{\underline{G}}^{SS} = \underline{G}^{\circ} + \underline{G}^{\circ}V\underline{G}^{SS} \qquad (16)$$

V being the perturbation matrix. Therefore, we obtain in the end

$$\bar{G}^{AA} = G^{AA} + \underline{\underline{G}}^{AA}\bar{\underline{\underline{H}}}^{AS}(\underline{\underline{1}}-\underline{\underline{G}}^{\circ}\underline{\underline{V}})^{-1}\underline{\underline{G}}^{\circ}\bar{\underline{\underline{H}}}^{SA}\bar{\underline{\underline{G}}}^{AA} \qquad (17)$$

This expression is especially interesting because normally $\underline{\underline{G}}^{\circ}$ is known; for instance, in a chain of N equal bonds like the bonds of polyacetilene, it is given by

$$G^{\circ}_{\mu\nu} = \frac{2}{N+1}\sum_k \frac{\sin \mu k\theta \sin \nu k\theta}{z - 2\cos k\theta} \qquad (18)$$

with $\theta = \pi/(N+1)$.

The matrix V is an 'induction-polarization' matrix; if it has non-zero elements only over a small number of sites, block polarization of eqn.(16) may lead to important simplifications. In the original Hückel method it would be strictly local,since it would result from corrections to the atomic parameters of the atoms linked to the 'adsorbed' molecule. In general, it will not only consist in corrections on all the matrix elements of the sites directly connected to the perturbing system, but will increase further and further in size along the iterations of an SCF procedure. The considerations of the preceding section apply to this case.

As is well known, formal treatments based on the Green matrix use the latter to derive all the properties of the system under study (in our case the electrons of A+S). For one thing, $\bar{\underline{\underline{G}}}^{AA}$ represents the Fourier transform of the probability amplitude for a state $|\mu\rangle$ to remain the same after a certain time t has elapsed

$$\frac{1}{2\pi i}\int_C \bar{G}^{AA}_{\mu\mu}\, e^{-izt}\, dz = \langle\mu|e^{-i\hat{H}t}|\mu\rangle \qquad (19)$$

where C is the usual circuit encircling the real axis from above; $\hat{H}$ is tha Hamiltonian operator corresponding to the matrix $\underline{\underline{H}}$ of eqn. (14). If $\bar{G}^{AA}$ has a pole on the real axis at $\varepsilon_{\mu\mu}$,i.e. has the form

$$\bar{G}^{AA}_{\mu\mu} = \frac{1}{z - \varepsilon_{\mu\mu}} \times \text{const.} + f(z) \qquad (20)$$

(with f(z) an analytical function of z) then, by Cauchy's integral formulas

$$\langle\mu|e^{-i\hat{H}t}|\mu\rangle = e^{-i\varepsilon_{\mu\mu}t}.\text{const.} \qquad (21)$$

and the state $|\mu\rangle$ is a stationary state (eigenstate) of the operator $\hat{H}$. We remind the reader that analytical continuation of $\bar{G}^{AA}_{\mu\mu}$ on the second Riemann sheet will yield a function which may have complex poles; these poles will have a negative imaginary part and will describe exponentially decaying states of the adsorbed impurity.

The charge-bond order matrix $\underline{\underline{P}}$ associated with $\underline{\underline{G}}$ is given (for doubly occupied orbitals) by

$$\underline{\underline{P}} = \hat{T}\underline{\underline{G}}\ , \quad \hat{T} = -\frac{2}{\pi}\int_{-\infty}^{\varepsilon_F} d\varepsilon \lim_{\text{Im}z\to 0^+} \text{Im}\ ; \qquad (22)$$

this expression makes it possible to write in particular the self--consistency condition in terms of the Green matrix

$$G = \{z\underline{\underline{S}} - \underline{\underline{H}}^{core} - \text{Tr}\left((\hat{T}\underline{\underline{G}})\underline{\underline{R}}\right)\}^{-1}, \qquad (23)$$

where $\underline{\underline{R}}$ is a supermatrix of two-electron integrals; the last term in parentheses being a matrix $\underline{\underline{X}}$ whose general term is

$$X_{\mu\nu} = \sum_{\rho,\tau} P_{\tau\rho}\left((\rho\tau|\mu\nu) - \frac{1}{2}(\rho\nu|\mu\tau)\right). \qquad (24)$$

5. COMMENT ON DYNAMICAL ASPECTS

The preceding brief mention of time-dependent aspects of local perturbations is intended as a hint to a novel and promising research field: the study of transport phenomena at the molecular level: in particular, the transmission of perturbations along polymer chains.

Such a transmission can take place according to two major types of mechanisms: electron and proton (ion) transfer. In quantum mechanical terms, it can be described as the evolution of a system

$$S_1PS_2 + e^- \text{ (or } H^+)$$

(where S_1 can be identified with an admolecule A) from an initial state

$$|0\rangle = |S_1^- PS_2\rangle \quad \text{or} \quad |(S_1H)^+ PS_2\rangle$$

to a final state

$$|(n+1)\rangle = |S_1PS_2^-\rangle \quad \text{or} \quad |S_1P(S_2H)^+\rangle$$

To illustrate briefly this problem let us consider a one dimensional chain of hydrogen bridges. This is a possibility in several polymers, especially polypeptides. Such a chain has a backbone typically represented by

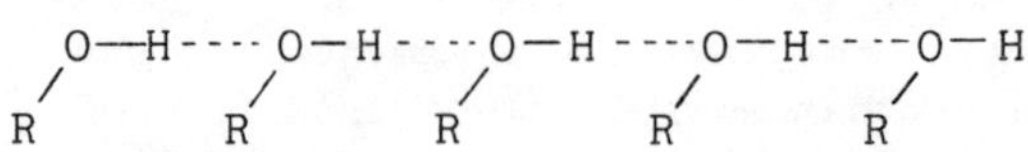

Addition of a proton to one of the alcoholic groups will give, say,

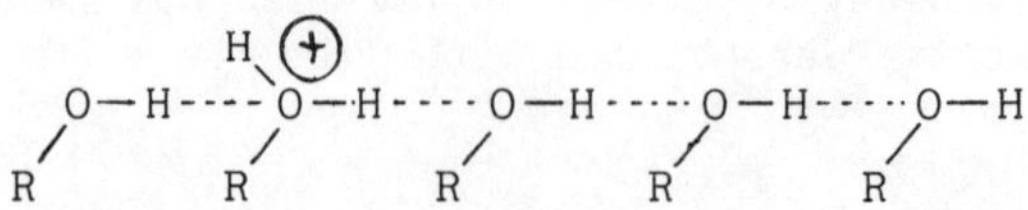

(I)

This chain may then undergo a "push-pull" rearrangement involving rotation of the additional OH bond and shift of the chain protons.

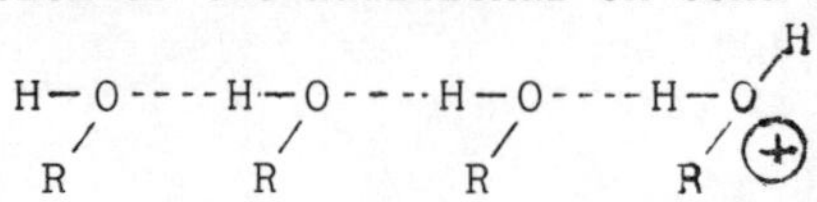

(II)

The terminal proton may then be yielded to acceptor S_2. Thus the local perturbation provided by the added proton, even if it extends very little away from a given site, may propagate far away from it.

The essentials of a theoretical treatment of this type of transition are as follows.

We want to know (a) the efficiency of the process, (b) its characteristic time. Both quantities are determined by the couplings (matrix element of the total Hamiltonian operator) between the states involved in the process. These states are: the states $|0$ and $|n+1\rangle$, the state $|1\rangle$ corresponding to the initial chain I, the state $|n$ corresponding to the final chain II, and the state $|j\rangle$ (J=2,...,n-1) where the ORH^+ ion is at the j-th oxygen atom.

If the couplings in question are only thermal, then a purely thermodynamical scheme applies, with the usual acrivation energies - and, of course, the choice between a stepwise and a concerted mechanism.

If the possibility of tunneling is taken into account, then the discussion becomes much more complicated and rests on the matrix elements of the temperature-dependent density operator exp(-$\hat{H}$/kT). The role of the local perturbation can then be illustrated by the considering only one proton and one hydrogen bridge:

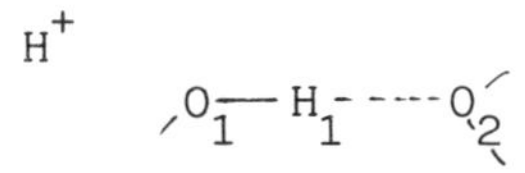

(III)

In the absence of an important influence of the additional proton, the 'channel' consisting in the shift of H_1 to O_2 is closed - viz. the coupling is very weak, say, because the left-hand well of the H-bridge double well is deeper than the right-hand one. On the other hand, the local perturbation induced by 'adsorption' of H^+ may modify the situation, so that the two systems

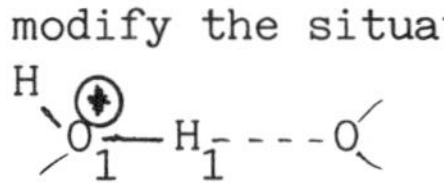

(IVa)

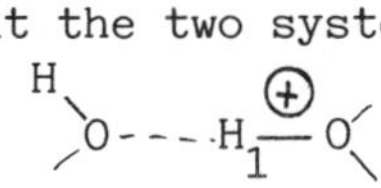

(IVb)

are close in energy and hence strongly coupled.

This situation makes the state corresponding to III (viz. the state $|0\rangle$ a quasi-stationary state with a finite life time. If the system in which it arises is a chain, the local perturbation will then be at site 2 after a certain time, give rise to the same effect, and hence move to site 3, even in the absence of thermal transitions. The combination of the tunnel and the thermal effect may indeed be responsible for a very fast propagation of the perturbation. A very striking example is anomalous conduction in water.

6. BIBLIOGRAPHICAL NOTE.

The local perturbation problem has a long history in solid-state physics. The 1969 paper by Newns on chemisorbed hydrogen is a classic in the field and may be consulted both for references to prvious work (in particular, Anderson's one) and as a fairly elementary application of Green matrices. Recent extensions with references to the general problems were provided with ref.s 1,2,3,13-15.

The method for σ electrons used here was proposed in ref.4 and the most recent review is given in re.5: here the interpretation of the atomic parameters as electronegativities is fully analized, following in ref.6. The MCF method was proposed in ref.7.

The electronegativity equalization principle seems to have been proposed by Sanderson in 1974 (ref.8) and has been used with varying success in theories of chemical bonds (cf. ref.9). The theoretical derivation of it in the case of donor-acceptor complexes has been given by Del Re in 1981 (ref.10).

A vary important source of SCF atomic data is provided by the Tables of Clementi and Roetti (ref.11).

The effect of a local perturbation on a polyene and its relat -

ionship with Friedel oscillations has been analyzed in ref.12. A generalization with use of the Green formalism is given in ref.15. Refs.2,3,13,14,15 contain the essential of the Green-matrix approach and additional references. For sec.5 cf. ref.16.

Acknowledgement.
These lecture notes include some background and some new results developed in the framework of a Research Project sponsored by the Italian National Council of Research and by the Italian Ministry of Education.

References.

1. D.M.Newns: Phys.Rev. 178, 1123 (1969)
2. J.Ladik,M.Seel: Phys.Rev. B 13, 5338 (1976)
3. G.Del Re,P.Otto,J.Ladik:Chem.Phys. 49, 321 (1980)
4. G.Del Re:J.Chem.Soc. 1958 ,4031
5. G.Del Re,G.Berthier,J.Serre: Lecture Notes in Chemistry,vol.13
6. M.T.Vandorffy: Acta Chim.Hung.(Budapest) 71,139 (1972)
7. P.Otto,J.Ladik: Chem.Phys.19, 209 (1977)
8. R.T.Sanderson, Educ.Chem. 11, 80 (1974)
9. R.G.Parr,R.A.Donnelly,M.Levy,W.E.Palke:J.Chem.Phys.68,3801 (1978
10. G.Del Re,J.Chem.Soc.(Far.Trans.2) 77, 2067 (1981)
11. E.Clementi,C.Roetti: At.Data and Nuclear Data Tables,14,177(1974
12. S.Canuto,J.L.Calais,O.Goscinski: J.Phys. B: At.Mol.Phys.14,1409 (1981)
13. G.A.Baraff,M.Schlüter:Phys.Rev.Lett.41,892 (1978)
14. J.Bernholc,N.O.Lipari,S.T.Pantelides:Phys.Rev.Lett.41,895 (1978)
15. M.Seel,G.Del Re,J.Ladik:Comp.Chem.1983 ,in press.
16. J.F.Nagle, M.Mille, H.J.Morowitz: J.Chem.Phys. 72, 3959 (1980)

ELECTRONIC LOCALIZATION AND DELOCALIZATION IN ORGANIC METALS AND SEMICONDUCTORS

Charles B. Duke
Xerox Corporation
800 Phillips Road, Bldg. 114
Webster, NY, 14580 (U.S.A.)

ABSTRACT. This paper is a brief review of three important developments which have occurred during the past decade in the theory of the electronic structure of organic materials. First, reliable and affordable semiempirical molecular orbital models of the electronic excitation spectra of large organic molecules and macromolecules have been developed and utilized extensively. Second, it has been recognized that under most circumstances disorder in solid-state polymers localizes the electronic excitations therein. Third, the collective behavior associated with quasi-one-dimensional systems has been studied intensively in numerous molecular solids, including pure and doped polyacetylene. These three developments, i.e., reliable MO theory, quantitative models of disorder, and the observation of collective phenomena in quasi-one-dimensional organic solids, have changed fundamentally both the conceptual foundations of the theory of electronic excitations in solid-state polymers and the design algorithms used to tailor these materials for specific applications.

I. INTRODUCTION

During the six years since the previous NATO school on the Quantum Theory of Polymers (1) three important developments in this topic have occurred. First, reliable and affordable molecular-orbital (MO) models of the electronic excitation spectra of large organic molecules (2) and polymers (3,4) have been developed and applied for a variety of purposes, including the design of xerographic photoreceptors (2,5) and battery electrodes (3,6). Second, the role of disorder in localizing electronic excitations

J. Ladik et al. (eds.), Quantum Chemistry of Polymers - Solid State Aspects, 325–335.

in polymers and molecular glasses has been recognized (7,8). Models describing the consequences of disorder have been utilized not only to interpret photoemission and UV absorption spectra from these materials but also to resolve important and long-unsolved practical problems such as the description of triboelectric charge transfer between polymers and both metals (9) and other polymers (10). Third, studies of the electrical properties of pure and doped polyacetylene have revealed evidence for the occurrence of a collective Peierls semiconducting ground state together with the unusual electronic excitation spectra which are implied by such an observation (11). The purposes of these lectures are to indicate the nature of each of these three developments and to provide references to the literature for those who wish to explore them in more detail.

II. ELECTRONIC STRUCTURE OF LARGE ORGANIC MOLECULES

The quantum theory of the electronic structure of infinite, periodic macromolecules is a well-developed and venerable topic which is reviewed often (1,12) and is well-known to the audience of this NATO school. The important aspect of this topic for our present purposes is the observation that ab initio (i.e., Hartree Fock plus configuration interaction) methods are simply too expensive to apply to the description of most of the large molecules (including polymers) of biological and technological interest. In addition, they fail to provide a quantitative description of UV absorption spectra without massive configuration-interaction analysis. Consequently, for quantum chemical calculations to be helpful in the design of polymeric materials of practical use in electronic applications (e.g., photoconductors (2,5), semiconductors (5), or battery electrodes (6)), some form of semiempirical model must be developed.

In the construction of semiempirical models it is useful to distinguish between ground-state properties (e.g., molecular geometries) and electronic excitation spectra. Typically models which describe one well describe the other poorly. Therefore in practical applications the molecular structure typically is obtained either experimentally or from one type of model (e.g., CNDO/2 and its extensions (13) or MNDO (14)) and then utilized as input to another model (e.g., CNDO/S3 (2,4) or VEH (2,6)) which is used to evaluate the electronic excitation spectra. Obviously this is not an optimal situation. It represents a useful compromise, however, between the demands of the problem at hand (i.e., the reliable evaluation of the electronic excitation spectra of large molecules), the limitations of the models, and the available computing power.

In this first lecture we consider the description of electronic excitation spectra by the CNDO/S3 molecular orbital model (2). It is a CNDO level model in which both the resonance integrals and coulomb integrals are obtained by empirical fits to the photoemission and UV absorption spectra of a few prototype molecules: benzene, substituted benzenes, pyrrole and furan, in particular (2,15). Typically low-binding-energy valence electron ionizations and low-energy $\pi \rightarrow \pi^*$ UV excited transitions are described to within an accuracy of 0.2eV using experimentally-determined molecular geometries. The model also has been applied to evaluate electron-molecular-vibration coupling constants in a variety of molecules (16). It has been utilized to interpret valence electron photoemission spectra (PES) and UV absorption spectra (UVA) of over 75 large organic molecules which include heteroatoms of O,S,N,F,Mg,Zn and Si (2,17). Thus, the successes and failures of the CNDO/S3 model are fairly well understood and documented, so that it can be utilized with reliability in the prediction of valence-electron PES and UVA.

The focus of our attention is the dependence of polymer electronic structure on molecular architecture. Since the subject is thoroughly reviewed elsewhere (18-20) the details of the analysis are not recapitulated here. Briefly, we begin by considering CHAIN ARCHITECTURES like polyethylene (PE) and polyacetylene (PA). PE is a one-electron insulator with a wide gap between its filled "bonding" orbitals and empty "antibonding" orbitals. PA, on the other hand, would be metallic if it exhibited only one CH group per unit cell. It is a Peierls insulator, however, because it dimerizes to form $(CH)_2$ unit cells, opening a small energy gap between orbitals which are bonding and antibonding across the short C=C units (4). This aspect of the molecular architecture of PA dominates its electronic excitation spectra.

We next consider RING ARCHITECTURES WITH SATURATED LINKAGES beginning with poly(p-phenylene) (21). Six sets of π orbitals occur for this macromolecule, one based on each of the six linearly independent orbitals of benzene. Three sets are occupied and three unoccupied. Emphasis is placed on the distinction between bonding (antibonding) π orbitals (which dominate the optical properties and charge transfer reactions) and non-bonding π orbitals (which dominate the valence-electron PES). As the saturated linkages between the phenyl moieties increase in length the interactions between these moieties diminish, leading to effectively non-bonding π excitations on the individual phenyl rings already for poly(p-xylylene) (22). Finally, the effects of heteroatoms in the ring are illustrated by poly(2,5-thienylene) (18,19) and polypyrrole (20,23). An important effect of the heteroatoms is their creation of frontier π orbitals which are very similar to those in PA, in spite of the difference between the molecular architecture of PA and that of polypyrrole and poly(2,5-thienylene).

We conclude with an examination of RING ARCHITECTURES WITH UNSATURATED LINKAGES. Analysis of poly(p-phenylene oxide) and poly(p-phenylene sulfide) (18,19) reveals how the p electrons on the chalcogen linkage groups can lead to extended π-electron orbitals even for highly non-planar conformations. A study of poly(p-phenylene vinylene) (20) is used to illustrate the occurrence of extended π-electron orbitals with unsaturated $(CH)_2$ linkage groups in contrast to the localized π orbitals resulting from the corresponding $(CH_2)_2$ linkage groups in poly(p-xylylene). Thus, the important consequence of unsaturated linkage groups is their generation of extended frontier orbitals between aromatic moieties even when these moieties are separated by large spatial distances and/or exhibit highly non-planar conformations.

In summary, the study presented in this first lecture of the influence of molecular architecture on the orbitals of ideal, periodic macromolecules has developed the insight required to understand the consequences of disorder for these rather different macromolecular species: the topic of the following lecture.

III. DISORDER AND LOCALIZATION: SMALL POLARONS IN POLYMERS AND MOLECULAR GLASSES

The study of disorder in solids has led, in the opinion of some (24), to a "revolution" in the sense of Thomas Kuhn in modern thinking about excitations in solids. In this view it is a fundamental fallacy to consider electronic excitations in polymers in terms of the traditional energy band theory of solids. Rather, such excitations in polymers should be regarded as localized entities which move through the solid via a series of incoherent "hops" (25). Moreover, the dynamics of this motion is that of small polarons because injected charges or excitons polarize the surrounding polymer medium (26,27). The illustration of the occurrence of this phenomena in three special cases, molecular glasses, pendant-group and molecularly-doped polymers (8,25-27), is the topic of this second lecture.

The local or extended nature of molecular-ion states in molecular solids is determined by a competition between fluctuations in the local site energies of these states (which tend to localize them) and the hopping integrals for inter-site excitation transfer (which tend to delocalize them). In order to define this fluctuation-induced localization concept more precisely, consider the model defined by the one-electron Hamiltonian

$$H = \sum_{i,\beta} \epsilon_i(\beta) a^+_{i\beta} a_{i\beta} + \sum_{i \neq i',\beta} t_{ii'} a^+_{i'\beta'} a_{i\beta}, \qquad (1)$$

in which i is a site index; β indicates the molecular orbital; the $t_{ii'}$ designate the intermolecular hopping integrals; and $a_{i\beta}$ is the annihilation operator for an electron occupying the orbital β at the site i.

The important new feature of polymers and molecular glasses, not present in periodic crystal lattices, is that the site energies, $\{\epsilon_i\}$ and the hopping integrals $\{t_{ij}\}$ form distributions. For example the probability of finding an eigenvalue ϵ_i with value ϵ is typically taken to be (8,26,27)

$$P(\epsilon) = (\pi\sigma^2)^{-1/2} \exp(-(\epsilon - \epsilon(\beta))^2/\sigma^2) \qquad (2)$$

in which σ^2 is the sum of a number of contributions associated with various types of disorder, and $\varepsilon(\beta)$ is an average eigenvalue associated with the molecular-ion state labelled by the orbital index β. Because both the $\{\varepsilon_i\}$ and $\{t_{ij}\}$ form distributions, the Hamiltonian (1) is specified statistically and hence can exhibit localized as well as extended engenstates (24,25).

The key question, of course, is what are the eigenfunctions and eigenvalues of Hamiltonian (1) given prescribed distributions of the $\{\varepsilon_i\}$ and $\{t_{ij}\}$. This question cannot be answered analytically at the present time. Some qualitative results have, however, been established numerically. The features of interest to us can be specified in terms of a mean site energy, $\bar{\varepsilon}$, the rms deviation from this mean, $\Delta = [\langle(\varepsilon_i - \varepsilon)^2\rangle_{Av}]^{1/2}$, and a mean hopping integral, t (28). Variations in the site energies from the mean (described by Δ) are referred to as "diagonal disorder" whereas analogous variations of the hopping integrals are called "off-diagonal disorder." Similarily, if these variations are caused by local time-independent fluctuations in composition or structure we speak of "static disorder" while if they are generated by the time-dependent thermal vibrations we employ the term "dynamic disorder." Injected charges are localized, i.e., they form molecular cations or anions within the solid, if

$$\Delta > czt \qquad (3)$$

in which c is a dimensionless number of the order of unity which depends both on the connectivity (i.e., dimensionality) of the system and on the extent of off-diagonal disorder. Typically

$c \sim 2.5$ for (isotropic) three-dimensional systems, 1.5 for two-dimensional systems, and zero for one-dimensional systems, although some confusion currently exists in the case of two dimensions (29). Inequality (3) is believed to be satisfied in molecular glasses and in certain aromatic polymers (7,8,25). In these cases electrons and holes injected into the material form local, molecular radical anions and cations, respectively, rather than extended mobile states like those characteristic of crystalline covalent semiconductors. Consequently, under these circumstances the electrons and holes are more accurately visualized as molecular ions in solution which move, however, by carrier hopping rather than by ionic diffusion.

Analyses of photoemission from polystyrene (26), poly(2-vinyl pyridine) (26), and poly(p-xylylene) (22) reveal that photoinjected holes are localized on individual phenyl moieties in these materials. In the case of polyacetylene (18,19), however, although π-electron hole states near the band edges are localized to eight or less carbon moieties, hole states near the center of the band are extended over twenty or more carbon species because of the large nearest-neighbor hopping integrals ($t \sim 2$ eV) in polyacetylene relative to polystyrene, poly(2-vinyl pyridine and poly(p-xylylene) ($t \sim 0.1$eV). Therefore we see that the localization of injected charges in polymers is related systematically to molecular architecture via the nature of the frontier orbitals discussed in lecture one (8,28).

Given the existence of molecular ions in polymers, their small-polaron dynamics is established by the interaction of the excess charge with those induced in the polymer matrix (8,26,27). For such a purpose the Hamiltonian of Eq. (1) must be extended to become

$$H = \sum_i \epsilon_i a_i^+ a_i + \sum_{i \neq j} (t_{ij} a_i^+ a_j + h.c.) + \sum_{i,n} g_{in} \hbar\omega_n a_i^+ a_i (b_n + b_n^+) + \sum_n \hbar\omega_n (b_n^+ b_n + 1/2). \qquad (4)$$

In Eqs. (4) a_i^+ designates the creation operator of an electron in an orbital of energy ϵ_i at site i, and b_n^+ that of a normal mode of vibration of energy $\hbar\omega_n$ coupled to the electronic state labelled by i via the dimensionless coupling constant g_{in}. The $\hbar\omega_n$ are the energies of the normal modes of vibration of the molecular medium in which we envision the sites to be embedded. The $\hbar\omega_n$ are obtained from the measured dielectric response of the polymer and the g_{in} are specified in terms of a classical electrostatics analysis of the interaction of a charge in a prescribed molecular orbital with the dielectrically active normal modes (26). This model has been applied successfully to interpret valence-electron photoemission and UV absorption spectra of polystyrene (7,26) and poly(2-vinyl pyridine) (26), as well as photoemission from a wide variety of molecular glasses (8,25). Transport data are less widely available, but the model is consistent with the limited data base currently in the literature (8,27). We conclude, therefore, that in molecular glasses (e.g., condensed ethyl benzene), pendant-group polymers (e.g., polystyrene) and molecularly-doped polymers (e.g., polycarbonate doped with trinitrofluorenone) the model defined by Eq. (3) with the $\{g_{in}\}$ and $\{\hbar\omega_n\}$ obtained from measured polymer dielectric response is both necessary and sufficient for the description of the small-polaron dynamics of charges and excitons injected into these materials. A thorough review of this topic has appeared in the literature (8), so we do not pursue it further here. The important aspect of the subject for us is the recognition and demonstration during the past five years that disorder does localize injected charges in polymers and that these charges exhibit small-polaron dynamics.

IV. DISORDER-INDUCED METALS: THE SEMICONDUCTOR-TO-METAL TRANSITION IN POLYACETYLENE

A third major innovation in polymer physics during the past decade is the recognition that because they are quasi-one-dimensional materials, some polymers exhibit collective semiconducting ground states(5,8). In contrast to the pervasive effects of disorder in almost all polymers, however, the occurrence of collective phenomena is uncommon. Specifically, it is characteristic of macromolecules like polyacetylene which have experienced a symmetry-lowering structural modification that introduces a semiconductor gap into what would have been a metallic electronic excitation spectrum. Moreover, in such materials the consequences of disorder can be dramatically different from those noted in the previous lecture. Thus, in this third and final lecture we examine the destruction of the collective semiconducting ("Peierls") ground state in doped polyacetylene by the interaction of the conduction electrons with a disordered array of donors or acceptors.

The context of the recent upsurge in interest in doped polyacetylene is the search for conducting polymers. In practice, such materials usually are prepared by dispersing particles or flakes of conducting material (e.g., carbon black or silver) in a polymeric binder to form a composite (5). At low concentrations of conducting filler, charge transport occurs via electron tunneling or hopping from one filler particle to another (30,31). At higher concentrations, $c > c_p \sim 30$ volume percent, conduction occurs via the formation of percolation pathways comprised of contacting filler particles within the polymer binder. A recent comprehensive review of the fabrication and uses of conducting polymers may be found in the literature (5).

The origin of the semiconducting behavior of pure polyacetylene is a lattice distortion of a symmetric $(CH)_n$ macromolecule, which would be metallic, into a dimerized configuration $(C_2H_2)_n$. Such distortions are thought to be inevitable in quasi-one-dimensional systems because of the accompanying gain in electronic energy (32). The resulting collective semiconducting state can be destroyed, however, by disorder. A model of the disorder-induced semiconductor-to-metal transition has been constructed by Mele and Rice (33). Since this model provides the only coherent interpretation of all of the available experimental data (34), it constitutes the basis of our discussion of the semiconductor-to-metal transition in doped polyacetylene.

Polyacetylene exhibits a doubly degenerate commensurate Peierls insulating ground state characterized by a bond-alternating geometry of the $(CH)_n$ backbone and a Peierls gap, E_p, in the electronic excitation spectrum. The introduction of charge transfer dopants initially induces domain walls (called "kinks" or (improperly) "solitons") between the two degenerate phases. For acceptor dopants like iodine, each kink is associated with an (induced) empty mid-gap electronic state, $E_k = E_p/2$, above the top edge of the valence band. The spatial extent of these kinks is about $10\underline{a}$, where $\underline{a}$ is the average distance between two carbon species along the backbone. Each charged kink is bound to the acceptor to which it has transferred its charge. As long as these kinks do not overlap (i.e., dopant concentrations are $c < c^* \sim \underline{a}/(20\underline{a})=0.05$), the Peierls state is insensitive to disorder and acts like a semiconductor of gap E_p with a suitable number of empty localized mid-gap states with spin 0 and charge +e. Consequently, the new feature of polyacetylene relative to the pendant-group polymers considered in the previous lecture is the distortion of the polyacetylene conformation by an injected charge. This distortion, in turn, induces a local mid-gap defect state initially independent of the disorder either inherent in the polymer or characteristic of the dopants. Moreover, this mid-gap state exhibits an unusual relationship between its spin and charge, i.e., charged states are spinless.

The introduction of localized defect states by injected charges is not peculiar to polyacetylene. Rather, it is a general feature of certain amorphous semiconductors (35). Indeed, amorphous semiconductors may be divided into two broad classes: disorder dominated and defect dominated. Pendant-group polymers and a-Si fall into the former class whereas doped polyacetylene and a-Se fall into the latter. From this more general perspective the quasi-one-dimensional nature of polyacetylene is relevant only because it affords the microscopic mechanism which causes injected charges to induce localized defect states in the Peierls gap. It is the defect states themselves which dominate the spectroscopic and transport properties of the material.

When $c > c^*$ in doped polyacetylene, isolated charged kink defects cease to be the ground state of the system. At these concentrations the defects interact, and in an ordered periodic system the lowest-energy state becomes a uniform incommensurate charge density wave (CDW) with its associated distorted geometry of the polyacetylene backbone. Disorder of the dopants destroys this CDW state, however, by pinning the induced charges on the backbone at the random positions of the impurities. This pinning destroys the order parameter, thereby eliminating both the one-electron gap and the regular bond alternation. In this limit polyacetylene becomes a "dirty" metal.

The predictions of the detailed analysis of Mele and Rice (33) are in remarkable correspondence with a variety of measurements on iodine-doped polyacetylene (34). First a sharp rise in the conductivity occurs at $c \sim 0.01$ associated with variable range hopping within (disorder-induced) electronic states in the gap just above the valence band edge. Then, in the range $0.05 \lesssim c \lesssim 0.1$ the magnetic susceptibility rises because of the breakdown of the Peierls state and the concomitant elimination of the gap. In particular, for $0.01 \lesssim c \lesssim 0.05$ the remarkable phenomenon of conductivity without comparable susceptibility is predicted, in excellent accord with both recent data and the early data (36) that originally stimulated the "soliton" concept. Therefore, doped polyacetylene affords an example in which disorder destroys a collective semiconducting ground state and produces an apparently "spinless" conductor as an intermediate stage characterized by only the partial breakdown of this state as reflected in a low but continuous distribution of electronic states in the Peierls gap. If the Mele-Rice model is correct, heavily doped polyacetylene is literally a disorder-induced metal.

V. SYNOPSIS

The past decade has been the scene of some remarkable advances in the understanding of the electrical properties of organic solids. Experimentally many new semiconducting and metallic organic materials have been prepared and studied, some of which have begun to achieve widespread practical application (5). From the perspective of theoretical modeling, reliable quantum chemical calculations for large molecules, radical improvements in the theory of disordered systems, and the proposal of quantitative models of collective semiconducting states in quasi-one-dimensional solids have conspired to yield quantitative interpretations of most of the available spectral and transport data on electrically active organics as well as useful guidance in the design of new materials. In these three lectures we have become acquainted with highlights of the advances in the theory of the electrical properties of polymers and molecular glasses, in particular. More extensive discussions of past accomplishments may be found in the literature cited. We can confidently anticipate, moreover, that further developments will continue unabated thereby rendering the subject of continuing interest in the future as well.

REFERENCES

1. J.M. André, J. Delhalle and J. Ladik, eds., Quantum Theory of Polymers, D. Reidel, Dordrecht, 1978, 365pp.
2. C.B. Duke, Int. J. Quant. Chem. Quant. Chem. Symp. 13, 267 (1979).
3. J.L. Brédas, R.R. Chance, R.H. Baughman and R. Silbey, Int. J. Quant. Chem. Quant. Chem. Symp. 15, 231 (1981).
4. W.K. Ford, C.B. Duke and A. Paton, J. Chem. Phys. 77, 4564 (1982).
5. C.B. Duke and H.W. Gibson, "Conductive Polymers," in Kirk-Othmer: Encyclopedia of Chemical Technology, D. Eckroth, ed., John Wiley, New York, 1982, Third Edition, Vol. 18, pp. 755-793.
6. J.L. Brédas, R. Silbey, D.S. Boudreaux and R.R. Chance, J. Am. Chem. Soc., in press.
7. C.B. Duke, Mol. Crystal. Liq. Cryst. 50, 63 (1979).
8. C.B. Duke, "The Electronic Structure of Semiconducting Polymers," in Extended Linear Chain Compounds, J.S. Miller, ed., Plenum, New York, 1982, Vol. 2, pp. 59-125.
9. T.J. Fabish and C.B. Duke, J. Appl. Phys. 48 4256 (1977).
10. C.B. Duke and T.J. Fabish, J. Appl. Phys. 49 315 (1978).
11. A.J. Heeger and A.G. MacDiarmid, "Solitons in Polyacetylene: A Summary of Experimental Results," in Physics in One Dimension, J. Bernasconi and T. Schneider, eds., Springer Series in Solid State Sciences, Springer-Verlag, Berlin, 1981, Vol. 23, pp. 179-193.
12. M. Kertesz, Adv. Quant. Chem. 15, 161 (1982).
13. J.A. Pople and D.L. Beveridge, Approximate Molecular Orbital Theory, McGraw Hill, New York, 1970, 214pp.
14. M.J.S. Dewar and W. Thiel, J. Am. Chem. Soc. 99, 4899 (1977).
15. N.O. Lipari and C.B. Duke, J. Chem. Phys. 63, 1748 (1975).
16. C.B. Duke, "Electron Interactions with Molecular Vibrations:

Their Nature and Consequences," in Synthesis and Properties of Low-Dimensional Materials, J.S. Miller and A.J. Epstein, eds., Ann. New York Acad. Sci., New York Academy of Sciences, New York, 1978, Vol. 313, pp. 166-178.
17. C.B. Duke, J. Vac. Sci. Technol. 15, 157 (1978).
18. C.B. Duke and A. Paton, in Conductive Polymers, R.B. Seymour, ed., Plenum, New York, 1981, pp. 155-169.
19. C.B. Duke, A. Paton and W.R. Salaneck, Mol. Cryst. Liq. Cryst.. 83, 177 (1982).
20. W.K. Ford and C.B. Duke, Mol. Cryst. Liq. Cryst. 93, 327 (1983).
21. W.K. Ford, C.B. Duke and A. Paton, J. Chem. Phys. 78, 4734 (1983).
22. C.B. Duke, R.W. Bigelow, A. Dilks, A. Paton, W.R. Salaneck and H.R. Thomas, Chem. Phys. Lett. 83, 255 (1981).
23. W.K. Ford, C.B. Duke and W.R. Salaneck, J. Chem. Phys. 77, 5030 (1982).
24. P.W. Anderson, "Lectures on Amorphous Systems," in Ill-Condensed Matter, R. Balian, R. Maynard and G. Toulouse, eds., North Holland, Amsterdam, 1979, pp. 159-261.
25. C.B. Duke, "Organic Solids: Traditional Semiconductors or Fermi Glasses?" in Festkörperprobleme, J. Treusch, ed., Advances in Solid State Physics, Vieweg, Braunschweig, 1982, Vol. 22, pp. 21-34.
26. C.B. Duke, W.R. Salaneck, T.J. Fabish, J.J. Ritsko, H.R. Thomas and A. Paton, Phys. Rev. B 18, 5717 (1978).
27. C.B. Duke and R.J. Meyer Phys. Rev. B 23, 2111 (1981).
28. C.B. Duke, Surf. Sci. 70, 674 (1978).
29. D. Weaire, "The Anderson Localization Problem," in Fundamental Physics of Amorphous Semiconductors, F. Yonezawa, ed., Springer Series in Solid State Sciences, Springer-Verlag, Berlin, 1981, Vol. 25, pp. 155-163.
30. E.K. Sichel, J.I. Gittleman and P. Sheng, Phys. Rev. B 18, 5712 (1978).
31. E.K. Sichel, Appl. Phys. Commun. 1, 83 (1981).
32. P.M. Chaikin, "Review of the Physics of Low-Dimensional Systems," in Synthesis and Properties of Low-Dimensional Materials, J.S. Miller and A.J. Epstein, eds., Ann. New York Acad. Sci., New York Academy of Sciences, New York, 1978, Vol. 313, pp. 128-144.
33. E.J. Mele and M.J. Rice, Phys. Rev. B 23,5397 (1981).
34. A.J. Epstein, H. Rommelmann, R. Bigelow, H.W. Gibson, D.M. Hoffman and D.B. Tanner, Phys. Rev. Lett. 50, 1866 (1983).
35. D. Adler, "Defects in Covalent Amorphous Semiconductors," in Fundamental Physics of Amorphous Semiconductors, F. Yonezawa, ed., Springer Series in Solid State Sciences, Springer Verlag, Berlin, 1981, Vol. 25, pp. 14-32.
36. W.P. Su, J.R. Schrieffer and A.J. Heeger, Phys. Rev. Lett. 42, 1698 (1979).

ELECTRONIC STRUCTURE OF HIGHLY CONDUCTING POLYMERS AND BIOPOLYMERS; SOLID STATE ASPECTS

Janos J. Ladik

Chair for Theoretical Chemistry, Friedrich-Alexander-University Erlangen-Nürnberg, D-8520 Erlangen, Egerlandstrasse 3, FRG
and Laboratory of the National Foundation for Cancer Research at the Chair for Theoretical Chemistry, University Erlangen-Nürnberg

ABSTRACT. - After a short review of the *ab initio* SCF LCAO crystal orbital method the negative factor technique for the determination of the density of states in a disordered chain is presented. After this the problem of electron correlation in polymers is discussed and the generation of correlated (quasi-particle) valence and conduction bands is reviewed.

The gap in alternating trans-polyacetylene is described on the basis of its quasi-particle band structure and the question of the Bloch-type conduction in DNA (either through doping or the possibility of intrinsic conduction due to charge transfer from the sugar rings to the nucleotide bases) is discussed. This is followed by a brief discussion of the electronic structure of disordered polypeptide chains.

Finally for the interpretation of the known biochemical activation mechanisms of human oncogens the different possible short- and long-range effects of carcinogen binding to DNA is summarized. The possiblity of soliton-generation in DNA through carcinogen binding and subsequent removal as one of the probable long-range effects is discussed in some detail.

1. INTRODUCTION

Polymers play an important role as plastics. Biopolymers like nucleic acids (DNA and RNA), proteins, polysacharides, lipids etc. have fundamental significance in life processes. In the last decade highly conducting polymers became objects of extensive experimental investigations being candidates for the discovery of new physical

J. Ladik et al. (eds.), Quantum Chemistry of Polymers - Solid State Aspects, 337–359.

phenomena and various attempts are made for their technical application (batteries etc.). To understand the different physical and chemical properties of polymers (which underlie in the case of biopolymers also their biological functions) one has to know their electronic structure.

To treat quantum mechanically highly conducting polymers or biopolymers one has to proceed stepwise.

1.) If the polymer is periodic (like polyacetylene, or homopolynucleotides or homopolypeptides) one can start with the calculation of their band structures using the *ab initio* SCF LCAO crystal orbital (CO) method /1/.
2.) As next step one has to take into account that nucleic acids or proteins, but also lightly or mediumly doped polyacetylene and polydiacetylene are aperiodic and therefore one has to apply appropriate methods for the treatment of disorder in quasi-one-dimensional chains/2/ using as input the results of periodic chain calculations (see below).
3.) Especially in the case of biopolymers it is important to take into account the effect of the environment (for instance surrounding water molecules and ions) on the band structure (in the case of a periodic chain) or on the level distribution (in the case of a disordered chain) by constructing an effective potential of the environment /3/.
4. Finally, one should take into account (using a good basis set) also the major part of the electronic correlation /4/.
5.) Having executed these steps for a polymer one is in a good position to compute different properties (electronic and vibrational spectra, transport- and magnetic properties etc.).
6.) In the case of biopolymers it is again especially important to treat interactions between chains (the genetic regulation of a cell depends first of all on DNA-protein interactions in the nucleoprotein complexes).

Since the unit cells in homopolynucleotides and homopolypeptides are in most cases fairly large, calculations performed until now are mostly only minimal basis *ab initio* band structure calculations. There are a few cases when aperiodicity in DNA and proteins were treated and a pilot calculation has been performed to treat the effect of the surrounding water on a cytosine stack /3/.

No correlation calculations have been performed until now on biopolymers, but such computations have been successfully executed in the cases of polymers with small unit cells (trans-polyacetylene and polydiacetylene; see below). The same holds for exciton spectra which have been successfully computed applying intermediate (charge transfer) exciton theory /5/ for the above mentioned two chains /6/. One should mention, however, that only the inclusion of the major part of correlation resulted in results in reasonable agreement with experiment. There is an early calculation on transport properties of periodic DNA models using simple tight binding (Hückel) band structures /7/. In the case of polyethylene ($(CH_2)_x$) a good basis set calculation with the inclusion again of the major part of the correlation even the mechanical properties of this system could be calculated obtaining results /8/ in good agreement with experiment. Finally, one should mention that there exist first calculations on the interactions between homopolynucleotides and a polyglycine in different conformations of the latter chain and in different relative positions of the interacting biopolymers /9/.

The good results obtained for chains with small unit cells make it probable that with further improved mathematical and numerical techniques and applying even larger computers, in the next few years the calculations on the electron structure and properties of biopolymers will reach the same level of sophistication as those for the above mentioned simple chains.

2. METHODS

2.1. Ab initio Crystal Orbital Method

If there is a translational (or more generally any periodic) symmetry in an infinite solid or polymers the infinite cyclic hypermatrix (which one obtains in any LCAO theory with periodic boundary conditions) can be brought with the aid of a simple unitary transformation into a block-diagonal form /1,10/. The order of these blocks is (in the ab initio case) equal to the number of basis functions in the unit cell. In this way the original hypermatrix equation splits into N+1 matrix equations if N+1 denotes the number of blocks (unit cells). Each such equation has an index which denotes the serial number of the matrix block to which the equation belongs. If $N \rightarrow \infty$ this serial number can be considered continous. Physically it is the vector $\vec{k}$ (or one of its components in the one-dimensional (1D) case) of the reciprocal lattice /1,10/.

We do not give here a derivation of the ab initio SCF LCAO CO method (for this see /1,10/), only the final expressions are written down in the 1D case. Let us write down a crystal orbital in the form of a linear combination of Bloch orbitals.

$$\Psi_n(k,\vec{r}) = 1/\sqrt{N+1} \sum_{j=-N/2}^{N/2} e^{ikja} \sum_{s=1}^{m} c(k)_{n,s} \chi_s(\vec{r}-\vec{R}_{js}). \qquad (1)$$

Here N+1 is the number of unit cells, m the number of basis functions in it, a the elementray translation, $\chi_s(\vec{r}-\vec{R}_{js})$ the s-th atomic orbital (AO) centered at the s-th atom ($s \in S$) in the j-th unit cell (this atom has the position vector $\vec{R}_{js}$) and n is the band index. After a Ritz variational procedure one obtains a generalized eigenvalue equation

$$\underline{\underline{F}}(k)\, \underline{c}_n(k) = \varepsilon_n(k)\, \underline{\underline{S}}(k)\, \underline{c}_n(k) \qquad (2)$$

for the determination of the coefficients $c(k)_{n,s}$. In equ. (2) the overlap matrix $\underline{\underline{S}}(k)$ is the Fourier transform of the matrix blocks containing the overlap integrals between basis functions belonging to different cells,

$$\underline{\underline{S}}(k) = \sum_{q=-N/2}^{N/2} e^{ikqa}\underline{\underline{S}}(q); \left[\underline{\underline{S}}(q)\right]_{r,s} = \langle \chi_r^o | \chi_s^q \rangle \quad (3)$$

(in the second equation of (3) the subscripts are basis function and the superscripts cell indices). Similarly one obtains /1,10/

$$\underline{\underline{F}}(k) = \sum_{q=-N/2}^{N/2} e^{ikqa} \underline{\underline{F}}(q) \quad (4)$$

The elements of the matrices $\underline{\underline{F}}(q)$ are defined according to the detailed derivation /1,10/ as

$$\left[\underline{\underline{F}}(q)\right]_{r,s} = \langle \chi_r^o | \hat{H}^N | \chi_s^q \rangle + \sum_{u,v,=1}^{m} \sum_{q_1,q_2=-N/2}^{N/2} \underline{\underline{P}}(q_1 - q_2)_{u,v} \times$$

$$\times \left[\langle \chi_r^o(1) \chi_u^{q_1}(2) | 1/r_{12} | \chi_s^q(1) \chi_v^{q_2}(2) \rangle - 1/2 \langle \chi_r^o(1) \chi_u^{q_1}(2) | 1/r_{12} | \chi_v^{q_2}(1) \cdot \chi_s^q(2) \rangle \right] \quad (5)$$

Here $\hat{H}^N$ stands for the one-electron part (kinetic energy + interaction with all the nuclei) of the chain and the elements of the charge-bond order matrix $\underline{\underline{P}}(q_1 - q_2)$ are defined as a generalization of the definition given by Coulson as

$$\underline{\underline{P}}(q_1 - q_2)_{u,v} = 2\, a/2\pi \int_{-\pi/a}^{\pi/a} \sum_{h=1}^{n^*} c_{h,u}(k) c_{h,v}(k) e^{ika(q_1-q_2)} dk \quad (6)$$

where n^* denotes the number of filled bands. (In the case of a metal with partially filled band(s), one has to integrate over k only for the filled part(s) of the Brillouin zone.)

Looking at equ.-s (2)-(6) it becomes obvious that they represent nothing else but Roothaan's SCF LCAO equations for a closed shell system /11/ generalized for an infinite chain with periodic boundary conditions. One has to solve them at a number of k points to be able to construct the matrices $\underline{\underline{P}}(q_1-q_2)$ for the next iteration step. Using appropriate numerical techniques usually it is enough to consider 6-8 k-points between 0 and $\pi/2$ $[\varepsilon(k) = \varepsilon(-k)]$ to obtain consistent results. The SCF procedure does not converge, however, in most cases so easily as in the corresponding molecules, especially with a larger basis set. The main reason for this is that in actual calculations one has to interrupt the summations (3) and (4) after a limited number of neighbors (more accurately to keep charge neutrality and to obtain reasonably reliable results one has to go until different numbers of neighbors for different types of intercell interaction integrals; for details see /12/) and the error caused by this procedure is strongly amplified in cases when the matrix $\underline{\underline{S}}(k)$ has some very small eigenvalues /13/. (For the elimination of $\underline{\underline{S}}(k)$ in equ. (2) one uses (similarly to the molecular case) again Löwdin's symmetric orthogonalization procedure.)

Finally, it should be mentioned that the formalism described here is valid also for the case of a repeated combined symmetry operation (for instance helix operation). As group theoretical considerations show it in this case 1.) one has to put the nuclei into the right positions by moving from one cell to the next and 2.) one has to rotate correspondingly also the basis functions /14/.

2.2. Negative Factor Counting Method for the Treatment of Disorder in Quasi-Onedimensional Systems

To determine the level distribution (density of states, DOS) in an aperiodic polymer chain one can apply the negative factor counting (NFC) method /2/. According to this method if we write for the disordered chain a Hückel determinant which is tridiagonal due to the fact that only first neighbors interactions are taken into account

$$|\underline{\underline{H}}(\lambda)| = \begin{vmatrix} \alpha_1-\lambda & \beta_2 & 0 & 0 \ldots & 0 \\ \beta_2 & \alpha_2-\lambda & \beta_3 & 0 & \\ 0 & \beta_3 & \alpha_3-\lambda & \beta_4 & \\ \vdots & & & \ddots & \\ 0 & & & \beta_N & \alpha_N-\lambda \end{vmatrix} = 0, \quad (7)$$

this can be easily transformed into a didiagonal form with the help of successive Gaussian eliminations. Therefore the determinant

$$|\underline{\underline{H}}(\lambda)| = \prod_{i=1}^{N} (\lambda_i - \lambda) \quad (8)$$

can be rewritten as

$$|\underline{\underline{H}}(\lambda)| = \prod_{i=1}^{N} \varepsilon_i(\lambda), \quad (9)$$

where the diagonal elements of the didiagaonal determinant are given by the simple recursion relation

$$\varepsilon_i(\lambda) = \alpha_i - \lambda - \beta_i^2/\varepsilon_{i-1}(\lambda), \qquad i = 1,2,3,\ldots,N, \tag{10a}$$

$$\varepsilon_1(\lambda) = \alpha_1 - \lambda \,. \tag{10b}$$

Comparing equ.-s (8) and (9) it is easy to see that for a given λ value, the number of eigenvalues smaller than λ ($\lambda_i < \lambda$) has to be equal to the number of negative $\varepsilon_i(\lambda)$ factors in equ. (9) /2/. (Calculations of all the eigenvalues λ_i for a long chain ($N = 10^4$ or 10^3) is impossible, but the computation of $\varepsilon_i(\lambda)$ factors with the help of Equ.-s (10a) and (10b) is very fast.) By giving λ different values throughout the spectrum and taking the difference of the number of negative $\varepsilon_i(\lambda)$'s belonging to consecutive λ values, one can obtain a histogram for the distribution of eigenvalues (density of states) of $\underline{\underline{H}}$ for any desired accuracy.

For actual calculations one has to make a band structure calculation for each component of the disordered chain assuming that it is periodically repeated. Then the values α_i (diagonal elements of $\underline{\underline{H}}$) can be determined from the positions of the bands of the components (the middle point or weighted middle points of the bands) and the off-diagonal elements β_i from the widths of the bands. It should be pointed out that the NFC in this simple form gives only the level distribution belonging to one band (for instance, valence band or conduction band) of the disordered chain.)

In the case of disordered quasi-1D systems the NFC method can also be applied for the case of an arbitrary number of orbitals per site either in an <u>ab initio</u> form (Day's and Martino's paper in ref. /23/ which contains also the detailed derivation) or in a semiempirical, for instance, extended Hückel form /15/; namely, one can show that if one has the secular determinant instead of a tridiagonal in a triblockdiagonal form

$$|\underline{\underline{F}} - \lambda \underline{\underline{S}}| = \begin{vmatrix} \underline{\underline{A}}_1 - \lambda \underline{\underline{S}}_1 & \underline{\underline{B}}_2 - \lambda \underline{\underline{Q}}_2 & & \underline{\underline{0}} & \\ \underline{\underline{B}}_2^{\dagger} - \lambda \underline{\underline{Q}}_2^{\dagger} & \underline{\underline{A}}_2 - \lambda \underline{\underline{S}}_2 & \underline{\underline{B}}_3 - \lambda \underline{\underline{Q}}_3 & & \\ \underline{\underline{0}} & \underline{\underline{B}}_3^{\dagger} - \lambda \underline{\underline{Q}}_3^{\dagger} & \underline{\underline{A}}_3 - \lambda \underline{\underline{S}}_3 & \underline{\underline{B}}_4 - \lambda \underline{\underline{Q}}_4 & \\ & & & \ddots & \\ & & & & \underline{\underline{A}}_N - \lambda \underline{\underline{S}}_N \end{vmatrix} = 0, \tag{11}$$

one can obtain for it the expression

$$|\underline{\underline{F}} - \lambda \underline{\underline{S}}| = \left(\prod_{i=1}^{n} s_i\right)\left(\prod_{j=1}^{n} (\lambda_j - \lambda)\right) = \prod_{i=1}^{N} \left(\prod_{k=1}^{l_i} u_{ik}(\lambda)\right); n = \sum_{i=1}^{N} l_i \,. \tag{12}$$

Here the s_i are the eigenvalues of the matrix $\underline{\underline{S}}$ and the λ_j are the roots of the generalized eigenvalue equation

$$\underline{\underline{F}}c_j = \lambda_j \underline{\underline{S}}c_j \ . \tag{13}$$

Further $u_{ik}(\lambda)$ stands for the kth eigenvalue of the matrix

$$\underline{\underline{U}}_i(\lambda) = \underline{\underline{A}}_i - \lambda \underline{\underline{S}}_i - (\underline{\underline{B}}_i^T - \lambda \underline{\underline{Q}}_i^T)\underline{\underline{U}}_{i-1}^{-1}(\underline{\underline{B}}_i - \lambda \underline{\underline{Q}}_i)\left[\underline{\underline{U}}_1(\lambda) = \underline{\underline{A}}_1 - \lambda \underline{\underline{S}}_1\right] \tag{13a}$$

and l_i is the dimension of the ith diagonal matrix block. The matrices $\underline{\underline{U}}_i(\lambda)$ can be diagonalized easily for a given value of λ. The number of the negative $u_{ik}(\lambda)$'s has again to be equal to the number of eigenvalues λ_j which are smaller than the chosen value. Changing λ, the whole spectrum can again be scanned and the density of states curve of the disordered system (in this case taking into account all the bands) can be obtained in this way to any desired accuracy.

To build up the matrix blocks occurring in equ. (5), one has to perform <u>ab initio</u> calculations both for the different units (to construct the diagonal blocks) and for their different clusters to construct the off-diagonal blocks (for instance, in a first neighbors interactions approximation in a binary disordered chain one has to calculate the AA, AB and BB clusters, where A and B are the two different units).

In connection with the NFC method in its matrix blocks form finally it should be mentioned that using the inverse iteration technique /16/ one can compute the wave function belonging to any particular energy level of the disordered system (which can be determined with any desired accuracy with the help of the NFC method). Keeping the part of electronic density (charge-bond order matrix) arising from the core levels and lower lying valence levels (which is known from the cluster calculations) constant one can iterate until self-consistency the physically interesting part (which lies near to the Fermi level) of the matrix $\underline{\underline{P}}$. In this way one obtains an approximate SCF solution for the whole disordered system /17/.

2.3. Treatment of Correlation in Polymers

For the calculation of the correlation energy per unit cell in the ground state of a polymer (either conductor or an insulator) one can use any size-consistent method (perturbation theory /18/, coupled cluster expansion /19/, electron pair theories /20/, etc.). In the case of insulators one can Fourier transform the delocalized Bloch orbitals into site semilocalized Wannier functions (WF-s) and perform the excitations between Wannier functions belonging to near lying sites /4/. (For the generation of optimally localized Wannier functions see /21/.) This procedure is, however, not possible in the case of metals, because no Wannier functions can be constructed from the Bloch states belonging only to the filled part of the valence band.

In the case of metals (but also in the case of insulators) one can take a grid in k in each band and perform the excitations between the chosen k values. (One can take either an equidistant grid in k, or one can also apply a non-equidistant grid in which case one can subdivide the band according to the DOS distribution; for details see the first reference of /4/.) To be able to perform the four index transformation necessary for any correlation calculation one can expand the LCAO coefficients $c(k)_{n,s}$ occurring in equ. (1) in a simple trigonometric series /22/.

$$c(k)_{n,s} = c^{(o)}_{n,s} + \sum_{j=1}^{N} c^{(j)}_{n,s} \cos(jk) \quad . \tag{14}$$

Actual calculations have shown that usually it is enough to go in this series until N = 2 or 3 /12,22/. In this way in any matrix element for a double excitation

$$\langle IJ \| AB \rangle \equiv \langle \Psi_i(k_1,\vec{r}_1)\Psi_j(k_2,\vec{r}_2) | 1/r_{12}(1-\hat{P}_{12}) | \Psi_a(k_3,\vec{r}_1)\Psi_b(k_4,\vec{r}_2)\rangle \tag{15}$$

the k-dependent parts of the Bloch functions (see equ. (1)) can be brought before the integration and so the integrals over the AO-s have to be calculated only once. (In (15) the band indices i and j denote filled and a and b unfilled bands, respectively, and $\hat{P}_{12}$ is the exchange operator.) This procedure made it possible to take a rather dense grid for all the three independent k-values in (15) ($k_1 + k_2 \equiv k_3 + k_4$).

As first step second order Moeller-Plesset perturbation theory (MP2) /20/ has been applied for different chains with small unit cells (see point 3). The MP2 correction to the Hartree-Fock energy can be written with the help of (15) in the simple form

$$E_2 = - \sum_{I,J,A,B} \frac{|\langle IJ \| AB\rangle|^2}{\varepsilon_A+\varepsilon_B-\varepsilon_I-\varepsilon_J} \tag{16}$$

where the summations over the composite indices I, J etc. means summations over band indices (i,j, etc.), k-values and spins.

It should be pointed out here that in the case of polymers with a larger molecule as unit cell (like TCNQ and TTF stacks, nucleotide base stacks or homopolypeptides with a larger amino acid residue as unit cell) it is necessary to localize further the WF's or Bloch orbitals into different parts of the molecule which is repeated in the periodic polymer. For that purpose one has to investigate how is possible to combine the Fourier transformation leading to WF's with the usual localization procedures /24,25/. (Localization studies on the four nucleotide bases have shown that both the Boys /24/ and the Edminston-Ruedenberg /25/ procedure have given excellent localization for the filled as well as for the virtual MO-s of the molecules /26/. One can hope only in this way to be able to select the important excited configurations in a way that their (for the correlation calculation) necessary number remains still low enough for an actual calculation.

A further difficulty arises from the fact that neither the WF's nor the wave functions obtained from a localization inside a molecule are solutions of the canonical Hartree-Fock equations. Therefore, any method treating correlation (perturbation theory, coupled cluster expansion, etc.) has to be reformulated for this case. This work has been done in the case of perturbation theory for the localization inside a molecule /27/ and it is in progress for the coupled cluster method /28/.

Besides the total energy one can correct in the band structure at least the physically interesting conduction and valence bands of a polymer by taking into account correlation effects. Namely, one can define following Takeuti's /5/ and Kunz and coworkers /29/ electronic polaron model (see also /22/) quasiparticle (QP) energy levels in the conduction and valence band, respectively,

$$\varepsilon_c^{QP}(k_c) = E^{(N+1)}(k_c) - E^{(N)} = A(k_c) \ , \tag{17a}$$

$$\varepsilon_v^{QP}(k_v) = E^{(N)} - E^{(N-1)}(k_v) = -I(k_v) \ . \tag{17b}$$

Here $E^{(N+1)}(k_c)$ is the total energy per unit cell of the chain with an extra electron in the conduction band at level k_c, $E^{(N)}$ is the total energy per unit cell of the ground state and $E^{(N-1)}(k_v)$ stands for the total energy per unit cell of the system with one electron missing from the level k_v of the valence band. Per definitionem $A(k_c)$ is the electron affinity (the energy gained by putting the extra electron into the conduction band at level k_c) and $I(k_v)$ the ionization potential (the energy needed to ionize an electron from level k_v of the valence band), respectively.

One can approximate the exact total energies $E = E_{HF} + E_{corr}$ as /22/

$$E = E_{HF} + E_2 \tag{18}$$

where E_2 is the correlation correction in the approximation of MP2. Substituting (18) and the expressions (Koopmans' theorem)

$$\varepsilon_c^{HF}(k_c) = E_{HF}^{(N+1)}(k_c) - E_{HF}^{(N)} \tag{19a}$$

$$\varepsilon_v^{HF}(k_v) = E_{HF}^{(N)} - E_{HF}^{(N-1)}(k_v) \tag{19b}$$

into equ.-s (15a) and (15b) one obtains

$$\varepsilon_c^{QP}(k_c) = \varepsilon_c^{HF}(k_c) + E_2^{(N+1)}(k_c) - E_2^{(N)} \tag{20a}$$

$$\varepsilon_v^{QP}(k_v) = \varepsilon_v^{HF}(k_v) + E_2^{(N)} - E_2^{(N-1)}(k_v) \tag{20b}$$

Substituting into equ.-s (20) the corresponding MP2 expressions (see equ. (16) we can write /22/

$$E_2^{(N)} = \sum_I \sum_J{}' \varepsilon_{IJ}^{(N)} = \sum_{I\neq V} \sum_{J\neq V} \varepsilon_{IJ}^{(N)} + \sum_{I\neq V} \varepsilon_{IV}^{(N)} \tag{21a}$$

$$E_2^{(N-1)}(k_v) = \sum_{I\neq V} \sum_{J\neq V}{}' \varepsilon_{IJ}^{(N-1)V} \tag{21b}$$

$$E_2^{(N+1)}(k_c) = \sum_I^{(c)} \sum_J^{(c)}{}' \varepsilon_{IJ}^{(N+1)C} = \sum_I^{(c)} \sum_J^{(c)} \varepsilon_{IJ}^{(N+1)C} + \sum_I \varepsilon_{IC}^{(N+1)C} \tag{21c}$$

Here the prime at the summation over J means that the states I = J have to be excluded, the pair correlation contributions $\varepsilon_{IJ}^{(N)}$ are defined as

$$\varepsilon_{IJ}^{(N)} = - \sum_{AB} \frac{|\langle IJ \| AB\rangle|^2}{\varepsilon_A + \varepsilon_B - \varepsilon_I - \varepsilon_J} \tag{22}$$

V stands for the state v, k_v and C for the state c,k_c, respectively. Further in equ. (21c) the summations over I and J contains also (with the exception of the last sum of the right hand side of (21c)) the extra occupied conduction bands state c . Substituting equ.-s (21) with the definition (22) into equs.-s (20) one obtains after some calculation the final expressions /22/

$$\varepsilon_C^{QP}(k_c) = \varepsilon_C^{HF}(k_c) + \Sigma_C^{(N+1)}(e) + \Sigma_C^{(N+1)}(h), \tag{23a}$$

$$\varepsilon_V^{QP}(k_v) = \varepsilon_V^{HF}(k_v) + \Sigma_V^{(N)}(e) + \Sigma_V^{(N)}(h). \tag{23b}$$

Here the electron and hole self-energies Σ are defined as

$$\Sigma_C^{(N+1)}(e) = \sum_I \varepsilon_{IC}^{(N+1)C} \tag{24a}$$

$$\Sigma_C^{(N+1)}(h) = \sum_I^{(c)} \sum_J^{(c)}{}' \varepsilon_{IJ}^{(N+1)C} - \sum_I \sum_J{}' \varepsilon_{ij}^{(N)} \tag{24b}$$

$$\Sigma_V^{(N)}(e) = \sum_{I\neq V} \varepsilon_{IV}^{(N)} \tag{24c}$$

$$\Sigma_V^{(N)}(h) = - \sum_{I\neq V} \sum_{J\neq V}{}' (\varepsilon_{IJ}^{(N)} - \varepsilon_{IJ}^{(N-1)V}) \tag{24d}$$

From equ.-s (24) it is easy to see the physical meaning of the self energies: $\Sigma_C^{(N+1)}(e)$ describes the new pair correlations formed between the extra electron in state C and all the other electrons and $\Sigma_C^{(N+1)}(h)$ gives the reduction of the pair correlations due to the fact that no scattering of the electrons into these newly occupied states is possible. Similarly $\Sigma_V^{(N)}(e)$ expresses the increase of pair correlations caused by the new scattering possibility to the empty state V and $\Sigma_V^{N}(h)$ gives pair correlations between the other electrons and the one in the state V. From equ.-s (24) and from this discussion follows that $\Sigma_C^{(N+1)}(e)$ and $\Sigma_V^{(N)}(e)$ are negative while $\Sigma_C^{(N+1)}(h)$ and $\Sigma_V^{(N)}(h)$ are positive. As the detiled derivation shows it /12,22/

$$\left|\Sigma_C^{(N+1)}(e)\right| > \left|\Sigma_C^{(N+1)}(h)\right| \tag{25a}$$

but

$$\left|\Sigma_V^{(N)}(e)\right| < \left|\Sigma_V^{(N)}(h)\right| \quad . \tag{25b}$$

Therefore the quasi particle gap is always smaller than the HF one. Further considerations /12,22/ indicate also that the band widths of the conduction and valence bands, respectively, are smaller in the QP (correlated) description, than in the HF case.

3. DISCUSSION OF SELECTED RESULTS

3.1. The Gap of Alternating Trans-Polyacetylene

If one performs a minimal (STO-3G) basis calculation for alternating trans-polyacetylene (PA) using bond distances determined from geometry optimization with this base, one obtains a gap of 8.91 eV /22/. This is more than four times the experimental value of $\sim$2 eV which is the position of the first peak in the absorption spectrum of pure trans PA /30/. (It should be mentioned that in trans PA there is no exciton band /6/ and therefore one can take this value as the gap value.) If one performs the geometry optimization with this basis, but takes the MP2 correction for correlation in the total energy the gap still remains 8.28 eV /22/.

If Table I we reproduce the results of Suhai /12,22/ for the gap of alternating trans PA using HF + MP2 optimized bond distances and different basis sets.

TABLE I. The quasi particle (QP) gap of alternating trans PA calculated in the HF + MP2 level and the Hartree-Fock (HF) gap using different basis sets (in eV-s)

Basis	ΔE_{HF}	ΔE_{QP}
STO-3G	8.28	7.67
STO-6-31G	4.93	4.13
STO-6-31G**	4.28	2.98

As we can see from the Table the minimal basis gap of 8.28 eV reduces to 4.28 eV if one applies the STO-31G** (double ζ + polarization functions both on the carbon and hydrogen atoms). If one calculates the QP gap at the MP2 level using the method described in part 2.3 the gap reduces only by about 0.6 eV in the case of the minimal basis, but it decreases by 1.3 eV using the STO-6-31G** basis. As detailed correlation calculations /3,12,22/ have shown it, to obtain with MP2 about 75 per cent of the correlation energy and correspondingly a larger reduction of the gap one has to use 1.) a good basis and 2.) to take into account in the double excitations $I \rightarrow A$, $J \rightarrow B$ all the filled and unfilled bands obtained in the given basis.

The value ΔE_{QP}=2.98 eV is still by $\sim$ 1 eV larger, than the experimental one /30/. This discrepancy can be explained, however, by 1.) the missing part of correlation (which would reduce according to the estimate of Suhai /22/ the theoretical gap of trans PA to $\sim$2.5 eV), by 2.) the neglect of relaxation effects (the N+1 and N-1 particle states were calculated in the formalism given in 2.3 with the help of wave functions obtained for the N particle state) and by 3.) the neglect of phonon polaron effects /22/. It should be mentioned that to take into account relaxation effects one could use the so-called "open-shell" SCF LCAO crystal orbital method published already in 1975 /10/.

3.2. The Possibilities of Bloch-type Conduction in DNA

3.2.A. <u>Conduction through doping</u>
Calculating with an STO-3G basis /31/ or with Clementi's minimal

basis set /32/ the band structures of the four nucleotide base (A,G,T and C) stacks in the conformation of B DNA /33/ we have found for the widths of the valence and conduction bands, respectively, values between 0.3 and 0.8 eV /34,35/. These widths lie in the same range as those of the valence band of a TTF stack (0.3 eV) and the conduction band of a TCNQ stack (1.2 eV) /36/ obtained with the aid of the STO-3G basis /31/. The TCNQ - TTF mixed molecular crystal is, due to the ∼ 0.6 e internal charge transfer (CT) from TTF to TCNQ, highly conducting. Since the physically interesting bands in this system (the valence band of TTF from which the CT occurs and the conduction band of TCNQ to which the CT takes place) have about the same widths as the nucleotide base stacks, one would expect that by appropriate doping (which is in the case of DNA by no means a trivial task) double stranded DNA with periodic DNA sequences (the base pair A-T repeated, or G-C repeated) could be made well conducting. Experiments along these lines are in progress.

In connection with this problem it should be mentioned that though native DNA is aperiodic, a mathematical statistical analysis of the experimentally determined DNA fragments sequences /37/ show that these sequences are by no means random, but there is a strong preference to have the same unit repeated several times /38/, and in some cases even twenty - fifty times /37/. In the light of these findings the possibility of Bloch-conduction through doping in the partially periodic base pair stacks becomes interesting also for the case of native DNA.

3.2.B. Is periodic DNA an intrinsic conductor?

The ab initio band structure calculations performed for the three homopolynucleotides (sugar-phophate-nucleotide base together forming the unit cell) polycytidine /34/, polyadenilic acid and polythymidine /35/ show a charge transfer of ∼ 0.2 e from the sugar part of the nucleotide to the corresponding nucleotide base. Though in the case of a better basis set the value of the transferred charge would most probably decrease somewhat, it still would remain non-negligible (∼ 0.15 e). The inspection of the band structures of these systems shows that the bands originating from the sugar-phosphate chain and those from the base stack are still distinguishable /34,35/.

The band structures of these polynucleotides show that these composite systems are still insulators (completely filled valence band) with a gap of ∼ 12 eV /34,35/. This very high gap excludes the possibility of another population of the bands by taking away a charge of ∼ 0.2 e from the highest lying sugar-phosphate-type band (which is the n^{*}-1 th band, the highest filled band being a base stack-type band /34,35/) and putting it to the lowest lying base stack-type band.

On the other hand we have seen in the case of $(CH)_x$ that by the application of a good basis the quasi particle gap becomes 3.0 eV instead of the minimal basis value of 8.3 eV /22, and point 3.1

in this article/. Further a recent 4-31 G (double ζ) split valence basis calculation for a cytosine stack resulted in a reduced gap of $\sim$6.8 eV /39/(instead of the $\sim$ 12 eV minimal basis calculation). Therefore one can expect that a 4-31 G** + MP2 calculation for nucleotide base stacks would give quasi particle gaps of $\sim$ 5 eV. Such calculations are in progress for the base stacks, but most probably cannot be executed in the next few years for the homopolynucleotides .

Considering these estimated reduced gaps one can raise again the question whether one would not obtain a deeper total energy per unit cell for a homopolynucletide if one would take into account the sugar - base charge transfer by the population of the bands (see above) in the course of the iteration procedure. If this would be the case one could expect that the periodic sugar-phosphate chain of DNA and the periodic segments of the nucleotide base stacks would show an intrinsic Bloch type conduction. This conduction would be of course perturbed by the aperiodic effective field acting on the sugar-phosphate chain due to the different base pairs (in the overwhelming non-periodic parts of the DNA sequences) and in all cases by the different water and ion structures around the base pairs /40/ - and in the case of nucleoproteins - by the neighborhood of different amino acid residues at different parts of the chain. Though this aperiodicity effects act in the direction of conductivity by hopping mechanism the possibility of Bloch-type conduction cannot be excluded even in this case (the mechanism of conduction if free charge carriers are present depends first of all on the relative position of the mobility edge with respect to the Fermi level). This interesting problem certainly deserves further investigations.

3.3. The Effect of Aperiodicity on the Electronic Structure of Multi-Component Polypeptide Chains

The homopolypeptides polyglycine and polyalanine have valence and conduction band widths, respectively, of $\sim$ 0.5 eV and $\sim$ 1.0 eV in the MINDO/3 CO approximation /41/ (the corresponding <u>ab initio</u> SCF LCAO CO band widths are larger by about a factor of 2 /42/). The valence band of polyalanine is shifted by about half of the band width (by 0.27 eV /41/) to higher (less negative) energies as compared to the valence band of polyglycine. This shift is enough to produce a band splitting in the periodic glyalaglyala... chain with a gap of $\sim$0.25 eV and the two new bands have widths of about 0.17 eV only /41/. This result is already quite surprising and contradictory to the chemical intuition, since alanine differs from glycine only by the substitution of an H atom by a CH_3 group.

Simple NFC calculations of a glycine chain with alanine impurities show already at the 5 per cent alanine case different peaks and gaps in the DOS spectrum of the valence band of glycine.

At 30 per cent alanine concentration the original band becomes completely destroyed and one can speak only about a distribution of levels in the region of the valence bands of polyglycine and polyalanine (for further details see the paper of Seel at ref. /2/).

In a subsequent paper /43/ the NFC technique in its more sophisticated matrix block form has been applied to mixed glycine-serine chains. The necessary matrix blocks (single glycine and serine units as well as gly-gly, gly-ser, gly-gly-ser clusters) were computed again in the MINDO/3 approximation to enable comparisons with the results of CO calculations performed for the corresponding periodic chains /41/.

The comparison of the results obtained by direct CO and by NFC calculation of the DOS-s show even in the case of the more complicated periodic poly(gly-gly-ser) chain excellent agreement (the band limits and the peaks in the DOS-s curves agree within 0.002 a.u. $\approx$ 0.05 eV which was the step length applied in the NFC procedure). Therefore one can expect that also in the case of disordered chains the NFC method in its matrix block form provides highly reliable results. For quasi-1D-systems there seems to be no other method which can compete with this technique. For 2- and 3D disordered systems unfortunately there is no way to write down a tridiagonal (or triblockdiagonal) matrix and therefore the method is not applicable (for further details see the paper of Ladik at ref. /2/).

In the case of a random chain of 38 gly and 42 ser units (nearly 50 - 50 per cent composition) one obtains again a DOS-s histogram which is completely different from the DOS-s curve of a periodic poly(gly-ser) chain. In this case, however, (in contrary to the gly-ala case) the conduction bands of gly and ser as well the valence bands of both systems merge into much broader bands, than those of the components. This example shows that the problem of disorder in proteins (as in DNA) is a very intricate one. The very fact of disorder does not mean necessary the exclusion of the possibility of Bloch-type conduction. This depends first on the position of the Fermi level (is it within a region of continously distributed energy levels, or does it fall into a gap?) and on the localization properties of the states. In other words if the Fermi level falls into such a region of states which are still delocalized (the Fermi level is above the mobility edge) Bloch-type (coherent) conduction is possible, while in the opposite case (the Fermi level is below the mobility edge) only hopping-type conduction occurs /44/. For this reason the investigation of localization properties of the states in a disordered system like a protein chain is of utmost importance for which a Green matrix technique has been developed /44,45/.

4. POSSIBLE MICROPHYSICAL MECHANISMS OF THE ACTIVATION OF ONCOGENS THROUGH CARCINOGENS

After the discovery of oncoviruses and oncogens in plants and animals in the last few years about twenty-five human oncogens have been discovered. These oncogens show the following three basic biochemical mechanisms (at least until now these three mechamisms are known). In one case a simple base substitution (the codon GGC has been changed to GTC) transforms a protooncogene to an active oncogene which causes human EJ bladder carcinoma /46/. This change which has been determined through the sequencing of important part of the gene in normal and tumor cells /46/ corresponds to a single glycine-valine substitution in the protein coded by this gene. Most probably the substitution of glycine by the more bulky valine inactivates an enzyme (or changes its function) which virtually changes the regulation of the cells in which this change occurs.

In the second case a gene occurring in normal cells becomes overactivated (without any change in its sequence) by the binding to its end a new control element (the so-called long terminal repeat (LTR)) /47/. The slicing out of LTR (which is of viral origin) from another part of DNA and its insertion to DNA at the end of the oncogene happens in an enzymatic way /47/.

The overproduction of the protein coded by the overactivated oncogene (by binding to it the LTR) obviously changes so much again the regulation of the cells which leads again to malignant transformation /47/.

The third mechanism is similar to the second one. In this case a whole normally functioning gene gets sliced out from one chromosome region and becomes inserted into another "more active" chromosomal region /48/. This DNA transposition (performed with the help of appropriate enzymes) causes again the overproduction of the protein coded by this gene. This results again in such changes of cell regulation /48/ which results at the final end in the development of a tumor.

In the case of $\begin{smallmatrix}G\\G\\C\end{smallmatrix} \longrightarrow \begin{smallmatrix}G\\T\\C\end{smallmatrix}$ point mutation (first mechanism) one has to observe that there is no way to get a G $\longrightarrow$ T base substitution in the same strand of DNA with the help of the usual Watson-Crick tautomeric shift-mutation mechanism /49/. On the other hand if during a duplication procedure in the complementary strand of DNA instead of C an A molecule is build in (A can form two hydrogen bonds with G), in the consequent duplication instead of G the nucleotide base T can be incorporated into the original strand.

Under normal conditions the described procedure is not possible, because both G and A are purine bases and therefore for a G-A base pair there would not be enough space in the double helical structure of B DNA. On the other hand if the double helix becomes through the binding of a chemical carcinogen (especially through the binding of a bulky one) in the neighborhood of the critical $\begin{matrix}G\\G\\C\end{matrix}$ codon distorted enough, the formation of the unusual G-A base pair can be easily visualized.

In a previous paper /50/ we have reviewed the most probable local (as well as long range) effects caused by carcinogen binding to DNA and/or proteins. Besides changes in charge distribution of the DNA constituents (including charge transfer) and changes of the vibrations in the neighborhood of the carcinogen binding (as well as breaking of bonds or forming of new bonds) first of all conformational changes caused by binding of carcinogens can be expected /50/. In this way though the $\begin{matrix}G\\G\\C\end{matrix} \rightarrow \begin{matrix}G\\T\\C\end{matrix}$ cancer causing point mutation cannot be explained by the usual tautomeric shifts (Watson-Crick mechanism /49/) in the base pair, it is still understandable on the basis of local (short-range) effects caused by carcinogen binding.

This is, however, not the case for the second and third above mentioned mechanisms. The transposition of LTR sequence in DNA (2. mechanism) or the transposition of a whole gene to another chromosomal region is explained on the biochemical level by enzyme action /47,48/. To understand, however, on the physico-chemical, how the binding of another DNA sequence (the LTR) to the end of a gene influences strongly its regulation, one has to suppose that DNA-protein interactions are essentially dependent on the sequences of the neighboring DNA segments. (The overactivation of the oncogene means most probably that it is blocked by a protein in a smaller fraction of time, than under normal conditions). In the same way for the transposition of longer DNA sequences (at mechanism 2 of LTR, at mechamism 3 of a whole gene) into another location one has to assume that the DNA-protein interactions of these DNA segments have been weakened in an extent (the DNA segments become deblocked) that the necessary enzymatic reactions could take place. Since the experiments described in ref.-s /47/ and /48/ have been performed with cancer cells and the corresponding normal cells, they do not contain any information for the start of these processes.

The most plausible assumption is also in the cases of mechanisms 2 and 3 that they were initiated by carcinogens and/or radiations which have besides local also long-range effects. In ref. /50/ a number of possibilities for long-range effects of carcinogen bindings on DNA-protein interactions have been reviewed. (Change of the strength of DNA-protein interaction in a longer sequence caused by charge transfer which can influence strongly the polarization and dispersion forces between the two chains. Long-range effect of carcinogen binding on the tertiary structure which results again in a change of DNA-protein interaction. Additional aperiodicity caused by carcinogen binding influencing DNA-protein interactions, etc. (for details see /50/)).

In the present paper we should like to elaborate only on one possibility, only shortly mentioned in /50/: long-range effects of carcinogens through formation of solitons in DNA. Let us assume that a bulky carcinogen, like the ultimate of 3,4-benzpyrene is bound to a nucleotide base. Certainly in the neighborhood of the attached carcinogen the structure of DNA, first of all the structure of the stacked base pairs will become strongly distorted. With this conformational change a change in the electronic structure of DNA will be coupled because of course the electronic interaction due to the stacking of the base pairs is strongly dependent on the relative position of the superimposed base pairs. In this way a non-linear change (conformational change coupled with electronic structure change) takes place at the site and neighborhood of the carcinogen binding.

In vitro a covalently bound ultimate carcinogen would remain attached to DNA for an indefinitely long time. Not this is the case, however, in vivo where for instance repair enzymes can remove the carcinogen within a few hours /51/. Until the carcinogen sits at its binding site the above described non-linear change will remain localized to the neighborhood of this site. After the removal of the carcinogen, however, it seems rather probable that the system will not relax immediately (the original conformation will not be restored instantaneously),because for this bulky molecular constituents of DNA (together with the water and ion structure surrounding them) have to be moved. On the other hand it is well known that 1.) solitons have several orders of magnitude longer life times and 2.) they can travel as solitary waves in an extended system /52/. Therefore one can postulate that after the removal of a carcinogen from DNA the non-linear (but previously local) change caused by its binding can travel through rather large distances along the chain causing a long-range effect which may effect along a larger segment of DNA its interaction with a protein molecule .

To test this hypothesis one can start by writing down a Hamiltonian for the soliton. Generalizing the theory of Su, Schrieffer and Heeger (SSH) /53/ which was developed for the non-linear change caused by a link in polyacetylene, we can write

$$H = H_{el} + H_{conf} + H_{el-conf} \tag{26}$$

Here we take as first (and rough) approximation for the description of the overlapping π electrons of the stacked basis H_{el} in the form (Hückel-type, or tight-binding approximation)

$$H_{el} = -t_o \sum_{n,s,\alpha} (C^+_{n+1,s,\alpha}C_{n,s,\alpha} + C^+_{n,s,\alpha}C_{n+1,s,\alpha}) \tag{27}$$

where n is the site index (which base), s($\pm$ 1/2) stands for the spin and α = 1,2 indicates the strand in the DNA double helix. Further the C^+-s and C-s are creation and annihilation operators, respectively, and t_o is the hopping integral for the undistorted chain. One should point out immediately that H_{el} is uncomplete because it does not contain electron-electron interaction terms. In a more complete formulation this has to be taken into account. Work along these lines is in progress.

To describe the conformational changes of the stacked bases one has to introduce three variables (in contrary of the one variable of the polyacetylene case /53/). We can denote by $Z_{n,\alpha}$ the shift from its equilibrium position along the Z axis (the main axis of the double helix) of the n-th base in the α-s chain and by $\Phi_{n,\alpha}$(following Krumhansl and Alexander /55/) the rotation of this basis (again measured from its equilibrium position) in the plane perpendicular to the Z axis. Finally, the angle $\Theta_{n,\alpha}$ measures the tilting of the base (thus Θ is the angle between the plane of the displaced base and the plane perpendicular to the Z axis). Assuming that the motion of the base described by these three variables can still be treated in the harmonic approximation, we can write

$$H_{conf} = 1/2 \sum_{n,\alpha} \Big[K_1(Z_{n+1,\alpha} - Z_{n,\alpha})^2 + K_2(\Phi_{n+1,\alpha} - \Phi_{n,\alpha})^2 + K_3(\Theta_{n+1,\alpha} - \Theta_{n,\alpha})^2 + M_{n,\alpha}(\dot{Z}^2_{n,\alpha} + \dot{\Phi}^2_{n,\alpha} + \dot{\Theta}^2_{n,\alpha}) \Big] \qquad (28)$$

Here K_1, K_2 and K_3 are the force constants belonging to the three variables Z, Φ and Θ and $M_{n\alpha}$ is the mass of the n-th base in the α-th strand ($M_{n\alpha}$ of course has only four different values).

Finally, for the coupling of the conformational change with the electronic structure change we have to introduce the modified hopping integral $t_{n+1,\alpha;n,\alpha}$ as

$$t_{n+1,\alpha;n,\alpha} = t_o - \Big[\beta_1(Z_{n+1,\alpha} - Z_{n,\alpha}) + \beta_2(\Phi_{n+1,\alpha} - \Phi_{n,\alpha}) + \beta_3(\Theta_{n+1,\alpha} - \Theta_{n,\alpha}) \Big] \qquad (29)$$

where β_1, β_2 and β_3 are the electron-base displacement (phonon) coupling constants. One should point out that equ. (29) is a straightforward generalization of equ. (2.2) of SSH /53/. Using the expression /30/ we can write for the non-linear electron-displacement coupling term

$$H_{el\text{-}conf} = \sum_{n,s,\alpha} (t_{n+1,\alpha;n,\alpha} - t_o)\,(C^+_{n+1,s,\alpha} C_{n,s,\alpha} + C^+_{n,s,\alpha} C_{n+1,s,\alpha}) \qquad (30)$$

Following SSH /53/ and Krumhansl and Alexander /55/ respectively, one can substitute the Hamiltonian (26) with definitions (27) - (30) (after taking expectation values of the electron-operators with the ground state many electron function) into the classical canonical equations of motion of Hamilton. For acutal numerical calculations one has to choose a set K_i and β_i (i = 1,2,3) values or one can try to determine them on the basis of quantum mechanical potential surface calculations.

The solutions of the classical equations of motion will then derive the time evolution of the solitary wave generated at the site of carcinogen binding. In this way one can learn about the life time and the range of the travelling non-linear distortion caused by carcinogen binding and subsequent releasing as a function of the parameter values. Obviously such results will have besides their physical significance a profound biological importance.

Finally, it should be mentioned that in our model we have neglected 1.) in contrary to Krumhansl and Alexander /55/ the coupling distortions of the stacked bases with the changes in the conformation of the sugar rings (the sugar puckering), 2.) non-linear terms in the motion of the bases and the sugar rings and 3.) the coupling to the environment (water molecules and ions). In a more sophisticated description which is in progress /56/ these effects together with electron-electron interaction terms should be included in the Hamiltonian. Further one has to point out that equ.-s (26)-(30) as they stand provide only three uncoupled poly-acetylene soliton problems in which only the parameter values (K_i and β_i) are different. One can, however, supplement equ.-s (28)-(29) with terms which would provide couplings between the coordinates Z_n, ϕ_n and Q_n, respectively.

ACKNOWLEDGMENT

The author would like to express his gratitude to the NATO Scientific Affairs Division for making possible to organize an Advanced Study Institute on the quantum theory of polymers. He is very much indebted to Professors J. Cizek, E. Clementi, T.C. Collins, G. Del Re, W. Forbes, M. Lax and P. Otto as well as to Dr.-s G. Corongiu, R. Day, B. Gazdy, M. Seel and S. Suhai for the continuous cooperation and for many fruitful discussions. He is further indebted to the "Kraftwerk Union AG" for providing free computer time and to the "Volkswagen" Foundation as well as to the "Fond der Chemischen Industrie" for financial support.

REFERENCES

/1/ Del Re, G., Ladik, J. and Biczó, G., Phys. Rev. 155 (1967) 997; André, J.-M., Gouverneur, L. and Leroy, G., Int. J. Quant. Chem. 1 (1967) 427 and 451.

/2/ Dean, P., Roy. Soc. London, Ser. A 254 (1960) 507; ibid 260 (1961) 263; Dean, P., Rev. Mod. Phys. 44 (1972) 127; Seel. M., Chem. Phys. 143 (1979) 103, Day, R.S. and Martino F., Chem. Phys. Lett. 84 (1981) 86; Day, R.S. and Ladik, J., Int. J. Quant. Chem. 21 (1982) 917; for a recent review see: Ladik, J., Int. J. Quant. Chem. 23 (1981) 1073.

/3/ Ladik, J., Lecture at Int. Symp. on Quantum Biology and Quantum Pharmacology, Palm Coast, FL (1981) unpublished; Otto, P., Ladik, J., Corongiu, G., Suhai, S., and Förner, W., J. Chem. Phys. 77 (1982) 5026.

/4/ Ladik, J. in "Recent Advances in the Quanum Theory of Polymers", ed. André, J.-M., Brédas, J.-L., Delhalle, J., Ladik, J., Leroy, G. and Moser, C., Springer Verlag, Berlin-Heidelberg-New York, 1980, p. 155; Suhai, S. and Ladik, J., J. Phys. C.: Solid State Phys. 15 (1982) 4327; Ladik, J., Int. J. Quant. Chem. 23 (1983) 1073.

/5/ Takeuti, Y., Progr. Theor. Phys. (Kyoto) 18 (1957) 421; Progr. Theor. Phys. (Kyoto) Suppl. 12 (1975) 75.

/6/ Suhai, S., J. Chem. Phys. (submitted).

/7/ Suhai, S., J. Chem. Phys. 57 (1972) 5599.

/8/ Suhai, S., J. Polymer Sci., Polymer Phys. 21 (1983) 000.

/9/ Otto, P., Clementi, E., Ladik, J. and Martino, F., J. Chem. Phys. (submitted); see also the contribution of Otto, P. in this volume, p. 361.

/10/ Ladik, J., in "Electronic Structure of Polymers and Molecular Crystals", ed. André, J.-M. and Ladik, J., Plenum Press, New York-London, 1975, p. 23.

/11/ Roothaan, C.C.J., Rev. Mod. Phys. 23 (1951) 69.

/12/ Suhai, S., in "Quantenmechanische Untersuchungen an quasieindimensionalen Festkörpern", Habilitation Thesis, Erlangen, 1983; J. Chem. Phys. (submitted).

/13/ Suhai, S., Bagus, P. and Ladik, J., Chem. Phys. 68 (1982) 467.

/14/ Ukrainski, I.I., Theor. Chim. Acta, 28 (1975) 139; Blumen, A. and Merkel, Ch., Phys. stat. sol. 83 (1977) 425.

/15/ Kertész, M. and Göndör, Gy., J. Phys. C 14 (1981) 851.

/16/ Wilkinson, J.H., "The Algebraic Eigenvalue Problem", Clarendon Press, Oxford.

/17/ Gadzy, B., Seel, M. and Ladik, J., Chem. Phys. Lett. (submitted).

/18/ Moeller, C. and Plesset, S., Phys. Rev. 46 (1934) 618.

/19/ Cizek, J., J. Chem. Phys. 45 (1966) 4256; Cizek, J. and Paldus, J., Int. J. Quant. Chem. 5 (1975) 359; Adv. Quant. Chem. 9 (1975) 105.

/20/ Ahlrichs, R. and Kutzelnigg, W., J. Chem. Phys. 48 (1968) 1819; Meyer, W. ibid 58 (1973) 1017.

/21/ Ladik, J. and Suhai, S., in Specialists' Report on Theoretical Chemistry, ed. Thomson, C., Royal Soc. of Chem. Burlington House, London, 1981, p. 49; Suhai, S., J. Chem. Phys. (submitted).
/22/ S. Suhai, S., Phys. Rev. B27 (1983) 3506.
/23/ Day, R.S. and Martino, F., Chem. Phys. Lett. 84 (1981) 86.
/24/ Boys, S.F., Rev. Mod. Phys. 32 (1960) 306.
/25/ Ruedenberg, K., Rev. Mod. Phys. 34 (1962) 326; Edminston, C. and Ruedenberg, K., ibid 35 (1963) 457.
/26/ Cizek, J., Förner, W. and Ladik, J., Theor. Chim. Acta (submitted).
/27/ Kapuy, E., Csepes, Z. and Kozmutza, C., Int. J. Quant. Chem. 23 (1983) 981.
/28/ Förner, W., Cizek, J. and Ladik. J., to be published in J. Chem. Phys.
/29/ Kunz, A.B., Phys. Rev. B6 (1972) 606; Devreese, J.T., Kunz, A.B. and Collins, T.C., Solid State Comm. 11 (1972) 670; Mickish, D.J., Kunz, A.B. and Collins, T.C., Phys. Rev. 139 (1974) 4461; Pantelides, S.T., Mickish, D.J. and Kunz, A.B., Phys. Rev. B10 (1974) 2602.
/30/ Etemad, S., Heeger, A.J., Lanchlan, L., Chung, T.-C. and MacDiarmid, A.G., Mol. Cryst. Liq. Cryst. 77 (1981) 43.
/31/ Hehre, W., Stewart, R.F. and Pople, J.A., J. Chem. Phys. 51 (1969) 265.
/32/ The basis used in the integral package IBMOL-5.
/33/ Arnott, S., Dover, S.P. and Wonacott, A.J., Acta Cryst. B28 (1969) 2192.
/34/ Ladik, J. and Suhai, S., Int. J. Quant. Chem. QBS7 (1980) 181.
/35/ Otto, P., Clementi, E. and Ladik, J., J. Chem. Phys. 78 (1983) 4547; see also the contribution of Otto, P. in this volume, p. 361.
/36/ Suhai, S. and Ladik, J., Phys. Lett. A77 (1980) 25.
/37/ See for instance: Dayhoff, M.O., Schwartz, R.M., Chen, H.R., Barker, W.C., Hunt, L.T. and Orcott, B.C., "DNA", Vol. 1, Mary Ann Liebert Inc., 1981, p. 51.
/38/ Gentleman, J.F., Shadbolt-Forbes, M.A., Hawkins, J.W., Ladik, J. and Forbes, W.F., Mathematical Scientist (submitted).
/39/ Suhai, S. and Ladik, J., to be published.
/40/ For a review see: Clementi, E., "Computational Aspects for Large Chemical Systems", Lecture Notes in Chemistry, Vol. 19 Springer Verlag, Berlin-Heidelberg-New York, 1980; Corongiu, G. and Clementi, E., Biopolymers 20, (1981) 551.
/41/ Suhai, S., Kaspar, J. and Ladik, J., Int. J. Quant. Chem. 17 (1980) 995.
/42/ Suhai, S., Collins, T.C. and Ladik, J., Biopolymers 18, 899 (1979).
/43/ Day, R.S., Suhai, S. and Ladik, J., Chem. Phys. 62 (1981) 165.
/44/ For details see F. Martino's paper in this volume, p. 279.

/45/ Day, R.S. and Martino, F., J. Phys. C. 14 (1981) 4247; Gazdy, B., Day, R.S., Seel, M., Martino, F. and Ladik, J., Chem. Phys. Lett. 88 (1982) 220.
/46/ See for instance: Santos, E., Tronick, S.R., Aaronson, S.A., Pulciani, S. and Barbacid, M., Nature 298 (1982) 343; Tabin, C.J., Bradley, S.M., Bargmann, C.I., Weinberg, R.H., Pagageorge, A.G., Scolnick, E.M., Dhar, R., Lowy, D.R. and Chang, E.H., Nature 300 (1982) 143.
/47/ See for instance: Chang, E.H., Furth, M.E., Scolnick, E.M. and Lowy, D.R., Nature 297 (1982) 479.
/48/ See for instance: Rechavi, G., Givol, D. and Canaani, E., Nature 300 (1982) 607.
/49/ Watson, J.D. and Crick, F.H.C., Nature 171 (1953) 737, 964; Watson, J.D.,"The Double Helix", Atheneum, New York, 1968.
/50/ Ladik, J., Suhai, S. and Seel, M., Int. J. Quant. Chem. QBS5 (1978) 135.
/51/ Weinstein, B.A. (personal communication).
/52/ See for instance: Davydov, A.S. and Kislusha, N.I., phys. stat. sol. 59 (1973) 465; Davydov, A.S., Phys. Scripta 20 (1979) 387.
/53/ Su, W.P., Schrieffer, J.R. and Heeger, A.J., Phys. Rev. B4 (1980) 2099.
/54/ Ladik, J. and Martino, F., to be published.
/55/ Krumhansl, J.A. and Alexander, D.M., in "Structure and Dynamics: Nucleic Acids and Proteins", ed. Clementi, E. and Sarma, R.H., Adenine Press, New York, 1983, p. 61.
/56/ Ladik, J., to be published in Int. J. Quant. Chem.

LARGE SCALE AB INITIO BAND STRUCTURE CALCULATIONS OF POLYNUCLEOTIDES AND POLYPEPTIDES

Peter Otto

Chair for Theoretical Chemistry, Friedrich-Alexander-University Erlangen Nürnberg, D-8520 Erlangen, Egerlandstrasse 3, FRG
and Laboratory of the National Foundation for Cancer Research at the Chair for Theoretical Chemistry, University Erlangen-Nürnberg

ABSTRACT. - Results of *ab initio* LCAO Hartree-Fock crystal orbital calculations are reported for single and double stranded periodic B-DNA models. For polycytosine, the results of model computations to investigate the effect of water molecules on its electronic structure are discussed. The presence of the water molecules, whose positions have been determined recently by a Monte Carlo simulation technique causes significant band shifts. Furthermore Monte Carlo computer experiments are presented for the systems of water molecules and positive ions enclosing single helix fragments of periodic polynucleotides in the B-DNA conformation. The identification of the positions and orientations for the water molecules in the first hydration shell is obtained from hydrogen and oxygen atoms probability distribution maps. The positions of the sodium ions in solution are given by the statistical analysis.

Finally the results are discussed of the theoretical investigation of the interaction between periodic single and double stranded DNA models and polyglycine in various conformations. The interaction energies are calculated with the help of the mutually consistent field method together with perturbation theoretical expressions. For all DNA-polyglycine complexes the structure is optimized with respect to the total energy of the combined system. It turns out that the complex , in which the agreement between the helical symmetry of B-DNA and polyglycine is realized, leads to the most stable configuration.

J. Ladik et al. (eds.), Quantum Chemistry of Polymers - Solid State Aspects, 361–392.

1. INTRODUCTION

Due to the central role of DNA and proteins in biochemistry and biophysics the computation of the electronic structure of periodic polymers built from nucleotide bases, base pairs, nucleotides and amino acids, respectively, had been of high interest since about twenty years. Early calculations of the band structure of DNA related periodic polymers have been performed with the crystal orbital (CO) method on the basis of different semiempirical levels (1). Recently the results of *ab initio* Hartree-Fock CO (2, 3) band structure calculations for the four nucleotide base stacks (4-6), the two Watson-Crick base pair stacks (6), the sugar-phosphate chain (4,5) and the three nucleotides cytidine (4,5), adenylic acid and thymidine (6) have been reported. These computations represent a significant progress but the following improvements are required for a more accurate description of the electronic structure of real DNA and its transport properties:

- i) The application of larger basis sets is necessary, since it has been shown recently that the energy gap between the valence and conduction band of a cytosine stack (7) is drastically decreased going from a minimal to a valence-split basis set. Further decrease of this quantity can be expected by including polarization functions.
- ii) The inclusion of correlation energy will further reduce the energy gap. This has been demonstrated impressively with calculations on polyacetylene (8).
- iii) Real DNA is aperiodic and the degree of aperiodicity will most probably determine the mechanism of conductivity.
- iv) In the living cell DNA is surrounded by water molecules and positive ions. In addition large parts of the genetic information in DNA is blocked due to complex formation with proteins.

For the presentation of methods and discussion of numerical results concerning points i) - iii) see the contribution of J. Ladik in this Volume.

In Section 2 the results are presented for the energy band structure computations on periodic DNA related polymers. Furthermore the Monte Carlo computer experiments are described leading to the water and counter-ion structure of periodic single helix B-DNA models.

The experimental evidence is increasing that the electrical properties of proteins play an essential role in their biological functions (9). Since the early suggestions of Szent-Györgyi (10) and of Laki (11) concerning the possibility of semiconduction in proteins and its relation to cancer (12) a number of theoretical investigations has been devoted to the determination of the energy band structures (13) and of the possible pathways for electron delocalization (14) in these systems.

Of special interest in molecular biology, however, are the complexes formed between DNA and proteins, the so-called nucleohistones. The change in the interactions between these macromolecules caused by chemical carcinogens, binding to DNA and/or proteins may be one of the determinant step in the mechanism of tumor development (15). These structural and energetical alterations of the binding properties in nucleohistones can be investigated only after having a detailed knowledge for the unperturbed macromolecular complex.

We started our investigations on this very complex problem with the computation of the interactions between the polypeptide backbone and DNA. In Section 3 the numerical results for very simple models of the macromolecular components are presented and discussed in the light of further improvements.

2.1. THE EFFECT OF HYDRATION ON THE ENERGY BAND STRUCTURE OF POLY-CYTOSINE

For a cytosine stack model calculations have been performed to investigate the effect of water molecules on its electronic structure (16). In this first approach the water structure of a cytidine unit in aperiodic B-DNA obtained with the help of Monte Carlo computer experiments (17) has been used. This choice of the water structure, of course, does not give a realistic description for cytosine,but we get in a relatively simple model calculation an order of magnitude estimate of the effects. The bound water molecules act on the electronic system of the stacked nucleotide bases as an external field. By a proper representation of the electron distribution of the water molecules through point charges one can easily calculate the matrix elements of this perturbing field in terms of the Bloch functions of the finite polymer and include in this way its effect on the band structure.

TABLE 1. The most important four energy bands of a cytosine stack calculated without and with hydration.(All energies in eV.)

Without hydration			With hydration		
E_{min}	E_{max}	δE	E_{min}	E_{max}	δE
8.421	8.697	0.276	7.026	7.331	0.305
5.916	6.738	0.822	5.608	6.473	0.865
-5.739	-4.872	0.867	-6.592	-5.679	0.913
-7.762	-7.386	0.376	-8.529	-8.110	0.419

From Table 1 we see that the potential of the hydration shell stabilized somewhat the base stack. All bands are moving downwards and also the total energy is lowered in the presence of the water molecules by about 8 eV per elementary cell. The band shifts are quite different and are varying between 0.3 and 1.4 eV. Though no dramatic changes can be observed in the band structure as a whole it should be pointed out that the hydration together with positive ions could play a very important role in the determination of the conductive properties of DNA. It will be, therefore, very important to repeat these computations using nucleotide units (sugar-phosphate-base) as a more realistic model for B-DNA.

As it becomes obvious from the above these investigations can be divided into four steps:

i) Compute the energy band structure of the unperturbed periodic polynucleotide,

ii) determine the water and ion structure for this macromolecule,

iii) find the representation of the electronic distribution of the water molecules by point charges and finally

iv) repeat the energy band structure calculation of the polynucleotide including the perturbation matrix due to the surrounding water molecules and positive ions.

2.2. THE ELECTRONIC STRUCTURE OF DNA RELATED PERIODIC POLYMERS

Results are reported of *ab initio* Hartree-Fock CO calculations using a minimal atomic basis set for the single stranded periodic B-DNA models of cytosine (C), thymine (T), adenine (A) and guanine (G) stacks and two polynucleotides with adenylic acid (ASP) and thymidine (TSP) as repeating unit, respectively. Further the energy band structures of two poly(base pairs), poly(adenine-thymine), (A-T), and poly(guanine-cytosine), (G-C), representing a simple model of B-DNA double helix are discussed.

The formalism of the *ab initio* Hartree-Fock CO method first proposed about 15 years ago (2,3) has been discussed in details elswhere (1,4) (see also the contribution of J. Ladik in this Volume). It has to be noted that in the case of a helix one steps from unit to unit, not with a simple translation but with a combined symmetry operation (translation along the helix axis and rotation around it). Therefore, one has to rotate both the atoms in the elementary cell and also the atomic orbitals centered on these atoms. This means that, having chosen the z axis along the main axis of the helix, the atomic orbitals with components in the xy plane have to be rotated (18). Second neighbors' interactions have been included with a correct electrostatically balanced cutoff (19) of the different types of integrals.

For the structural parameters of the helix and the geometry of the molecules in the unit cell the experimental data of B-DNA (20) have been taken. The coordinates for the hydrogen atoms have been added with the aid of a special computer program (21) making use of recently reported bondlengths and bond angles for the hydrogen atoms in the four bases (22) and in the sugar unit (23). In the study of the polynucleotides, a sodium ion has been attached to each phospahte group;in this way these chains attain electroneutrality. The position of these added counter-ions was assumed to correspond to the one determined from quantum-mechanical computations on the interaction of one Na+ ion and one sugar-phosphate fragment (24). This assumption, however, is only an approximation since the Na+ ion in a chain of periodic polynucleotides interacts not only with one sugar-phosphate unit but also with the entire DNA chain (25). In all calculations a minimal basis set (4s functions for hydrogen atoms, 7s and 3p functions for the second row elements, and 9s and 6p primitive Gaussian functions for the atoms of the third row) has been applied.

TABLE 2. The limits and widths of the two highest filled and two lowest unfilled bands of the four nucleotide base stacks (in eV).

System	Band	E_{min}	E_{max}	δE	MO	Gap
	$n^{*}+2$	4.200	4.406	0.206	3.894	
	$n^{*}+1$	1.697	2.517	0.820	1.638	
PolyC	n^{*}	-9.526	-8.737	0.789	-9.945	10.434
	$n^{*}-1$	-11.493	-11.422	0.071	-11.905	
	$n^{*}+2$	3.441	4.833	1.392	3.778	
	$n^{*}+1$	1.228	1.538	0.310	1.585	
PolyT	n^{*}	-11.248	-10.717	0.531	-10.985	11.945
	$n^{*}-1$	-12.201	-11.646	0.555	-12.332	
	$n^{*}+2$	3.292	3.619	0.327	3.026	
	$n^{*}+1$	2.444	2.184	0.260	1.924	
PolyA	n^{*}	-10.117	-9.677	0.440	-10.369	12.121
	$n^{*}-1$	-11.731	-11.574	0.157	-12.036	
	$n^{*}+2$	3.680	3.886	0.206	3.393	
	$n^{*}+1$	2.568	3.279	0.711	2.429	
PolyG	n^{*}	-9.466	-8.696	0.770	-9.488	11.264
	$n^{*}-1$	-11.754	-11.612	0.142	-12.312	

Table 2 contains the characteristics of the two highest filled and the two lowest unfilled bands of the four nucleotide base stacks, while Table 3 gives the same information for the double stranded poly(base pairs) (poly(A-T) and poly(G-C)). In columns 3 and 4 we report the lowest and highest value for each band. In column 5 we give the bandwidth. Column 6 contains the energy of the corresponding molecular orbital obtained via Hartree-Fock computations of the molecular system representing the unit cell. Finally, in the last column we report the difference between the energy minima of the $n^{\times}+1$ band and the energy maxima of the valence band $n^{\times}$ (headed Gap). The correspondence between the individual molecular levels and the bands is always unambiguous. Although the symmetry is broken in the periodic chains as a result of the stacked arrangement of the units, one can still define quasi-π-type bands which are located mainly around the Fermi level.

TABLE 3. The characteristics of the two highest filled and the two lowest unfilled bands of poly(G-C), poly(A-T), poly(ASP) and poly(TSP) in (eV). For comparison the Table also contains the location of the molecular orbitals of the base pairs.

System	Band	E_{min}	E_{max}	δE	MO	Gap
	$n^{\times}+2$	2.750	3.020	0.270	2.111	
	$n^{\times}+1$	1.716	2.011	0.295	1.536	
Poly(A-T)	$n^{\times}$	-10.516	-9.996	0.520	-10.137	11.712
	$n^{\times}-1$	-11.046	-10.464	0.581	-10.985	
	$n^{\times}+2$	4.143	4.823	0.680	3.521	
	$n^{\times}+1$	1.691	2.411	0.720	1.606	
Poly(G-C)	$n^{\times}$	-10.939	-10.099	0.840	-9.292	11.790
	$n^{\times}-1$	-11.952	-11.252	0.700	-11.886	
	$n^{\times}+2$	3.816	4.118	0.302		
	$n^{\times}+1$	2.646	2.926	0.280		
Poly(ASP)	$n^{\times}$	-9.824	-9.514	0.310		12.160
	$n^{\times}-1$	-11.664	-11.520	0.144		
	$n^{\times}+2$	3.882	5.225	1.343		
	$n^{\times}+1$	1.989	2.309	0.320		
Poly(TSP)	$n^{\times}$	-10.099	-9.799	0.300		11.788
	$n^{\times}-1$	-11.569	-11.068	0.501		

Comparing our results in Table 2 with the previously reported *ab initio* CO calculations (4,5) for which the STO-3G (26) minimal atomic basis set had been applied we find that the bandwidths of both the valence (denoted by n^x) and the conduction bands (denoted by n^x+1) are in good agreement. The positions of these bands and of the molecular orbitals are shifted by a constant amount of about 4.3 eV towards deeper energies due to the more flexible minimal atomic basis set used in these computations. Therefore, both calculations predict a gap between the highest occupied and lowest unoccupied bands in all cases more than 10 eV.

Little is available in the literature to discuss our energy band structures for the more complex periodic DNA models reported in Table 3. For poly(A-T) and poly(G-C), the features of the bands, their positions and their bandwidths can be related to those of the corresponding bases from which they originate. Equivalent findings have been obtained with the help of the semiempirical PPP and CNDO/2 CO calculations, respectively. In the last column (headed Gap) the smallest possible excitation energy of the poly(base pairs) is given. In poly(A-T) the highest filled and the lowest unfilled levels originate from poly(A) and poly(T), respectively, and therefore the first transition is of interbase type. In the case of poly(G-C) our results indicate again an intertype transition from G to C.

With the help of Mulliken's population analysis, the amount of transferred charge from the sugar phosphate chain to the adenine or thymine chain has been computed using the results of the poly (ASP) and poly(TSP) chain calculations. The amount of 0.212 and 0.190e per molecule pair is transferred to A and T, respectively. In the case of poly(CSP) (in this computation a hydrogen atom has been attached to each phosphate group thus the polymer chain is neutral) a charge transfer of 0.187e from the sugar phosphate unit to C has been calculated (4,5). Recently, the internal electronic charge transfer in B-DNA has been the subject of detailed computational investigations (27). The different molecular fragments were chosen in such a way that each base is in a field similar to the one experienced in the macromolecule; the fragments consist of one phosphate group, two sugar residues and the base bound to one of the sugar units. From these molecular calculations a charge transfer from the sugar units to the bases is obtained confirming the results of our CO computations. A quantitative comparison, however, shows a decrease of about 20 per cent for poly(ASP) and poly(TSP) relative to the corresponding values of the molecular models. The above difference may be due to the unit to unit interactions which are considered up to second neighbors in the periodic macromolecule. Another reason for the above discrepancy is that the interacting bases are different in the two computations. The main point, however, is confirmed, namely different methods yield a notable charge transfer from the sugar to the base.

The widths of the valence and conduction bands for poly(ASP) and poly(TSP) have the same order of magnitude as in the corresponding base stacks, indicating that the possibility for Bloch-type conduction in these systems exists if free charge carrier are generated in them. The gap is in all cases more than 10 eV. It is known that a Hartree-Fock calculation gives a too large gap for conduction, therefore, the results do not rule out the possibility of intrinsic semiconduction in DNA. Increasing the base set would lead to smaller values for the gap within the Hartree-Fock approximation. The calculated charge transfer suggests the possibility of creation of free charge carriers in these systems; it is known that in a closed-shell restricted Hartree-Fock CO computation only completely filled bands can be obtained even if there is considerable charge transfer; therefore, the above suggestion requires further studies with different formalisms.

2.3. MONTE CARLO COMPUTER EXPERIMENTS ON THE WATER AND SODIUM ION STRUCTURE OF PERIODIC MODELS OF THE B-DNA SINGLE HELIX

In the last years the problem of the proper theoretical treatment of solvation effects in DNA has been solved to a great extent (17). In a number of publications the interaction energy between water molecules and the nucleotide bases (28), base pairs (29), single (30), and double helices (31) of DNA has been calculated. Recently, also the complete solvent structure of a B-DNA double helix fragment with 12 base pairs and the corresponding sugar and phosphate units has been determined (32). In these Monte Carlo simulations, 447 water molecules have been included and their interaction energies and probability distributions (at a temperature of 300 K) have been calculated. In recent publications the Na+ ion structure of B-DNA at different humidities, ionic concentrations and temperature has been presented (33).

In the following section the results of Monte Carlo computer experiments are reported for the systems of water molecules enclosing single helix fragments of periodic polynucleotides in the B-DNA conformation, which are built up from 30 sugar-phosphate-base units and 30 Na+ ions. The number of water molecules surrounding these polymers in a cylindrical volume are 1410 for poly(ASP), 1326 for poly(GSP), 1365 for poly(TSP) and 1347 in the case of poly(CSP). The computations are carried out at a simulated temperature of 300 K. The identification of the positions and orientations for the water molecules in the first hydration shell is obtained from hydrogen and oxygen atoms probability distribution maps. These bound water molecules can be partitioned into clusters strongly interacting with the phosphates, the sugars, the bases and the counter-ions, respectively. From the statistical analysis we also obtain the positions of the sodium ions in solution.

We expect different positions of the Na+ ions relative to the sugar-phosphate-base unit in the presence of water molecules than have been assumed in the CO calculations described in 2.2. We also expect that the most stable arrangement of the sodium ions relative to the DNA helix will strongly depend on the nature of the base. The selected single helix fragments are composed of 30 bases (three full B-DNA turns). Since in our DNA model the atoms have translational symmetry turn to turn (a combined symmetry operation is effective base to base), we impose this symmetry constraint on the water molecules and on the counter-ions. More than 400 water molecules and ten Na+ ions are placed within the middle section of the cylindrical volume. The water molecules and the ions are further constraint (boundary conditions) to be within a cylinder co-axial to the periodic DNA fragment with a diameter of 27 Å and a height of 101.4 Å. In our cartesian system the z-axis is along the height of the cylinder. The energy data we report in the Tables refer to the DNA single helix model, the water molecules and the ions in one third of such a sample.

The relevant interaction energies are obtained with atom-atom pair potentials constructed in such a way as to accurately reproduce the intermolecular interaction energies computed by <u>ab initio</u> methods for the complex solute-water (34-38). For the water-water potential we use a CI-type potential (39,40). For the Na+ ... B-DNA fragment we use the same (7s, 3p)-type basis set but with inclusion of the basis set superposition correction (41).

The Monte Carlo technique, originally proposed by Metropolis (42) as used in this work, represents a modification previously described (43). Starting from some initial distribution of solvent molecules, a statistically meaningful number of configurations is generated according to a Boltzmann-weighted algorithm. To reach equilibrium we have considered 700,000 to 900,000 water and Na+ configurations; the data analyzed below were obtained by computing 500,000 additional configurations.

The entire study is condensed in a number of Tables. The use of these Tables is examplified by considering in detail some of the trends emerging from these experiments. The analysis is completed by graphical representations which show the statistical distributions of the oxygen and hydrogen atoms of the water molecules.

In Table 4 we summarize the average data obtained from the four experiments. For each system we present the average number N of water molecules and the average interaction energy of the water molecules bound to specific groups of atoms or subunits of DNA and to the Na+ ions. The decision, wether a water molecule is bound is made on the basis of its position and orientation relative to a given atomic site. In addition the data are presented for those water molecules not included in the first hydration shell: these are referred to "groove" water molecules (44,45).

TABLE 4. The average number of water molecules and the average interaction energies (in Kj/mole).

Site/Group	Poly(ASP) N	E	Site/Group	Poly(GSP) N	E
NH_2	3.07	-87	NH_2	3.57	-107
PO_4	6.66	-82	PO_4	6.78	-84
Base	5.05	-85	Base	7.07	-84
Groove	34.17	-77	Groove	30.52	-72
Na+	3.44	-33	Na+	2.68	-18

Site/group	Poly(TSP) N	E	Site/Group	Poly(CSP) N	E
			NH_2	3.69	-70
PO_4	6.38	-72	PO_4	6.66	-84
Base	3.76	-79	Base	5.41	-82
Groove	33.37	-59	Groove	31.13	-74
Na+	3.52	-8	Na+	3.95	-33

In Table 5 the average positions relative to the atoms of DNA are given for the individual Na+ ions. For each of the ten counter-ions the average is obtained from all the Monte Carlo steps (excluding those before equilibration). In the second column we give the average distance of the ion from the helix axis. In the third cloumn (headed I.-n.n.) we present the DNA atom which is nearest to the sodium ion; the atom is denoted by its standard name (for example O1P) and by the number of the nearest base it is linked to. In the forth column we give the distance to the first nearest atom of DNA. In the fifth column we specify the second nearest neighbor, II.-n.n., and its average distance from the sodium ion is given in column six. In the last column we present the number of water molecules in the first solvation shell of each Na+.

TABLE 5. The average position for sodium ions (all distances are given in Å).

I	R	I.-n.n.	dist.	II.-n.n.	dist.	N(w)
		Poly(ASP)				
1	9.40	O2P (A 1)	2.15	C2' (A 2)	2.89	2.98
2	11.10	O2P (A 2)	2.96	O3' (A 3)	3.01	3.21
3	8.26	C2' (A 4)	2.81	O2P (A 3)	3.40	3.72
4	8.23	C2' (A 5)	2.92	O2P (A 5)	3.38	4.00

continued on next page

Table 5 continued.

5	4.66	N7	(A 5)	2.12	C8	(A 6)	2.15	2.85
6	4.83	N7	(A 6)	2.11	C8	(A 7)	2.15	2.08
7	8.37	O2P	(A 7)	2.37	C2'	(A 8)	2.67	3.04
8	8.89	O2P	(A 8)	2.48	C2'	(A 9)	2.68	4.00
9	7.19	O2P	(A 9)	2.59	C2'	(A10)	2.82	3.00
10	10.72	O2P	(A10)	3.12	C3'	(A11)	3.13	3.68

Poly(GSP)

I	R	I.-n.n.		dist.	II.-n.n.		dist.	N(w)
1	4.76	N9	(G 2)	2.05	C8	(G 2)	2.05	2.00
2	4.92	C8	(G 3)	2.17	N7	(G 2)	2.18	3.00
3	4.79	N7	(G 3)	2.13	C8	(G 4)	2.14	2.42
4	4.91	C8	(G 5)	2.08	N9	(G 5)	2.10	2.00
5	3.86	O6	(G 5)	2.27	C6	(G 5)	2.77	3.00
6	4.63	C8	(G 7)	2.21	N7	(G 6)	2.16	2.79
7	6.78	C2'	(G 8)	2.73	C8	(G 8)	2.77	2.58
8	4.80	N7	(G 8)	2.14	C8	(G 9)	2.14	2.97
9	7.10	C2'	(G10)	2.47	O2P	(G 9)	3.19	3.00
10	4.87	C8	(G11)	2.06	N9	(G11)	2.15	3.00

Poly(TSP)

I	R	I.-n.n.		dist.	II.-n.n.		dist.	N(w)
1	6.86	C5M	(T 1)	2.48	C6	(T 2)	3.26	3.00
2	6.86	C5M	(T 2)	2.36	C6	(T 3)	3.32	4.00
3	8.35	O2P	(T 3)	2.38	C2'	(T 4)	2.66	3.00
4	8.16	O2P	(T 4)	2.45	C2'	(T 5)	2.60	3.00
5	8.10	O2P	(T 5)	2.66	C2'	(T 6)	2.70	3.91
6	8.45	O2P	(T 6)	2.36	C2'	(T 7)	2.66	3.99
7	8.15	O2P	(T 7)	2.57	C2'	(T 8)	2.66	3.00
8	7.84	C5M	(T 8)	2.56	C2'	(T 9)	2.78	4.00
9	8.38	O2P	(T 9)	2.53	C2'	(T10)	2.65	3.64
10	7.53	C5M	(T10)	2.47	C2'	(T11)	3.06	3.00

Poly(CSP)

I	R	I.-n.n.		dist.	II.-n.n.		dist.	N(w)
1	5.39	C6	(C 1)	2.58	C5	(C 2)	2.70	3.92
2	5.33	C5	(C 2)	2.34	C6	(C 3)	2.72	4.66
3	8.73	C2'	(C 3)	2.68	O2P	(C 4)	2.90	4.00
4	6.64	C6	(C 4)	2.79	C5	(C 5)	2.76	4.00
5	5.46	C6	(C 5)	2.65	C5	(C 6)	2.73	3.99
6	5.47	C6	(C 6)	2.62	C5	(C 7)	2.71	4.00
7	5.95	C6	(C 7)	2.58	C5	(C 8)	2.80	4.00
8	8.21	C2'	(C 8)	2.90	O5'	(C 9)	3.36	3.88
9	9.81	O1P	(C 9)	2.50	C3'	(C10)	3.03	2.58
10	8.62	C2'	(C10)	2.61	O2P	(C11)	2.80	2.91

In Table 6 we summarize the energetic data for the four Monte Carlo experiments, presenting the water-water, water-DNA, ion-water, ion-DNA and ion-ion average interaction energies, E(W-W), E(W-DNA), E(I-W), E(I-DNA) and E(I-I), respectively.

TABLE 6. Average interactions energies (Kj/mole).

	Poly(ASP)		Poly(GSP)	
E(W-W)	-26.608	+0.028	-25.127	+0.048
E(W-DNA)	-22.651	+0.109	-25.506	+0.037
E(I-W)	-1457.8	+1.5	-1599.3	+5.5
E(I-DNA)	-2828.8	+2.3	-3133.5	+4.8
E(I-I)	1094.4	+1.5	1309.9	+4.5
	Poly(TSP)		**Poly(CSP)**	
E(W-W)	-25.655	+0.044	-24.901	+0.050
E(W-DNA)	-9.665	+0.072	-26.185	+0.033
E(I-W)	-1481.2	+1.6	-1599.4	+3.6
E(I-DNA)	-2794.5	+0.7	-2807.9	+0.7
E(I-I)	1131.7	+0.9	1172.6	+1.8

We have determined the probability to find oxygen (or hydrogen) atoms of water at a given R-value measured from the z-axis for the phosphate group, the sugar residue and the base for each of the periodic polynucleotide (46). In Fig. 1 we report the distribution of water molecules around the phosphate groups (atomic sites O1P, O2P, O3' and O5'). In Fig. 2 we present the distribution of the water molecules solvating the bases. In Fig. 3 the results of the water molecules are shown, which are bound to the sodium ions. Finally, in Fig. 4 the distributions of the water molecules bound to the atoms of the DNA helix are graphically visualized. In these Figures the probability distribution for the hydrogen atoms is given as a dotted line and the one of the oxygen atoms is drawn as a solid line.

Comparing the total water distribution around the phosphate groups in Fig. 1 we observe a basically similar pattern for (GSP), (TSP) and (CSP) with well-defined maxima and minima. In the case of (ASP) we see a very distinct peak at R=7Å and a set of peaks that can be approximately associated to an asymmetrical Gaussian distribution with a maxima at about 10 Å. This finding indicates that the Na+ ions are located nearer to the free oxygen atoms of the PO_4 group in the case of (ASP), causing in this way a less ordered water structure.

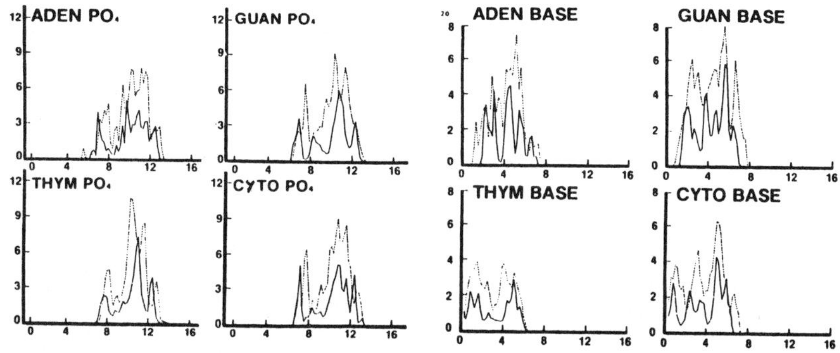

FIG. 1 and 2. The statistical hydrogen and oxygen distributions of water molecules bound to the PO_4 groups and the bases, respectively.

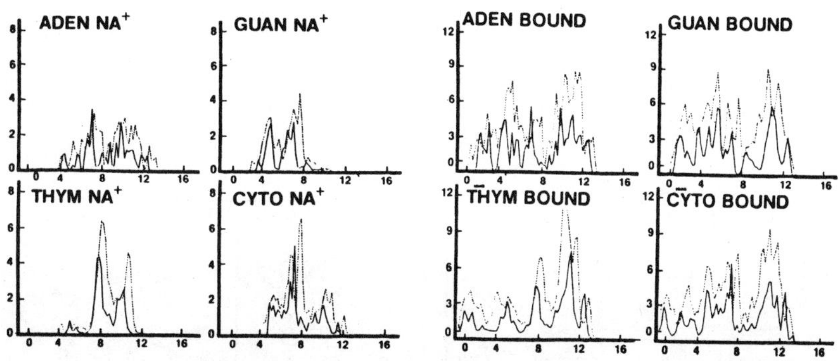

FIG. 3 and 4. The statistical hydrogen and oxygen distributions of water molecules bound to the sodium ions and the nucleotide units, respectively.

Due to the fact that in the single helix those atoms can be solvated, which are involved in the formation of hydrogen bonds between the bases in the DNA double helix, the average number of water molecules bound to the bases is about twice as large in the single-stranded cases than in the double-stranded samples (33). G and C having three atomic sites available for hydrogen bonds have the largest number of bound water molecules (7.07 and 5.41). T is less solvated than A, both having two additional atomic sites. The methyl group and the sugar residue are near the atoms O4 and O2 in T and sterically hinder the approach of water molecules to these atoms.

The main difference between the solvation of the amino groups in the bases (A, G and C) is that in the case of C the water molecules also approach this hydrophilic group from the inside of the helix. The different chemical properties of the four bases are expressed most obviously in the structural pattern of the sodium ions, which as a consequence is also reflected in the solvation of these ions. From Table 4 we see that the average number of water molecules per ion is 3.95 for CSP and is almost the same for ASP and TSP (3.44 and 3.52). In the case of GSP we obtained the relatively small average value of 2.68 per ion. From these numbers one can deduce that the ideal coordination number of the ions is reached in different ways. In GSP the ions have in the average three nearest neighbor atoms belonging to DNA, while in CSP four water molecules are needed to complete the coordination number to six. From Fig. 3 it can be recognized that the solvation pattern of the ions is similar for TSP and GSP on one side and ASP and CSP on the other side. Two intensive peaks are observed, separated by about 3 Å in the first case. For the second group a highly structured spectra within the region of 4 to 12 Å distant from the helix axis can be seen.

In Figures 5 and 6 we report the average positions of those water molecules which are bound to one turn of the four polynucleotides, respectively, (projected into the xy and yz planes).

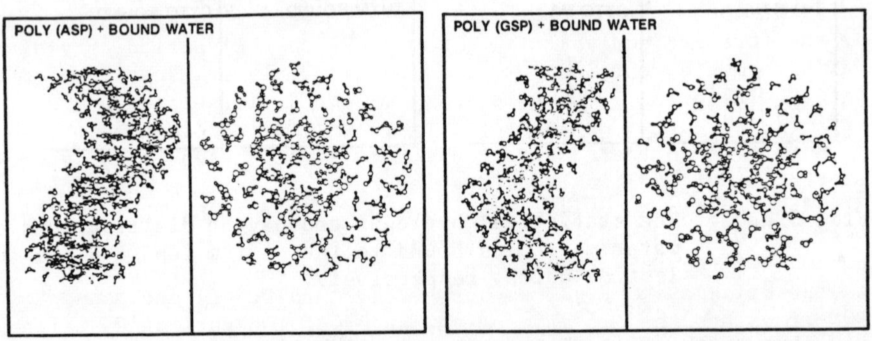

Fig. 5. The average positions of the water molecules bound to ASP and GSP projected into the yz (left) and xy (right) planes.

The most direct way to analyze the ion structure is provided by projections of the statistical distributions obtained from the Monte Carlo data in the xy and yz planes. The graphical representations are presented in Figures 7 and 8, respectively. A counterion corresponds to each "spot"; the size of the spot provides a

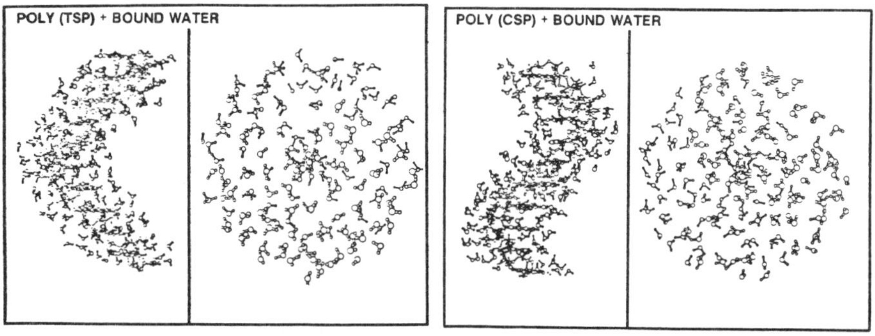

FIG. 6. The average positions of the water molecules bound to TSP and CSP projected into the yz (left) and xy (right) planes.

measure of the ion mobility. The mobility is large in the x,y directions and relatively small in the z direction. From both graphical representations one obtains a rather clear information about the location of the ions relative to the DNA single helix and to each other. It is still difficult to recognize if the individual ions are located in the same environment of the DNA. We have applied the inverse rotational symmetry operation to the atoms of the DNA and to the average positions of the ions. The projections of these new atomic positions are shown in Fig. 9. We are aware of the fact that this "unwinded" conformation of the DNA helix does not describe properly all geometrical relations among the atoms and that this information becomes worse with increasing distance from the helix axis. On the other hand we can easily recognize wether the positions of the ions are equivalent. In the ideal case of periodicity of the helix formed by the ions we would observe only one spot (projection into the xy plane). The more spots we can distinguish the higher will be the extent of aperiodicity in the helix formed by the ions.

According to their average distances from the helix axis and the nearest atoms (see Table 5) the ten ions per turn in ASP can be divided into three groups. The farthest placed ions (R ~ 11 Å) are near the O2P atom of the PO_4 group and are linked to the atoms C3' and O3' of the sugar ring. The ions of the second group are closer to the helix axis (R-values between 7.2 and 9.4 Å) and connect the PO_4 group and the C2' atom of the sugar of two nucleotide units. Finally, the ions of the third group have moved close to the base (R ~ 4.7 Å) and are located near N7 and C8 of two purine rings.

In the case of GSP all ions have moved away from the PO_4 group towards the bases. Some of the ions (I2, I3, I6 and I8) "link" two bases via the atoms N7 and C8. The ions (I1 and I4) are in the plane of the base near to the atoms C8 and N9. The ion (I5) closer to

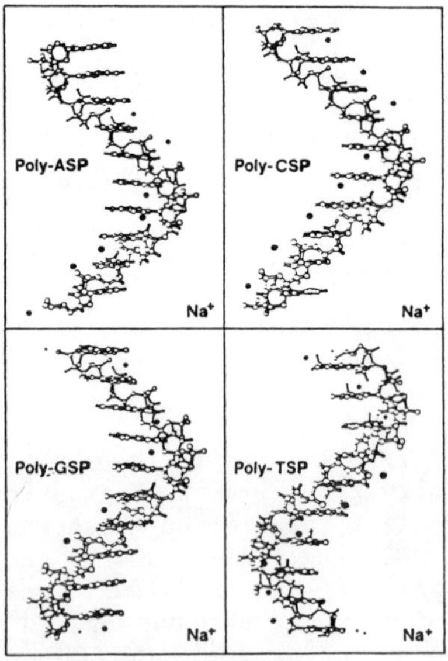

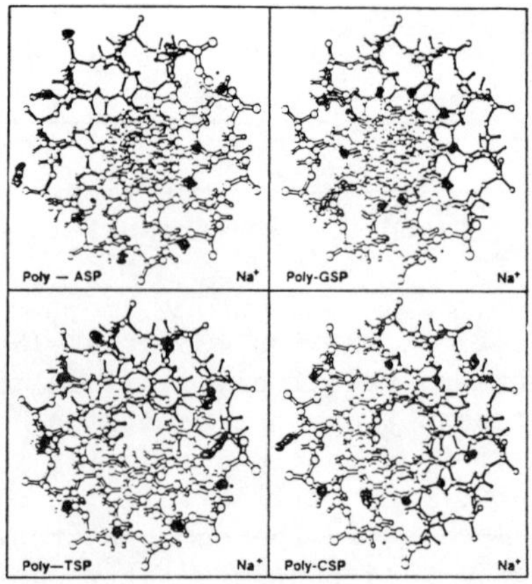

FIG. 7 and 8. The projection of the ten counter-ions into the yz (left) and into the xy (right) planes.

the helix axis by about 1 Å interacts with the π-electron system of the carbonyl group (C6=O6).

Two geometrical arrangements of the ions occur in solvated TSP. For all ions the two nearest neighbor atoms of DNA belong to two nucleotide units. The ions (I1, I2, I8 and I9) are positioned near the carbon atom C5 of the methyl group and either C6 of the base or C2' of the sugar. In the latter case R is larger by about 0.8 Å. The second group of ions with R-values of 8.2 Å is bound to O2P and C2'. In TSP the sterical effect of the methyl group prevents the ions to approach the bases.

In CSP one can distinguish between two groups of ions. One group with distances between 5.3 and 6.0 Å and another one with distances larger than 8.2 Å from the helix axis. The first class of ions is linked to the π-bonds (formed by C5 and C6) of two cytosine residues. The ions of the second group are located near to the PO_4 group.

The conclusions drawn from the results listed in Table 5 are confirmed by looking at Figure 9. In ASP we find the largest degree of spreading in the positions of the ten ions whereas in TSP the average positions of the ten ions relative to DNA show only small deviations.

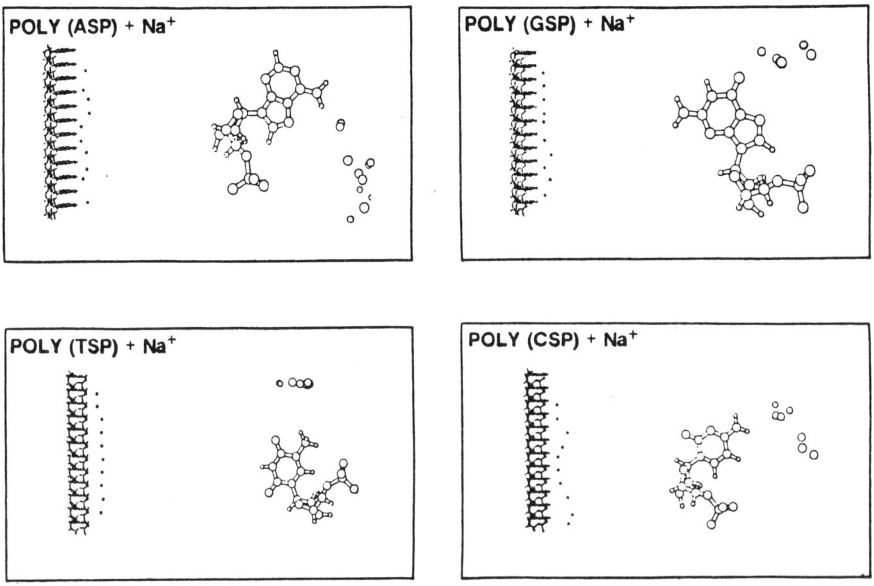

FIG. 9. The "unwinded" configuration of the polynucleotides together with their Na+ ions. Left insert: projection into the yz plane; right insert: projection into the xy plane.

In Table 6 the average interaction energies between the different components are summarized. One can see that with increasing distance of the ions from the helix axis the average total energy decreases, mainly due to the large decrease in the attractive energy contribution with DNA. This term overcompensates the gain in energy based on the considerably reduced ion-ion repulsion. The ion-water term follows the same behavior as the ion-ion contribution but the changes are much smaller.

It is known (47) that by increasing the relative humidity above 90 per cent, the added water molecules exhibit a bulk-water behavior. Therefore, the samples of more than 400 water molecules per turn of the DNA single helix is sufficiently extended to describe in a meaningful way not only the first hydration shell, but also the water molecules enclosed in the grooves.

The statistical results for the water and ion structures shall be used to determine the electric field which is acting on the polymer chain. As the discussion above has shown the problem is much more difficult due to the missing periodicity (from nucleotide to nucleotide subunit) in the position of the ions and, therefore, also in the geometrical arrangement of the solvating water molecules. A more detailed analysis of the obtained structure of the hydration shell is required, to find out wether it is possible to determine average positions of the water molecules in such a way that the periodicity is retained going from one subunit to the next one.

2.4. REPRESENTATION OF THE WATER MOLECULES BY POINT CHARGES

Until now the effect of the water molecules and the ions on the energy band structures of the periodic homopolynucleotides has not been investigated. Therefore , in this Section only a short review will be given of the method to determine the point charge representation of the hydration shell, as it has been used in the study of the effect of hydration on the electronic structure of a cytosine stack (see Section 2.1).

The water molecules forming the hydration shell can be divided into clusters by inspection of their geometrical positions. The clusters are chosen in such a way that the intracluster interaction should be larger than the intercluster ones. As a next step _ab initio_ calculations have to be performed for each water cluster. The resulting canonical Hartree-Fock orbitals are transformed to localized ones. To find a suitable representation of the electron distribution of the water clusters by point charges first the electrostatic potential $V_1^N(r_{i1}^N)$ for each localized orbital ψ_1^N of the Nth cluster at a given set of positions r_{i1}^N has to be determined. Each of these point sets is located on spheres with different radii around the center of charge of the given orbital. The radii cover the short medium and long range regions of the molecular potential field.

As a next step the resulting potentials are to be fitted using point charges potentials by minimizing the deviation of the two electrostatic fields in the least-square sense (48):

$$F_1^N = \sum_{i=1}^{m_1^N} \sum_{j=1}^{M^N} \left| \frac{q_{i1}^N}{r_{i1}^N - r_{j1}^N} - V_1^N(r_{i1}^N) \right|^2 = \min$$

$$1 = 1, 2, \ldots, n_N^*. \qquad (1)$$

M^N stands here for the number of point charges representing one localized orbital, m_1^N is the number of points for which the two potentials have been calculated, and n_N^* is the number of occupied orbitals in the Nth water cluster. q_{i1}^N and r_{i1}^N stand for the charge and location of the resulting point charges, respectively, and are determined by solving the system of nonlinear equations resulting from the minimum condition in eq. (1). The values $m_1^N = 4$ have been used which give satisfactory accuracy according to previous experiences gained by application of the above described procedure to the problem of intermolecular interactions (48).

3. THEORETICAL STUDY ON THE INTERACTION BETWEEN POLYGLYCINE AND PERIODIC B-DNA MODELS

3.1. MODELS FOR B-DNA AND POLYGLYCINE

Interactions between proteins and nucleic acids are of central importance in molecular biology. Acid proteins, basic proteins form complexes with single and double stranded polynucleotides, called chromatines, which form the fundamental structure of the chromosomes. A structure conserving and stabilizing function has been attributed to the basic proteins, however, it appears that they are responsible for the more general function of gene repression (49).

Despite the vast amount of chemical and physical experiments the information is still not sufficient to explain the detailed structure of the nucleohistones, with the exceptions of certain DNA-repressor protein complexes. Some general tendencies have been found: the α-helix conformation of the polypeptides is predominant in histones (from ORD experiments (50)); lysine-rich histones mainly occupy the major groove of B-DNA (51), while protamines are bound to DNA in the minor groove (52). Homopolylysine interacts cooperatively with A- and T-rich sequences of DNA and these complexes are stabilized either by shortening of the DNA turn or by superfolding (53). The preferred binding sites of homopolyarginine are G- and C-rich sequences in DNA (54).

The level of detailed knowledge is much higher for certain repressor proteins and their interactions with DNA. The structures of four polypeptides that specifically bind to DNA have been recently determined, namely, the <u>lac</u> repressor protein (55,56), the <u>cro</u> repressor protein from bacteriophage λ (57), the catabolite gene activator protein from <u>Escheria coli</u> (58) and the amino-terminal fragment of the CI repressor protein from bacteriophage λ (59). In addition to the primary polypeptide structure the sequence of the base pairs of DNA are known to which the relatively short polypeptides are bound. Furthermore, structural data about the conformation of the peptide backbone, strands of α-helices alternating with antiparallel ß-sheets, are available.

One of the main difficulties in the treatment of such complicated systems consists in the choice of the model. One can think on two different approaches, according to the state of structural knowledge. On one side we have the "basic" nucleohistone complexes and on the other side the well-defined repressor protein-operator systems. In the first case we have to assume certain conformations based on informations of the polypeptide backbone, then we have to adjust the side groups and finally to optimize this geometry relative to the DNA structure by calculating the interaction energy (60-62). For those cases where the fundamental structural properties are known, there remains the still formidable task to fit both macromolecules stereochemically together to form an energetically stable complex (57,63). Recent progress in model building for complementary structures between flexible strands has been achieved with the help of computer programs for model building and graphical representation (64) combined with the theoretical computation of interaction energies between the macromolecules (65,66).

We have started our study with the investigation of the interaction between the polypeptide backbone, in different conformations and periodic single and double helices using the structural parameters of B-DNA. We expected from the results the answers to several questions: i) does there exist a stable complex between the polypeptide backbone and DNA, and if so ii) which conformation is the most favorable one for the polypeptide and iii) can the interaction be a dominant factor to stabilize the macromolecular complex.

The periodic single stranded DNA helices have been represented by the four base stacks (C, T, A and G), poly(ASP) and poly(TSP). The Watson-Crick base pairs G-C and A-T and the dinucleotide ASP-TSP as unit cell of the macromolecule have been chosen to build up the models of the DNA double helices.

Polyglycine has been taken as the peptide component assuming a diglycyl unit in the elementary cell. The following conformations, for which the structural parameters are given in Table 7, have been applied: the fully extended form of polyglycine (PGFE), the α-helix (PGαH) and four helices with radii of 15 Å (radius of B-DNA helix 10.5 Å). These helices have been constructed with the help of computer programs for the calculation of helical parameters (67). The turn lengths of the selected polyglycine helices are 101.4 , 67.6, 33.8 and 8.45 Å and correspond to the ratios turn length (PG)/turn length (B-DNA) = 3, 2, 1 and 0.25, respectively. In the following the abbreviations PG101, PG67, PG33 and PG8 are used to denote these helices.

TABLE 7. The structural parameters of the six investigated polyglycine models and the definition of the supercell.

Model	Helical parameters a_{trans} (Å)	α_{rot}	Dihedral angles ψ_1	φ_1	ψ_2	φ_2	Super cell[1)], Nb. of subunits PG	DNA
PGFE[2)]	3.380	180.00	0	0	0	0	10	10
PGαH	6.450	197.59	123	132	123	132	11	20
PG101	5.310	18.65	180	180	200	180	19	30
PG67	3.581	19.03	140	240	300	60	19	20
PG33	2.039	21.76	200	180	320	40	17	10
PG8	0.604	25.48	0	20	200	140	55	10

1) Supercell is the smallest translationally symmetric subunit.
2) In PGFE the subunit is the glycyl residue.

In Table 8 the energy parameters of the highest filled and the lowest unfilled bands are summarized for the polyglycine models. The *ab initio* band structure calculations using a minimal atomic basis set (7s/3p) have been performed in the second neighbors' interactions approximation. From the total energy per unit cell (column 6 in Table 8) it follows that the extended conformation is the most stable one. PGαH is less stable by about 6 kcal, followed by PG33, PG67, PG8 and PG101. The interaction between two subunits being separated by the distance of one turn length, however, is not taken into account in the CO calculation. This effect which might be important in the case of PG8, has been estimated by computing the electrostatic part of this energy with the aid of point charges. The values given in Table 8 show only a very small attractive contribution in cases PG33 and PG67. The repulsive interaction of about 3 kcal per unit for PG8, however, is not large enough to balance the difference in the total energy between PG101 and PG8.

TABLE 8. The characteristics of the valence and conduction bands for the six polyglycine models. The last two columns contain the difference in the total energy per unit cell with respect to PGFE and the interaction energy between subunits of one helix separated by one turn

Model	Band	E_{min}	E_{max}	δE	Gap	E_{tot}	E_{stab}	E_{intra}
		(eV)				(au)	(kcal)	
	n^*+1	3.624	4.169	0.545				
PGFE	n^*	-12.548	-12.184	0.364	16.353	-412.10270	0.0	
	n^*+1	3.479	3.949	0.470				
PGαH	n^*	-12.071	-12.047	0.024	15.996	-412.09319	5.97	
	n^*+1	3.636	3.659	0.023				
PG101	n^*	-11.757	-11.742	0.015	15.401	-412.00464	61.55	
	n^*+1	3.737	4.108	0.371				
PG67	n^*	-11.418	-11.223	0.195	15.331	-412.07371	18.20	-0.004
	n^*+1	3.719	4.064	0.345				
PG33	n^*	-11.693	-11.522	0.171	15.586	-412.08079	13.75	-0.023
	n^*+1	3.893	4.189	0.296				
PG8	n^*	-9.912	-9.589	0.323	13.778	-412.02935	46.04	2.990

3.2. CALCULATION OF INTERACTION ENERGIES

In principle those methods, which are used to compute intermolecular interaction energies can be generalized to compute the interactions between macromolecules. The superchain method, however, where the elementary cell of the combined system is formed with the reference cells of the interacting polymers, can only be applied when the translation vectors of both macromolecules are parallel and of equal length. This method has to be ruled out for the case of DNA interacting with polyglycine.

Another possibility consists in the application of the perturbation theoretical energy expressions, modified for interacting macromolecules (68). Its usage would cause computational problems due to the large number of atoms in our model systems.

In this work we have used this version of the mutually consistent field method (MCF) (48) which has been developed to treat the interactions between polymer chains (69). Each subsystem is computed in the potential field of the partner system. The Coulomb potentials of the elementary cell of one chain, represented by a point charge distribution which are fitted to the Hartree-Fock Coulomb potential are included in the one-electron part of the Fock matrix of the other chain and vice versa (for more details see Section 2.4). The procedure of taking into account the effect of the mutually polarization is repeated until consistent solutions are obtained for the charge distributions. Computing finally the interaction between these point charge representations, one obtains the electrostatic and the polarization energy contribution together.

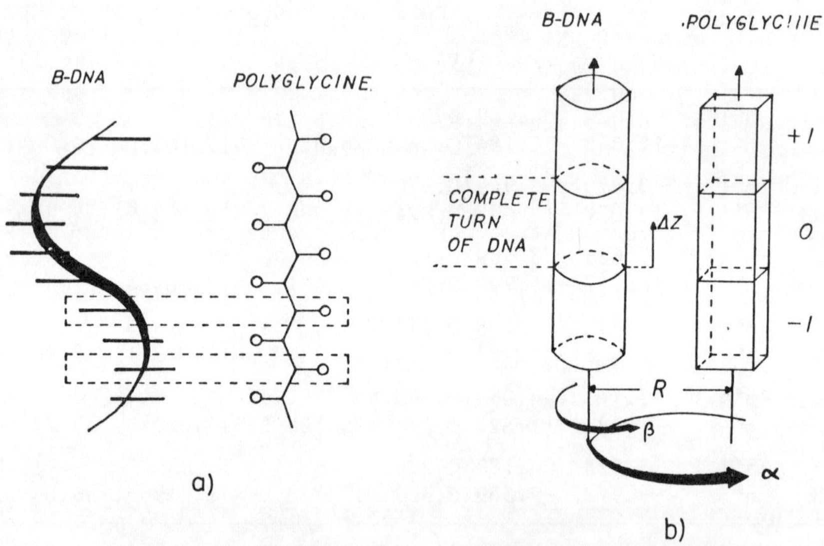

FIG. 10. A schematic graphic of the complex PGFE and DNA to demonstrate the occurrence of aperiodicity within each of the interacting chains (a); a graphical representation of the degrees of freedom used to determine the most stable DNA-polyglycine structure (b).

Both one-dimensional periodic polymers become aperiodic as soon as they are interacting. As can be seen from the schematic illustration in Fig. 10a each unit cell of one chain experiences a different effective potential of the partner chain. As a consequence the electronic charge distribution will be different in each unit cell of both macromolecules. We have considered this effect with the help of the MCF method. The translationally symmetric common unit cell contains ten nucleotide and ten glycyl units in the case of the DNA-PGFE complex. A starting configuration of the complex is chosen. Then the band structure calculation is repeated for each of the ten glycyl units, assuming a periodic potential arising from the atoms of the DNA helix, which are located in the space of a disk whose height is given by the length of the internal translation vector of PGFE. In the same way the different charge distributions are determined for the DNA component. Finally, one complete turn of DNA and polyglycine is built up from the point charge distributions of the different subunits, respectively.

The exchange and charge transfer energy contributions have been computed with the help of the perturbation theoretical expressions (70), according to the procedure described in (71). The dispersion energy has been approximated using London's formula and the atomic polarizabilities und ionization energies given by (72).

The most stable configuration of both polymers has been computed as a function of i) the rotation of DNA around its helix axis, ii) the rotation of DNA around the PG axis, iii) the shift of DNA along its axis and iv) the distance between both polymer axes (see Fig. 10 b). Of course, for the helical models of polyglycine (the helix axes of both chains coincide) only two degrees of freedom remain, namely i) and iii). The computed interaction energy refers to the supercell of the complex (which is the smallest translationally symmetric fragment of the complex). The energy is given by the sum of the interactions between the reference supercells ΔE^{00}, and the interactions between the reference supercell with the lower and upper neighboring supercells of the partner system, ΔE^{0-1} and ΔE^{0+1}

$$\Delta E = \Delta E^{00} + \Delta E^{0+1} + \Delta E^{0-1} + \Delta E^{+10} + \Delta E^{-10} =$$

$$= \Delta E^{00} + 2(\Delta E^{01} + \Delta E^{0-1}). \tag{2}$$

For the general term ΔE^{0I} in the expression we can write

$$\Delta E^{0I} = E^{0I}_{elst+pol} + E^{0I}_{exch} + E^{0I}_{ch.tr} + E^{0I}_{disp} . \tag{3}$$

It has to be mentioned that the mutually consistent band structure calculations which lead to the different charge distributions in the originally periodic polymer, have to be performed for different values of the degrees of freedom.

3.3. DISCUSSION OF THE RESULTS

In Table 9 the results of the calculated intèraction energies defined in Eq. (2) are summarized for PGFE and the periodic single and double stranded B-DNA models. Table 10 contains the same kind of informations for polyglycine in the chosen helical conformations interacting with poly(ASP) and poly(ASP-TSP), respectively.

TABLE 9. The contributions to the interaction energies between polyglycine (PGFE) and periodic single and double stranded B-DNA helices (energies in kcal/mole).

B-DNA Model	Polyglycine fully extended conformation		
	$(E_{elst+pol}+E_{exch}+E_{ch.tr})$	E_{disp}	ΔE
Poly(A)	-2.9	-2.4	-5.3
Poly(G)	-3.1	-2.0	-5.1
Poly(T)	-3.7	-4.4	-8.1
Poly(C)	-3.6	-3.9	-7.5
Poly(ASP)	-24.6	-1.7	-26.3
Poly(TSP)	-21.0	-1.7	-22.7
Poly(A-T)	-9.6	-6.9	-16.5
Poly(G-C)	-7.7	-5.8	-13.5
Poly(ASP-TSP)	-24.4	-2.8	-27.2

3.3.1. PGFE - DNA models

a. poly(A), poly(G), poly(T), poly(C)

It is not surprising that we obtain basically similar results for the four base stacks. The total interaction energies are small ranging from -5 to -8 kcal. The potential surface as a function of the parameters (α, ß, Δz and R, defined in Fig. 10b) show a large number of very flat minima. The equilibrium distance between the polymer axes is 9.5 Å. The analysis of the energy contributions per nucleotide base with PGFE shows that only one local attractive interaction is responsible for the binding.

b. poly(A-T), poly(G-C)

The binding properties with these double helix models are basically a superposition of the results for the corresponding single helices. The electrostatic part is somewhat larger than the sum because PGFE is differently polarized by the atoms of the double compared to those of the single helix. The two helix axes again are separated by about 9.5 Å.

c. poly(ASP), poly(TSP)

An essentially different situation occurs for the periodic polynucleotides which contain the polar PO_4 group. The total interaction energy is -26 and -23 kcal for poly(ASP) and poly(TSP), respectively. The potential hypersurface now has a single deep minima. The binding originates from one local electrostatic interaction between the $Na^+PO_4^-$ group and the polar carbonyl group of the neighboring glycyl unit. The interchain distance has been found to be 14.3 Å in both cases.

d. poly(ASP-TSP)

The number of atoms in the dinucleotide unit ASP-TSP is too large to perform CO calculations. Therefore, the point charge representation was generated from the charge distributions of the corresponding single helices. The total interaction energy is -27 kcal; the binding is effected again by one strong local interaction.

TABLE 10. The interaction energies between poly(ASP) and poly(ASP-TSP), respectively, with the different helical conformations of polyglycine. (in kcal/mole).

PG-helix	Poly(ASP) $(E_{elst+pol}+E_{exch}+E_{ch.tr})$	E_{disp}	ΔE
PGαH	-14.4	-1.1	-15.5
PG101	-4.8	-2.8	-7.6
PG67	-10.7	-1.7	-12.4
PG33	-55.3	-0.9	-56.2
PG8	-5.6	-3.2	-8.8
	Poly(ASP-TSP)		
PG101	-0.8	-5.1	-5.9
PG67	+21.9	-3.3	+18.6
PG33	-27.6	-4.0	-31.6
PG8	-2.4	-6.1	-8.5

3.3.2. PGαH and poly(ASP)

The relative configuration of the component polymers is similar to the case of PGFE. The calculated interaction energy, however, is only -15.5 kcal for the interchain distance of 15.9 Å. The difference is even higher when we consider that in the case of PGαH the supercell consists of two complete turns of DNA. More atoms are packed in the volume unit of PGαH than of PGFE. Therefore, more dipoles are oriented unfavorably with respect to the DNA subunits, decreasing thus the interaction energy with PGαH. Two local attractive interactions, separated by one turn length of DNA are basically responsible for the complex stabilization (see Fig. 11a).

3.3.3. PGxxx - DNA models

a. poly(ASP)

From Table 10 we see, that the stability of the complex formed with the single helix poly(ASP) depends strongly on the turn length of the PG helix. The binding energy is only -7.6 kcal for PG101, despite the fact that three complete windings of DNA now define the supercell. The complex with PG67 is stabilized by -12.4 kcal and an interaction energy of -8.8 kcal is calculated for PG8. The large value of -56.2 kcal computed for the binding energy with PG33 can be explained on the basis of the local binding analysis. In the cases of PG101 and PG8 all subunits contribute with almost equal amount but alternating sign to the total energy, leading to the small overall value. In PG67 the first three and the last four nucleotide units of the DNA super cell exhibit strong binding properties. About seven subunits in between, however, are repulsive relative to the neighboring glycyl units. In PG33 - poly(ASP) each of the ten ASP units contributes about -5 kcal to the binding energy. The polyglycine helix aranges itself in such a way relative to the DNA helix that both windings are parallel to each other as far as possible. Of course, with PG33 this condition can be optimally fulfilled, resulting in an almost equal surrounding for each diglycyl and ASP subunit, respectively. In Fig. 11 this property is visualized by projecting the atomic positions in the complex into the yz plane.

b. poly(ASP-TSP)

The interaction energies are drastically reduced in the complexes formed with the DNA double helix compared to the single helix. Only a weak binding is observed with PG101 (-5.9 kcal) and PG8 (-8.5 kcal), which originates mainly from the dispersion energy. For the complex with PG67 no stable configuration is found anymore. The "optimal" one is repulsive by about 19 kcal. Despite the appreciable reduction in stabilization, the PG33 complex is still rather stable (-31.6 kcal). The analysis of the subunit to subunit interaction shows again no remarkable local binding for PG101 and PG8. In the complex with PG67 one can distinguish between four regions with alternating interaction potential. The nonbonding terms, however, are larger in magnitude than the bonding ones. The binding energy with PG33 is approximately equal for each ASP-TSP cell. This finding confirms again the tendency of the polypeptide backbone to adjust itself to the helical structure of DNA as much as possible.

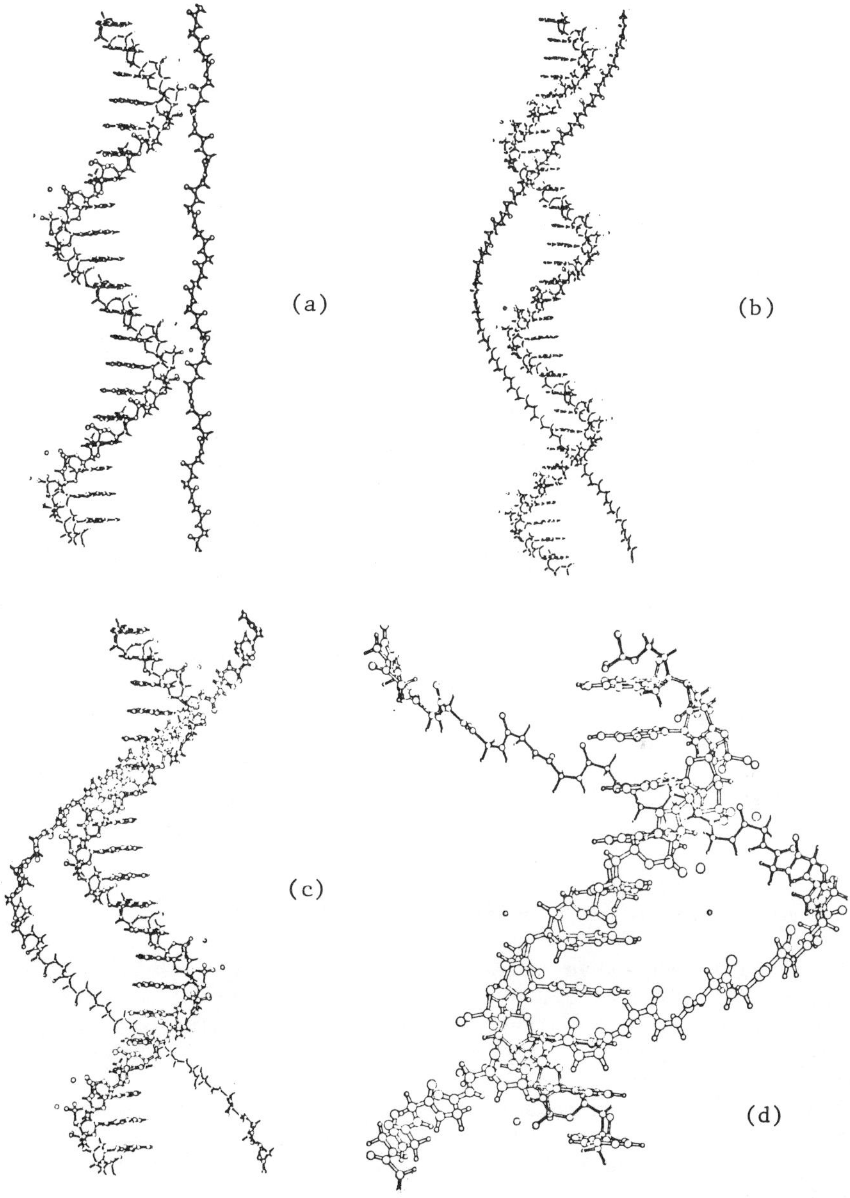

FIG. 11. The geometry of the most stable orientation of poly(ASP) to polyglycine in the supercell. (a) PGαH, (b) PG101, (c) PG67 and (d) PG33.

Despite several serious but well-defined simplifications and assumptions in the model building, general conclusions valid for the real system can be deduced. Mainly three approximations have been introduced: i) neglect of possible interactions between amino acid side chains and subgroups of DNA, ii) restriction of full periodicity in the polypeptide backbone and the DNA helix and iii) neglect of the effects of the solvation shell.

The PGFE conformation is appropriate thus that every second amino acid residue is directed towards the DNA helix and can interact with the base or PO_4 group, especially when a distortion as proposed by Feughelmann (73) takes place. A similar statement can be made for the α-helix conformation of polyglycine with the peptide helix being strictly parallel to the DNA helix axis. The local interaction per DNA turn is still smaller than for PGFE. Only every 4-5th amino acid side chain may interfere with suitable binding sites in DNA. Obviously better binding properties can be achieved by introducing a distortion into the linear polymer such that the α -helix follows to some extent the winding of the DNA helix.

The results strongly support the hypothesis that DNA is able to induce its helical symmetry on the conformation of the polypeptide chain, which is suggested by the existence of a 1:1 stoichiometry between synthetic basic homopolypeptide-DNA complexes. The polypeptide backbone of PG33, even in the absence of amino acid side groups, contributes essentially to the overall stability of the complexes. Additional links between subgroups of both chains, of course, will enforce the strength of the bonding.

One of the results of the Monte Carlo computer experiments (see Section 2.3) (74) has been that the position of the sodium ions depends on the nature of the base. In our model systems we have assumed the counter-ions bound to the free oxygen atoms of the phosphate group. Therefore, one can expect drastical changes in the stability and structure of those DNA-polyglycine complexes, where the binding is due to ion-dipole interactions between $PO_4^-Na^+$ and the glycyl unit (e.g. PGFE and PGαH). For the helix models of polyglycine the effect will be less dramatic, because the stability is based mainly on dipole-dipole interactions (75).

A more conclusive answer to all the problems mentioned above can only be given by further theoretical investigations. On one side one has to improve the model system (e.g. building up polypeptide helices which fit into the major groove of B-DNA) and to consider amino acid side chains. This last improvement has to go parallel with the introduction of chemical aperiodicity in the DNA helix. Furthermore the water molecules and ions have to be included in the computations. On the other hand it is absolutely necessary to develop new theoretical and computational concepts for future investigations to treat these complicated systems in an efficient way without loosing the reliability of numerical results.

ACKNOWLEDGMENTS

The author is greatly indebted to Professors J.Ladik and E.Clementi for helpful discussions and continuous interest. Thanks are due to Dr. G. Corongiu for stimulating discussions and for helping in the computational work. It is a pleasure to thank IBM Germany for a research grant and the IBM Computer Center for its collaboration. In addition I wish to thank the National Foundation for Cancer Research for partial support of these investigations.

REFERENCES

(1) Ladik, J. 1975, in "Electronic Structure of Polymers and Molecular Crystals", Eds. André, J.-M. and Ladik, J., Plenum Press, New York, p.663.
(2) Del Re, G., Ladik, J. and Biczo, G. 1967, Phys. Rev 155, p.997.
(3) André, J.-M., Gouverneur, L. and Leroy, G. 1967, Int. J. Quant. Chem. 1, p.427.
(4) Ladik, J. and Suhai, S. 1980, Int. J. Quant. Chem. QBS7, p. 181.
(5) Ladik, J. Suhai, S. and Seel, M. in "Physics of Polymers, Liquid Crystals and One Dimensional Systems", Eds. March, N.H. and Tosi, M.P., Plenum Press, New York, 1983, in press.
(6) Otto, P., Clementi, E. and Ladik, J. 1983, J. Chem. Phys. 78, p. 4547.
(7) Ladik, J. and Suhai S., to be published.
(8) Suhai, S. 1983, Phys. Rev. B27, p.3506.
(9) Szent-Györgyi, A. 1973, Bioenergetics 4, p. 535; Szent-Györgyi, A. 1973, Acta Biochim. Biophys. Acad. Sci. Hung. 8, p. 177; Pethig, R. and Szent-Györgyi, A. 1977, Proc. Natl. Acad. Sci. USA 74, p. 226; Bone, S., Lewis, T.F., Pethig, R. and Szent-Györgyi, A. 1978, Proc. Natl. Acad. Sci. USA 75, p. 315.
(10) Szent-Györgyi, A. 1941, Nature 148, p. 157.
(11) Laki, K. 1942, Studies from the Inst. Med. Chem. University Szeged 2, p.43.
(12) Szent-Györgyi, A. 1941, Nature 148, p. 157.
(13) Suhai, S., Collins, T.C. and Ladik, J. 1979, Biopolymers 18, p. 899.
(14) Suhai, S. 1974, Biopolymers 13, p. 1701.
(15) Ladik, J., Suhai, S. and Seel, M. 1978, Int. J. Quant. Chem. QBS5, p.35.
(16) Otto, P., Ladik, J., Corongiu, G., Suhai, S. and Förner, W. 1982, J. Chem. Phys. 77, p. 5026.
(17) Clementi, E. in "Computational Aspects for Large Chemical Systems; Lecture Notes in Chemistry", 1980, Springer-Verlag, Berlin, Vol. 19.
(18) Ukrainski, I.I. 1975, Theoret. Chim. Acta 30, p.139; Merkel, Ch. in "Elektronische Eigenschaften von Molekülkristallen" 1977, Thesis, Techn. University of Munich.
(19) Ladik, J. and Suhai, S. in "Theoretical Chemistry", 1981, Ed. Thomson, C., Specialists' Report (Royal Society of Chemistry, London), Vol. 4, p. 49.
(20) Fieldman, R. in "Atlas of Macromolecular Structure on Microfiche (AMSOM)", 1976, Document 13.2.1.1.1., (National Institute of Health, Bethesda, Washington.

(21) Hermans, J., Ferro, D.R., McQueen, J.E and Wang, S.C. 1976, in "Environmental Effects on Molecular Structure and Properties", Ed. Pullman, B., Reidel, Dordrecht, p. 459.
(22) Clementi, E., André, J.-M., André, C., Klint, D. and Hahn, D. 1969, Acta Phys. Acad. Sci. Hung. 27, p.493.
(23) Corongiu, G., Clementi, E. 1978, Gazz. Chim. Ital. 108, p. 273.
(24) Clementi, E. 1980, in "Lecture Notes in Chemistry", Vol. 19, Springer-Verlag, New York.
(25) Clementi, E. and Corongiu, G. 1981, Biopolymers 20, p. 2427; 21, Biopolymers, p. 763.
(26) Hehre, W., Stewart, R.F. and Pople, J.A. 1969, J. Chem. Phys. 51, p. 2657.
(27) Clementi, E. and Corongiu, G. 1982, Int. J. Quant. Chem. QBS9, p. 213.
(28) Scordamaglia, R., Cavallone, F. and Clementi, E. 1977, J. Am. Chem. Soc. 99, p. 5545.
(29) Clementi, E. and Corongiu, G. 1980, J. Chem. Phys. 72, p. 3979.
(30) Clementi, E. and Corongiu, G. 1979, Biopolymers 18, p. 2431; Clementi, E. and Corongiu, G. 1979, Int. J. Quant. Chem. 16, p. 897.
(31) Clementi, E. and Corongiu, G. 1979, Chem. Phys. Lett. 60, p. 175
(32) Corongiu, G. and Clementi, E. 1981, Biopolymers 20, p. 551.
(33) Clementi, E. and Corongiu, G. 1981, in "Biomolecular Stereodynamics", Ed. Sarma R.H., Adenine Press, New York, p.209.
(34) Clementi, E. 1976, in "Determination of Liquid Water Structure, Coordination Numbers for Ions and Solvation for Biological Molecules", Lecture Notes in Chemistry, Vol. 2, Springer-Verlag, Berlin.
(35) Clementi, E., Scordamaglia, R. and Cavallone, F. 1976, J. Am. Chem. Soc. 99, p. 5531.
(36) Ragazzi, M., Ferro, D. and Clementi, E. 1979, J. Chem. Phys. 70, p. 1040.
(37) Carrozzo, L., Corongiu, G., Petrongolo, C. and Clementi, E. 1978, J. Chem. Phys. 68, p. 787.
(38) Kistenmacher, H, Lie, G.C., Popkie, H. and Clementi, E. 1974, J. Chem. Phys. 61, p. 546.
(39) Yoshimine, M., Matsuoka, O. and Clementi, E. 1976, J. Chem. Phys. 64, p.1351.
(40) Popkie, H., Kistenmacher, H. and Clementi, E. 1973, J. Chem. Phys. 59, p. 1329.
(41) Corongiu, G. and Clementi, E., to be published.
(42) Metropolis, N., Rosenbluth, A.W., Rosenbluth, M.N. and Teller, A.H. 1953, J.Chem. Phys. 21, p. 1087.
(43) Romano, S. and Clementi, E. 1978, Gazz. Chim. Ital. 108, p.319.
(44) Corongiu, G. and Clementi, E. 1983, Biopolymers in press.
(45) Clementi, E. and Corongiu, G. 1981, Annales of the New York Academy of Sciences 367, p. 83.
(46) Lie, G.C., Yoshimine, M. and Clementi, E. 1976, J. Chem. Phys. 64, p. 2314.
(47) Texter, J. 1978, Prog. Biophys. Mol. Biol. 33, p. 83.

(48) Otto, P. and Ladik, J. 1975, Chem. Phys. 8, p. 192;
Otto, P. and Ladik, J. 1977, Chem. Phys. 19, p. 209;
Otto, P. 1978, Chem. Phys. 33, p. 407.
(49) Kleinsmith, L.J., Heidema, J. and Carroll, A. 1970, Nature 22, p. 1025; Teng, C.S., Teng, C.T. and Allfrey, V.G. 1971, J. Biol. Chem. 246, p. 3597.
(50) Hindley, J. 1963, Biochem. Biophys. Res. Commun. 12, p. 175;
Barr, G.C. and Butler, J.A.V. 1963, Nature 199, p. 1170.
(51) Zubay, G. and Doty, P. 1959, J. Mol. Biol. 1, p. 1.
(52) Wilkins, M.H.F., Zubay, G. and Wilson, R.H. 1959, Trans. Faraday Soc. 55, p. 497.
(53) Luzzati, V. and Nicolaieff, N. 1963, J. Mol. Biol. 7, p. 142.
(54) Tsuboi, M. and Matsuo, K. 1968, J. Mol. Biol. 15, p. 256;
Olins, D.E., Loins, A.L. and von Hippel, P.H. 1968, J. Mol. Biol. 33, p. 265.
(55) Leng, M. and Felsenfeld, G. 1966, Proc. Natl. Acad. Sci. USA 56, p. 1235.
(56) Adler, K., Beyreuther, K., Fanning, E., Geisler, N., Cronenborn, B., Klemm, A., Müller-Hill, B., Pfahl, M. and Schmitz, A. 1972, Nature 237, p. 322.
(57) Beyreuther, K., Adler, K., Geisler, N. and Klemm, A. 1973, Proc. Natl. Acad. Sci. USA 70, p. 3576.
(58) Anderson, W.F., Ohlendorf, D.H., Takeda, Y. and Matthews, B.W. 1981, Nature 290, p. 754.
(59) McKay, D.B. and Steitz, T.A. 1981, Nature 290, p. 744.
(60) Pabo, C.O. and Lewis, M. 1982, Nature, 298, p. 443.
(61) De Santis, P., Forni, E. and Rizzo, R. 1974, Biop1oymers 13, p. 313.
(62) Carter, C.W. and Kraut, J. 1974, Proc. Natl. Acad. Sci. USA 71, p.283.
(63) Church, G.M., Sussman, J.L. and Kim, S.H. 1977, Proc. Natl. Acad. Sci. USA 74, p. 1458.
(64) Ohlendorf, D.H, Anderson, W.F., Risher, R.G., Takeda, Y. and Matthews, B.W. 1982, Nature 298, 718.
(65) Nir, S., Garduno, R. and Rein, R. 1976, Polymer 18, p. 431;
Nir, S., Garduno, R., Rein, R., Coeckelenberg, Y. and MacElroy, R.D. 1977, Int. J. Quant. Chem. QBS4, p. 135.
(66) Rein, R., Garduno, R., Egan, T.J., Columbano, S., Coeckelenberg, Y. and MacElroy, R.D. 1978, Origin of Life 265.
(67) Columbano, S. and Rein, R. 1980, Computer Programs in Biomedicine 11, p. 3.
(68) Imamura, A., Suhai, S. and Ladik, J. 1982, J. Chem. Phys. 76, p. 6067.
(69) Ladik, J. 1975, Int. J. Quant. Chem. QBS9, p. 563;
Ladik, J. and Suhai, S. 1980, in "Molecular Interactions", Eds. Ratajczak, H. and Orville-Thomas, W.J., Wiley Ltd. New York, p. 151;
Ladik, J. 1980, in "Recent Advances in the Quantum Theory of Polymers", Eds. André, J.-M., Bredas, J.-L., Delhalle, J., Ladik, J., Leroy, G. and Moser, C., Springer-Verlag, Berlin, p. 155.

(70) Murrell, J.N., Randic, M. and Williams, D.R. 1965, Proc. Roy. Soc. A284, p. 566.
(71) Otto, P. 1979, Chem. Phys. Lett. 62, p. 538; Förner, W., Otto, P., Bernhardt, J. and Ladik, J. 1981, Theoret. Chim. Acta 60, p. 269.
(72) Kang, Y.K. and Jhong, M.S. 1982, Theoret. Chim. Acta 61, p. 41.
(73) Feughelmann, R, Langridge, R. Seeds, W.E., Stokes, A.R., Wilson, H.R., Hooper, L.W, Wilkins, M.H.F., Barcklay, R.K. and Hamilton, L.D. 1955, Nature 175, p. 834.
(74) Otto, P., Clementi, E. and Corongiu, G. to be published.
(75) Otto, P., Ladik, J., Clementi, E. and Martino, F. to be published.

INDEX OF AUTHORS

INDEX OF SUBJECTS